Insect-Plant Interactions

Volume II

Editor

Elizabeth A. Bernays, Ph.D.

Professor and Head
Department of Entomology
University of Arizona
Tucson, Arizona

CRC Press

Taylor & Francis Group
Boca Raton London New York

CRC Press is an imprint of the
Taylor & Francis Group, an **informa** business

First published 1990 by CRC Press
Taylor & Francis Group
6000 Broken Sound Parkway NW, Suite 300
Boca Raton, FL 33487-2742

Reissued 2018 by CRC Press

© 1990 by Taylor & Francis
CRC Press is an imprint of Taylor & Francis Group, an Informa business

No claim to original U.S. Government works

A Library of Congress record exists under LC control number: 88035918

Publisher's Note
The publisher has gone to great lengths to ensure the quality of this reprint but points out that some imperfections in the original copies may be apparent.

Disclaimer
The publisher has made every effort to trace copyright holders and welcomes correspondence from those they have been unable to contact.

ISBN 13: 978-1-138-50590-2 (hbk)
ISBN 13: 978-1-138-56032-1 (pbk)
ISBN 13: 978-0-203-71173-6 (ebk)

Visit the Taylor & Francis Web site at http://www.taylorandfrancis.com and the CRC Press Web site at http://www.crcpress.com

PREFACE

Insect-Plant Interactions is a series devoted to reviews across the breadth of the topic from cellular mechanisms to ecology and evolution. Articles are selected from areas of particular current interest or subjects that would especially benefit from a new review. It is hoped that the interdisciplinary selection in each volume will help readers to enter new fields of insect-plant interactions.

Volume 2 contains six very different articles. Courtney and Travis consider the way in which females choose host plants for their offspring, and how this may be altered by various selection pressures. They make the first attempt at a general model for host plant choice.

Two articles deal with learning: first a thorough review of the effect of experience on plant feeding species is given by Szentesi and Jermy who take some novel and challenging approaches to the topic, while Lewis and Lipani present a comparison of bees and butterflies in relation to nectar foraging.

Price and his collaborators discuss possible new ways to view herbivore population dynamics. They use species that feed on trees as examples and suggest that they fall into two general categories that involve different modes of oviposition. This leads into a discussion of the importance of natality and its importance, especially in the light of file table analyses where it is traditionally ignored.

Simpson and Simpson give a much needed review of how plant-feeding insects deal with the nutritional variation of their foods, and what the mechanisms are that allow them to compensate for low nutrient levels in general, or to switch diets as needed. Ecologists may find this subject enhances their view of field behavior.

Finally Rowell-Rahier and Pasteels review the subject of willow leaf beetles and the three trophic levels that interact in the selection of different protective strategies by beetles. The costs and benefits of different food choices are discussed.

THE EDITOR

Elizabeth Bernays, Ph.D., is a Professor and Head of the Department of Entomology, and Professor in the Department of Ecology and Evolutionary Biology at the University of Arizona, Tucson. After receiving a B.S. with honors in 1962 at the University of Queensland, Brisbane, Australia, she traveled in Europe and taught biology before studying for the M.Sc. and then Ph.D. at the University of London, England. Her Ph.D. was a study of the hatching mechanism in the desert locust involving morphology, physiology, and behavior.

From 1970 to 1983, Dr. Bernays was scientist, senior scientist, and then principal scientist in a British Government Research Institute involved with research on insect pests in developing countries. This insitute is now called the Overseas Development Natural Resources Institute. During this period, she worked in the laboratory on the physiological regulation of feeding behavior in grasshoppers and locusts, and on the effects of various plant compounds on behavior and physiology of plant-feeding insects. In the field, she worked on cassava pest biology in Nigeria and cereal resistant to insects in India. With Dr. R. F. Chapman she organized, in 1978, the Fourth International Symposium on Plant-Insect Interactions and edited the resultant book on that topic.

In 1983, Dr. Bernays moved to the University of California Berkeley where she worked on a variety of topics in plant-insect interactions and was Professor of Entomology and Professor of Zoology. In 1989 she moved to The University of Arizona where she continues her research.

In the field of plant-insect interactions, Dr. Bernays combines her physiological and behavioral experience with functional morphology and ecology, to provide a multidisciplinary approach. Her major interests concern host range in phytophagous insects; functional and causal mechanisms of host choice; variation and the role of learning chemical ecology of plants and insects; biological control of weeds and crop-plant resistance to insects.

Dr. Bernays has published widely in international journals of ecology, morphology behavior, and physiology and is an assistant editor for the *Journal of Chemical Ecology, Journal of Insect Behavior,* and *Entomologia Experimentalis et Applicata.* She has presented numerous papers at universities and institutes in the U.S. and Europe.

TABLE OF CONTENTS

Chapter 1
Insect Herbivore Population Dynamics on Trees and Shrubs: New Approaches
Relevant to Latent and Eruptive Species and Life Table Development 1
**Peter W. Price, Neil Cobb, Timothy P. Craig, G. Wilson Fernandes,
Joanne K. Itami, Susan Mopper, and Ralph W. Preszler**

Chapter 2
The Role of Experience in Host Plant Choice by Phytophagous Insects 39
Árpád Szentesi and Tibor Jermy

Chapter 3
Phenolglucosides and Interactions at Three Trophic Levels: Salicaceae-
Herbivores-Predators 75
M. Rowell-Rahier and J. M. Pasteels

Chapter 4
Learning and Flower Use in Butterflies: Hypotheses from Honey Bees.............. 95
Alcinda C. Lewis and Gloria A. Lipani

Chapter 5
The Mechanisms of Nutritional Compensation by Phytophagous Insects........... 111
S. J. Simpson and C. L. Simpson

Chapter 6
Mother Doesn't Know Best: Selection of Hosts by Ovipositing Insects 161
Steven P. Courtney and Travis T. Kibota

Index 189

1

Insect Herbivore Population Dynamics on Trees and Shrubs: New Approaches Relevant to Latent and Eruptive Species and Life Table Development

Peter W. Price, Neil Cobb, Timothy P. Craig, G. Wilson Fernandes, Joanne K. Itami, Susan Mopper, and Ralph W. Preszler
Department of Biological Sciences
Northern Arizona University
Flagstaff, Arizona

TABLE OF CONTENTS

I.   Introduction..2
II.   Life Table Analysis in Retrospect..4
        A.   The Sampling Problem...4
        B.   The Plot Problem...5
        C.   The Experimental Problem...7
        D.   The Mortality-Natality Problem.....................................8
III.   Latent Species — Galling Sawflies on Willows..............................8
IV.   Phylogenetic Constraints, Adaptive Syndromes, and Emergent Properties:
        Differences Between Latent and Eruptive Species...........................17
        A.   An Example of a Latent Species....................................18
        B.   An Example of an Eruptive Species.................................20
        C.   General Cases for Latent and Eruptive Species.....................25
V.   Reinterpretation of Life Table Studies....................................27
        A.   An Evolutionary Perspective.......................................27
        B.   Female Behavior...28
        C.   Resource Quantity and Quality.....................................28
        D.   Weather...29
        E.   Competition...29
        F.   Natural Enemies...29
        G.   Natality or Mortality?..30
VI.   Conclusions...30
Acknowledgments...32
References..32

I. INTRODUCTION

The development of life tables and their analyses so dominated the field of insect population dynamics in the 1960s that the approach may be regarded as a paradigm: the prevalent model in the field which frames the way in which we view the natural world. We examine, here, the validity of the life table approach, and what we have learned from several decades of its use. In 1989 the first life table for understanding natural populations of insects and their dynamics, developed by Morris and Miller,[113] was 35 years old (see Reference 37 for earlier life table developments). Analytical approaches for a series of life tables began to appear in 1959.[110,167] This method for studying natural populations became widely adopted, and was a force which developed enormous research energy in the 1960s. Insect population dynamics became a major focus in ecology, and a flush of books soon reflected this interest (e.g., References 24, 38, 111, 159, and 169).

With the advantage of hindsight we can now evaluate the contribution this research has made to understanding insect population dynamics. This is important because the approach is still used for studying major insect pests in agriculture and forestry, for investigating herbivores to be used in biological weed control projects, and for the better conservation of rare and endangered species (e.g., References 48, 61, 84, and 170). In addition, millions of research dollars were expended on the often large projects which undertook life table construction for many insect generations, so it is appropriate for planners and research directors to understand the quality of research yielded and its long-term impact on progress of understanding. Finally, the large body of literature on life tables and analysis has shaped the way in which we now view nature and where we place research emphasis.

We develop the argument that life table construction emphasized mortality in populations, and the method dictated the kinds of results, and conclusions reached, in insect population dynamics studies. The important influences of natality and female oviposition behavior in response to plant quality were frequently overlooked. Female oviposition behavior is phylogenetically constrained and acts as an ultimate factor profoundly influencing the adaptive syndrome of the species, and the proximate ecological factors that are commonly studied in insect life table analysis. Therefore, a radical shift in perspective is needed to place the evolutionary history of a species in a central causal position in order to understand the emergent ecological phenomena involved with population change. Such a shift also requires a revision of life tables to better evaluate natality effects, female oviposition behavior, and the role of plant quality. We use these approaches to identify the fundamental differences between latent and eruptive species of insect herbivores.

We emphasize in this review the gathering of the empirical data and the construction of life tables. We do not stress the theoretical basis of forest insect population dynamics which has been reviewed effectively in the recent past (e.g., References 8, 12, 15, 16, 18, 66, and 92).

We use an historical perspective to consider what we know about insect population dynamics, and what contributions the field of evolutionary ecology can make. We then discuss modifications or alternative approaches that might increase our understanding of population dynamics and foster new research directions or emphases. We argue that a synthesis of several areas of evolutionary ecology with a modified life table approach may change our perceptions of how insect population dynamics should be studied and the major factors driving population change.

Our perspective is colored by the organisms we have studied over the past 10 years. Of particular interest is the shoot-galling sawfly, *Euura lasiolepis*, which attacks arroyo willow, *Salix lasiolepis*, and the sawfly's relatives. This sawfly is hardly regarded as a pest; it is not eruptive or an outbreak species, and has shown remarkably stable populations over the last decade in our study sites. It is perhaps an apparent paradox that we claim to shed new light on pest insect population dynamics by having considered species with stable populations. However, we argue that a strong comparative approach among species with stable populations and those with eruptive populations can further our understanding of the mechanisms involved with both kinds of insect populations.

The eruptive and noneruptive species are only two ends of a continuum of variation in natural insect herbivore populations, but for simplicity we will concentrate on these extremes. Therefore, a few definitions are needed to clarify the terms we use.

A **latent species** refers to a species with typically latent populations, and here is defined as an insect herbivore species that remains at steady population densities, varying between one or two orders of magnitude, and usually incapable of increasing explosively to cause heavy damage to the host plant population (synonyms; nonoutbreak,[89,171] nonpest, noneruptive). The key characters are steady densities with low levels of damage to host plant populations. Species are not necessarily rare, or at low densities.

An **eruptive species** refers to a species with eruptive populations, and here is defined as an insect herbivore species that has both an endemic phase of low density and low damage to the host-plant population, and an epidemic phase in which populations become dense and damaging, with fluctuations ranging over three to five orders of magnitude (synonyms: cyclic,[17,118] irruptive,[171] outbreak,[171] pest).

Endemic — the low-density, nondamaging phase in an eruptive species often associated with populations locally distributed in pockets or foci in particularly favorable host plant groups or populations.

Epidemic — the high-density, damaging phase in an eruptive species frequently associated with either rapid spread of populations or an apparent spread resulting from rapid *in situ* increases from very low populations.

We prefer to use the terms latent and eruptive species for several reasons. Latent and eruptive are descriptors of species type and therefore more helpful than such designations as nonoutbreak vs. outbreak, or nonpest and pest. The term latent implies the potential to increase to high densities, a possibility we will discuss later, but in most places for most of the time we argue that such species are so constrained that the latent state will prevail. In our term eruptive we include all of the outbreak types recognized by Berryman and Stark[17] and Berryman,[13] so our use is much broader simply because we wish to emphasize extremes in population types, while they were concerned with classifying kinds of outbreaks. These terms also permit the classical epidemiological usage of endemic and epidemic for phases in an eruptive species' population change.

Use of the term **steady population density** in the definition of a latent species needs some clarification. We imply here changes in density of only one or two orders of magnitude between the highs and lows in population size over several years. This contrasts with many eruptive species which commonly show density changes of three to five orders of magnitude (e.g., References 100, 118). We do not imply stability in the long term, over decades, which is usually used in ecology to denote a stable equilibrium with low variance around an average population size.[95]

Important contributions to understanding differences between latent and eruptive species have been made recently by Rhoades,[148] Nothnagle and Schultz,[120] Mason,[89] Wallner,[171] and Myers.[118] Each had a slightly different focus with Rhoades being very general, Mason emphasizing latent forest Lepidoptera, Wallner contrasting latent and eruptive species, and Myers, and Nothnagle and Schultz concentrating on eruptive forest Lepidoptera. These publications mark an important surge in synthesis, and together with Barbosa and Schultz[8] may well mark a new phase in reaching generalizations about latent and eruptive species.

On a less positive note these authors do not claim any great success in reaching a synthesis. They reinforce the view that we still have much to learn, generalities are hard to achieve, and little really definitive understanding of insect herbivore population dynamics is available. Mason (Reference 89, p. 52) declared that "There is a surprising lack of basic biological information on the nonoutbreak defoliators, and virtually nothing is known about their population dynamics." Wallner[171] categorized nonoutbreak species as K-selected and rare (his Table 1) and the extreme form of outbreak species as r-selected and eruptive, but Nothnagle and Schultz[120] maintained that current knowledge was too incomplete to assess such a categorization. They said, "Although much is known about some irruptive pest species, the majority of the nonirruptive species remain virtually uncharacterized in any meaningful way. Hence, comparing the biological traits of pest and nonpest species is very difficult" (p. 60). Myers (Reference 118, p. 231) focused on disease as a major factor in cycles while admitting "I may simply be emphasizing the importance of disease because it is the least-known factor in studies of insect cycles." But in their penultimate sentence Nothnagle and Schultz (Reference 120, p. 74) concluded "that one of the most influential environmental factors is likely to be variation in host plant quality for young larvae." In fact we see little convergence in these studies toward a common theme, or a general synthesis on latent and eruptive species (even though general theory on insect population dynamics is progressing well, e.g., References 15, 18).

Perhaps the most tangible contribution we can make to this debate is that for 10 years we have studied a latent species and we believe we understand it well. Perhaps it is one of the best-studied latent species we have available for comparing with eruptive species. Therefore we approach the problem of differentiating between latent and eruptive species from an unusual perspective, which we feel increases our understanding of the mechanisms regulating insect population dynamics.

II. LIFE TABLE ANALYSIS IN RETROSPECT

The life table approach gave impressive impetus for the study of insect population dynamics when first adapted for natural insect populations by Morris and Miller.[113] This was reinforced by the development of formal analytical approaches—single factor analysis[110] and key factor analysis,[167] and analysis of determination.[115] Coupled with standardized sampling techniques and a firm statistical base for analyzing data,[109] a rigorous approach to the study of field populations was developed. These factors we believe sealed the future of life table construction and analysis as the major approach to understanding natural populations.

With the advantage of hindsight we contend that life table construction, in the way originally developed, predisposes the data for certain kinds of results, and precludes other kinds of phenomena from being easily recognized. In fact the methods dictate the kinds of results obtained. We will argue this point by a series of examples, which focus on problematical areas in the methods used. Our intention is not to belittle what we regard as impressive advances made using life table analysis, but rather to build constructively on a solid research base. However, recognizing limitations in a major force for understanding dynamics is an important prelude to further advances. This discussion attempts only to cover general emphases in the literature.

A. THE SAMPLING PROBLEM

Morris[109] stated "The fact that the variability in insect population between trees is so much greater than the variability within trees...may well prove to be a principle common to all forest insects" (p. 244). To cope with this large variation Morris set as an acceptable standard the requirement that sample sizes should be large enough to yield a standard error of 10% or less of the mean density of insects per sample unit. Working with 10 square feet of branch surface as a standard method of expressing spruce budworm density on balsam fir, he used Snedecor's[158] formula for calculating the number of trees needed relative to budworm density in order to sample a population adequately. At relatively high budworm densities perhaps only 35 trees were needed to sample egg masses (e.g., 170 egg masses per 10 square feet). But Morris's[109] Figure 14 (p. 276) shows that at low budworm densities tree sampling becomes excessive. "In 1948, population increased to three larvae and to 0.5 pupae or egg masses, indicating that required sample size would be about 40, 210, and 310 trees respectively. With a field establishment of some 20 workers, it is probable that satisfactory life tables could be developed for at least one plot at this level of population....The next increase gave populations of roughly 20 larvae and five pupae or egg masses, indicating a sample size that would permit the study of several plots" (p. 280). Because budworm reaches such low endemic densities Morris (Reference 109, p. 281) bemoaned the fact that "The possibility that the preparation of reliable life tables will be extremely difficult, if not impractical, at endemic levels is disappointing. Our knowledge of the epidemiology of any species will be incomplete until we know what factors are responsible for the maintenance of endemic populations over a period of years, and through what mechanism the population succeeds in 'escaping' from the endemic level."

With Morris's[109] publication the sampling problem was fully exposed, and resolved. Large samples had to be taken, requiring large teams of workers. Such samples were required several times per year to cover all the age intervals of an insect generation necessary for life table construction. Such efforts severely limited the number of sample plots which could be studied in any one year. Costs to study the population dynamics of any one insect pest were very high.

We have concentrated on Morris's[109] paper and used several quotations because these arguments profoundly influenced the sampling programs used in other studies. Sample sizes were frequently large (Table 1) and usually costly in time and money.

The consequences for what we learned about insect herbivore population dynamics were also profound. Expressing insect density as a mean of many tree samples obscured much of the crucial biology of the plant-herbivore interaction. Individual plant variation relevant to many herbivores, now recognized as important (e.g., Reference 39), was not quantified. And yet for many insect species, although perhaps not all, it was probably plant variation in genotype, environment, and phenotype which could account for the large variance in insect population densities within a population of trees. As a consequence plant quality was inadequately considered and frequently did not enter as a factor in a life table (Tables 1 and 2). Major exceptions involved studies on bark beetles where tree resistance by resin expulsion and hypersensitive responses were conspicuous (Table 3) and its variation in plant populations became a focal point in the development of general theory on bark beetle outbreaks (e.g., References 10, 11, 17, 141, and 143). Further consequences of the sampling problem are reflected in what we call the plot problem, the experimental problem, and the mortality problem.

B. THE PLOT PROBLEM

Sampling became such a demanding and expensive activity that only a small number of plots, at best, could be studied, and emphasis was placed on how populations changed within these plots. A strong spatial component in epidemiology was largely lost, even though we now know that analysis of gradients is a powerful comparative approach in ecology (e.g., References 79-81, and 188). In addition, plots were commonly placed in

Table 1
EXAMPLES OF SAMPLE SIZES USED FOR DEVELOPMENT OF LIFE TABLES

Insect species	Sample unit	Sample size	Source
1. *Diprion hercyniae*	Tree	150	119
2. *Pristiphora erichsonii*	Branch	200	67, 68
3. *Neodiprion swainei*	Tree	75	99
4. *Choristoneura fumiferana*	Branch	31—300	109, 111
5. *Lithocolletis blancardella*	Leaf cluster	400	126
6. *Recurvaria starki*	Branch tip	100	162
7. *Brachys tessellatus*	Leaf	>200	166
8. *Archips argyrospilus*	Leaf cluster	100	77
9. *Spilonota ocellana*	Leaf cluster	100	77
10. *Coleophora serratella*	Leaf cluster	100	77
11. *Lepidosaphes ulmi*	Leaf cluster	96	156
12. *Neodiprion fulviceps*	10 Branches/tree	10 trees	34
13. *Operophtera brumata*	2 Traps/tree	5 trees	168, 169
14. *Pikonema alaskensis*	Tree	6	64
15. *Lithocolletis salicifoliella*	25 Leaves/tree	10 trees	56
16. *Agromyza frontella*	Variable	24—36	102
17. *Taxomyia taxi*	454 g foliage/tree	3—11 trees	145

Note: The studies using large sample sizes are given first. Note that samples were pooled for life table construction so effects of plant variation were not evaluated.

Table 2
EXAMPLES OF LIFE TABLE STUDIES WHICH OMITTED OR INCLUDED EXPERIMENTAL WORK OR STUDIES ON PLANT EFFECTS ON INSECT POPULATIONS

Insect species	Experimental studies	Plant effects explored?	Source
1. *Diprion hercyniae*	Cocoon planting	No	119
2. *Pristiphora erichsonii*	No	No	67, 68
3. *Neodiprion swainei*	Insecticide treatment	No	99, 100
4. *Choristoneura fumiferana*	Extensive	Stand age, density, flowering, etc.	111
5. *Lithocolletis blancardella*	No	Superficially	126
6. *Recurvaria starki*	No	Resination and needle drop	162
7. *Brachys tesselatus*	No	No	166
8. *Archips argyrospilus*	No	No	77
9. *Spilonota ocellana*	No	No	77
10. *Coleophora serratella*	No	No	77
11. *Lepidosaphes ulmi*	No	No	156
12. *Neodiprion fulviceps*	No	No	37
13. *Operophtera brumata*	No	No	168, 169
14. *Pikonema alaskensis*	Predation on overwintering cocoons	Canopy level	64
15. *Lithocolletis salicifoliella*	Competition	Plant age	56
16. *Agromyza frontella*	No	No	102
17. *Chionaspis pinifoliae*	No	No	84

Note: Where plant effects were studied variation among plants was usually omitted because of pooling samples (see Table 1).

Table 3
**EXAMPLES OF LIFE TABLE STUDIES AND
OTHER STUDIES ON BARK BEETLES IN
WHICH NUMBERS OF ATTACKING FEMALES
WERE, OR COULD HAVE BEEN, USED TO
INITIATE THE LIFE TABLE, AND EFFECTS OF
PLANT RESISTANCE ON FECUNDITY COULD
BE EVALUATED**

Insect species	Host plant	Source
1. *Dendroctonus adjunctus*	*Pinus ponde-rosa*	23
2. *D. brevicomis*	*P. ponderosa*	36
3. *D. frontalis*	*Pinus* sp.	26, 63
4. *D. ponderosae*	*P. contorta*	2, 140, 141
5. *D. simplex*	*Larix laricina*	74
6. *Ips calligraphus*	*P. elliottii*	57, 58
7. *Scolytus scolytus*	*Ulmus procera*	9
8. *S. ventralis*	*Abies grandis*	10

high herbivore density sites, because sampling with some accuracy was feasible, and the collection of negative data in many samples at low densities was unproductive. Effort was concentrated on epidemic populations because the sampling approach precluded productive research on endemic populations. This is a case of the method dictating what can and cannot be studied in insect population dynamics.

C. THE EXPERIMENTAL PROBLEM

The apparent rigor of the sampling program and analysis of results seems to have beguiled many researchers into thinking that life table construction and analysis was sufficient to explain insect population dynamics, an end unto itself. The manpower and cost of obtaining the samples each generation also detracted from emphasis on more detailed mechanistic studies on interactions and interrelationships, what Morris[112] called process studies, usually involving experimental methods (Table 2). Even when experiments were employed they were not detailed or extensive enough to reveal mechanisms.

The lack of experimental work left the empirical population censuses and life tables as the only source of data for interpreting the causes of population change. Even though Morris (Reference 110, p. 587) noted "A key factor was defined as any mortality factor that has useful predictive value and no attempt has been made to establish cause and effect," many authors have ignored the caution. For example, a density-dependent factor identified by key factor analysis is commonly interpreted as a regulating factor. "The population is regulated at another season by the density dependent action of pupal predation" (Reference 168, p. 141). This is in spite of the evident alternative possibilities that density dependence simply shows that a factor responds to insect density rather than causes its change, or that insect density and the density-dependent factor covary in response to a third independent cause.

The problems with a purely correlative approach to understanding population dynamics have been explained in many essays and we do not need to go into details here (see References 43, 62, 83, 85, 101, 112, 114, 128, 152-154, and 161). The fact that correlation does not reveal causation is well understood, and the need for experimental studies to establish cause and effect is widely recognized. However, even in 1988 Myers (Reference 118, p. 224) could say "There is an unfortunate lack of experimental investigation of population cycles of forest Lepidoptera." The general lack of experimental studies coupled with life table construction has left us with very little mechanistic understanding of insect population dynamics (bark beetles excepted, e.g., Reference 142). This impotence in evaluating many studies is explored in more detail in the role of natural enemies by Price.[130]

D. THE MORTALITY-NATALITY PROBLEM

The life table emphasizes the loss of individuals from a cohort of eggs because of their death. The eggs laid by a population of females per sample unit (e.g., 10 square feet of balsam fir foliage for the spruce budworm) is the basis of the life table. After all, it can be argued that this is the start of the new generation. Therefore, mortality factors are relatively well documented, even though they are often lumped. For example 8 studies in 23 analyzed by Podoler and Rogers[125] include a complex of factors not separated into component parts.

Emphasis on mortality leaves natality factors less effectively documented. We contend that revealing modification of life table construction would initiate the generation with the eggs, or potential eggs, in the female insect, enabling the study of many important questions (e.g., Reference 127). Did females lay their full complement of eggs? If females were smaller than normal, was this owing to larval competition or to low food quality? Does competition among females for oviposition sites affect the size of the egg cohort (e.g., References 181, 183)? If sex ratio is skewed, what were the causes and consequences?

These kinds of questions are seldom answered directly in life table analysis, bark beetles excepted. This is because the original construction left questions of fecundity until the end of the life table (see Reference 113). The difference between the number of eggs expected, based on the number of females surviving at the end of the generation, and the actual number found, was interpreted as being caused by immigration or emigration of females to or from the study site. This was not monitored empirically, nor were alternative mechanisms adequately tested for gain or loss of eggs. As a result, frequently no hard evidence was obtained on this critical stage in the life cycle of the insect herbivore. In most studies using Morris and Miller's[113] format, natality effects are not mechanistically separated from the effects of migration. Again, an important exception is in studies of bark beetle population dynamics, which also assessed the potential fecundity of females attacking trees (Table 3).

These problems need to be overcome before life tables can be generated that realistically portray the dynamical properties of insect populations. From our studies on *Euura lasiolepis* and related sawflies we have learned that general approaches to life table construction may need modification, at least for some species, and that an experimental approach is central to evaluating cause and effect. Experiments include an examination of intra- and interplant variation in quality. Therefore, we will review what studies of the willow, herbivore, natural enemy system have taught us.

III. LATENT SPECIES — GALLING SAWFLIES ON WILLOWS

We have studied three species of sawfly in enough detail to know that the general patterns in plant-herbivore interactions and population dynamics are strong. These are the shoot galler *Euura lasiolepis* on *Salix lasiolepis* (References 27, 29-32, 127, 131, 133-135, 138, 172), the shoot galler *Euura exiguae* on *Salix exigua*,[132] and the bud galler *Euura mucronata* on *Salix cinerea* (References 136, 137, 149).

The general life history is that sawflies emerge in the spring, mate, and females

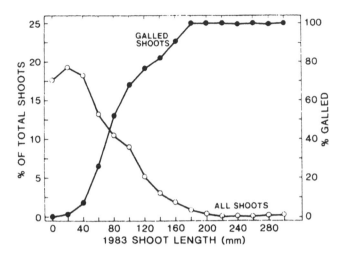

FIGURE 1.   The distribution of all shoots on ramets of *Salix lasiolepis*, and the distribution of attacks by *Euura lasiolepis*. Note that 100% of very long shoots are attacked but these represent a small proportion of those available. From Craig, T. P., Price, P. W., and Itami, J. K., *Ecology*, 67, 419, 1986. Copyright 1986 by the Ecological Society of America. Reproduced with permission.

oviposit into rapidly growing young shoots. Females cause gall development and larvae feed within the gall. The shoot gallers overwinter in the gall, spinning a cocoon in the gall and pupating in the spring. The bud galler emerges from the gall as a larva and spins a cocoon in the soil to overwinter.

The willow species are shrubs which readily spread vegetatively into clones covering several square meters. Thus we can distinguish between the genotype, or genet, for the whole clone, and individual stems, or ramets, which emerge from the ground.

All species most frequently attack the most vigorous plants in a population and the most vigorous shoots within a plant. Long shoots that are attacked with the highest probability are relatively rare in a population of shoots (References 27, 136) (Figure 1). Females show a preference for long shoots and attack shoots in order of their length.[30] Larvae survive better in long shoots than in short shoots so there is a clear link between female ovipositional preference and larval performance.[30,132,136,137] Oviposition scars make nodes unacceptable to females[28] and protect the node against further oviposition rather like a territorial defense using a deterrent pheromone (e.g., Reference 139). As a result females compete for high-quality oviposition sites and interference competition is an important determinant of the shoot quality a female can utilize.[31]

Female oviposition behavior is strongly influenced by shoot quality, which is best predicted by shoot length. In an experiment with a high-water treatment and a low-water treatment on potted plants the behavior of females resulted in major differences in the number of eggs deposited per plant. Two females in each cage had the capacity to lay a total of about 100 eggs, and we used this number as the start of the life table (Table 4). On well-watered plants females laid 72.5% of these eggs, but on low-water plants only 14.9% of eggs were laid. This was because of reduced gall initiation, and eggs were not laid in 60.8% of initiated galls.[127] In both life tables on high- and low-water plants, female oviposition behavior in response to plant quality was the most important factor causing loss of individuals to the population. This natality effect was of paramount importance. We have shown that this effect remains when experimental plants are uncaged and females can move between them[133] and that the experiments simulate effects seen on wild clones in high-water and low-water sites.[127]

Table 4
LIFE TABLES FOR POPULATIONS OF *EUURA LASIOLEPIS* ON EXPERIMENTAL PLANTS GIVEN A HIGH WATER TREATMENT AND A LOW WATER TREATMENT

x (Stage)	l_x (No. Alive at beginning of x)	d_xF (Factor responsible for d_x)	d_x (No. dying during x)	$100q_x$ (d_x as a % of l_x)
High water treatment				
Eggs in females	100	Failure to initiate gall	0.0	0.0
	100	Egg retention	27.5	27.5
		Total maternal response	27.5	27.5
Eggs in galls	72.5		0.0	0.0
First instar	72.5			
		Parasitism	0.0	0.0
		Host effects	9.9	13.6
		Total	9.9	13.6
Second instar	62.6	Host effects	0.0	0.0
Last instar	62.6			
Low water treatment				
Eggs in females	100	Failure to initiate gall	62.1	62.1
	37.9	Egg retention	23.0	60.8
		Total maternal response	85.1	85.1
Eggs in galls	14.9		0.0	0.0
First instar	14.9			
		Parasitism	0.0	0.0
		Host effects	8.1	54.5
		Total	8.1	54.5
Second instar	6.8	Host effects	0.3	3.9
Last instar	6.5			

Note: This illustrates a new emphasis on capturing female ovipositional behavior in the beginning of the life table.

In the same experiments the second most important effect was plant resistance to first-instar larvae.[127] In the high-water treatment, interval mortality of first-instar larvae caused by plants was 13.6% (100 qx, Table 4), but in the low-water treatment it was 54.5%. Such differences have been found in other experiments and in field clones in years after heavy and light precipitation.[133]

These are very clear results supported by several sets of experiments and field observations. They show that female behavior is critical in the population dynamics of this insect, and that plant resistance to first-instar larval establishment in galls is also a major factor. These roles of behavior and plant resistance have seldom entered into life table construction (Tables 1 and 2) except for bark beetles (Table 3).

Even though watering and fertilizing experiments using potted plants produced opposite effects on plant protein content and phenolic defenses relative to shoot length, females still attacked long shoots.[172] In the water experiment increased shoot length correlated with increased phenolics and decreased protein, while in the fertilizer experiment the opposite relationships were observed. Shoot length, therefore, is a better criterion for attack than nutrient status or general levels of chemical defense. However, the individual phenolic glycosides evaluated increased as shoot length increased[138] so these may act as cues to which females respond. Nevertheless, all indications are that shoot length and vigor at time of attack is the best predictor of attack and females respond to this, independent of the general state of plant nutrients and defensive chemicals.

We use each willow genet as the sampling and experimental unit. Each genet displays many ramets of different ages and vigors, presenting a complex array of resources of varying quality for sawflies.[27,18,132,136,137] Young ramets are more vigorous, produce long shoots, and vigor declines with ramet age. Thus, the pattern of attack by sawflies correlates positively with shoot length and negatively with ramet age (Figure 2). Heterogeneity of willow clone quality is increased by variation in water availability. Years with high winter precipitation result in better willow growth, as in water treatment experiments,[134] and wet sites result in better growth each year than dry sites. The three clones with sawfly densities of over 700 galls per 1000 shoots are all in much wetter sites than the other clones which line a temporary stream (Figure 2). Well-watered plants received more sawfly attack and larvae survive better than poorly watered plants.[134] Thus each willow clone is practically a unique resource for sawflies with its own genotype, and phenotype as influenced by age and environment. In addition, history of sawfly attack influences the effective vigor of ramets because galls cause "pruning back" of growth and more vigorous regrowth — an effect we call resource regulation.[27,149] We would fail to observe all this variation important to the sawflies if we lumped samples across clones as has been practiced commonly for life table construction.

Population stability is characteristic for each clone over the 8-year period we have sampled *Euura lasiolepis* (Figure 3). Densities typically vary over only two orders of magnitude, and positive correlations of densities in one year with any other year have always been significant (e.g., Reference 27). Note that densities of sawflies are high on some clones, and this latent species could not be considered as a rare insect. Limitation of population size is clearly defined by the generally small number of long shoots per unit area in which larvae can survive. High precipitation and resource regulation may modify these limits, but only slightly. The rapid feedback causing tight population regulation works through female competition for oviposition sites in high-density patches, and the difficulty in finding high-quality resources in slowly growing willow patches.

Natural enemies are limited by the plant to a large extent. Gall size limits access to one of the major parasitoids, *Pteromalus* sp., such that it is most successful on the less vigorous willow clones which produce smaller galls, but have low *Euura* populations. Thus, a negative spatial density-dependent relationship exists between *Euura* density and parasitism by *Pteromalus*.[131,134] The other major parasitoid is the larger ichneumonid, *Lathrostizus euurae*. This species is not limited by gall size but by gall toughness, with a short time of access to the host larvae which keeps levels of parasitism low.[32] These three-trophic-level interactions constrain the major parasitoids from becoming important mortality factors on the sawflies, and this feature is characteristic of the other species we have studied.[132,137]

The generality that these kinds of latent species attack longer shoots in a population of shoots can be extended further by more superficial studies on many other species. We have now estimated the relationship between shoot length class and probability of attack for 14 galling sawfly species and all show positive and significant patterns (Table 5). This sample includes shoot, petiole, bud, midrib and leaf gallers, and leaf folders; all the galling types known for the tenthredinid sawflies, as well as cases on the only two genera attacked, *Populus* and *Salix*. Other studies on sawfly communities support these patterns. Galling sawfly species tended to be positively correlated across host plants,[51] within host plants on longer shoots,[52] and on genotypes which produced longer shoots.[53] Outside the tenthredinid sawflies positive responses to shoot length have been

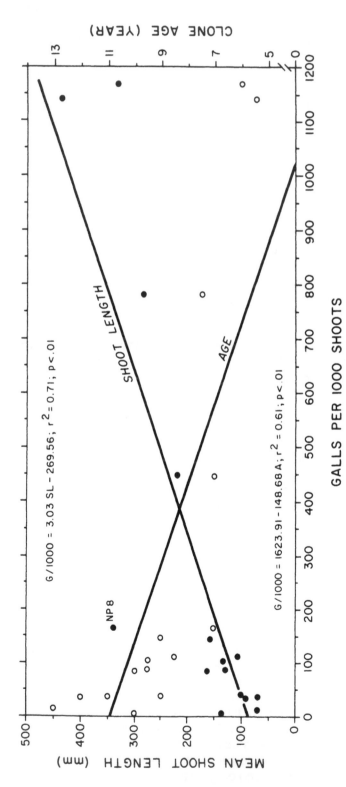

FIGURE 2. The relationships between mean shoot length per willow clone and clone age, and density of *Euura lasiolepis* galls in 1986, on 15 clones in Flagstaff, AZ. The unusual clone (NP8) with high mean shoot length but low attack is a very favorable clone for sawflies but a dense population was almost eradicated in 1981 by grasshoppers feeding heavily on galls, and the sawfly population has not recovered. When NP8 is removed the regressions account for 91% and 96% of the variance for shoot length and age, respectively. G/1000 = galls per 1000 shoots, SL = shoot length, A = age. Open circles are clone age and closed circles are for mean shoot length per clone.

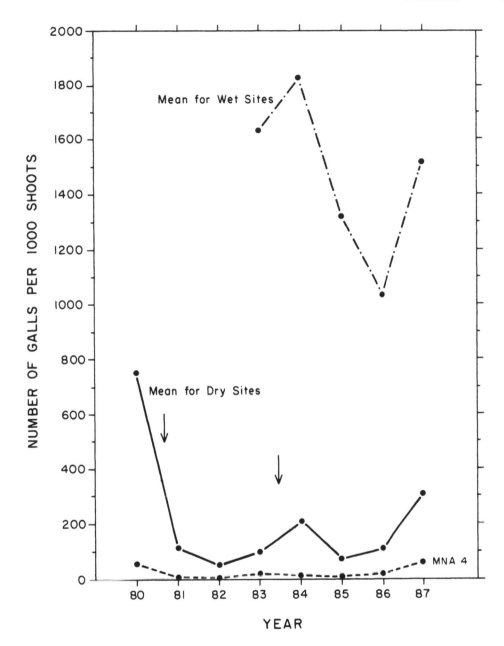

FIGURE 3. Populations of *Euura lasiolepis* on some representative clones in the Flagstaff, AZ area. Clones with the highest densities of sawflies are illustrated (mean of 3 clones in wet sites) as well as the population mean across the 12 clones in drier sites. Lowest density on the clones samples was on clone MNA 4, also illustrated. Arrows indicate winters with relatively low precipitation. Note that maximum differences in population density are relatively small. For wet sites the highest density is only 1.8 times higher than the lowest density. For dry sites the difference is 17.0 times.

documented for species in five other major taxa: cynipid gall wasps (Cynipidae), gall midges (Cecidomyiidae), aphids (Aphididae), moths (Lepidoptera), and weevils (Curculionidae) (Table 6). Therefore, the patterns we have documented for the species we have studied in detail may well be representative of many herbivore species.

We must emphasize the importance of long shoots and vigorously growing plants to these kinds of herbivores. Although such resources are relatively rare, oviposition is

Table 5
SPECIES OF TENTHREDINID SAWFLY SHOWING A POSITIVE, SIGNIFICANT RESPONSE TO SHOOT LENGTH

Insect species	Host plant species	Locality	r²	p
1. *Euura amerinae*	*Salix pentandra*	Joensuu, Finland	0.92	.01
2. *E. atra*	*S. alba*	Joensuu, Finland	0.91	.01
3. *E. exiguae*	*S. exigua*	Weber River, UT	0.51	.05
4. *E. lasiolepis*	*S. lasiolepis*	Flagstaff, AZ	0.92	.01
5a. *E. mucronata*	*S. cinerea*	Joensuu, Finland	0.95	.01
5b. *E. mucronata*	*S. cinerea*	Kiel, Germany	0.84	.01
5c. *E. mucronata*	*S. lapponum*	Kevo, Finland	0.62	.05
6. *Euura* petiole galler	*S. lasiolepis*	Flagstaff, AZ	0.63	.01
7. *Euura* midrib galler n. sp.	*S. exigua*	Lees Ferry, AZ	0.76	.01
8. *Euura* shoot galler n. sp.	*S. interior*	Urbana, IL	0.73	.01
9. *Phyllocolpa coriacea*	*S. cinerea*	Joensuu, Finland	0.86	.01
10. *P. excavata*	*S. pentandra*	Joensuu, Finland	0.51	.01
11a. *Phyllocolpa* leaf fold morph	*S. lasiolepis*	Flagstaff, AZ	0.87	.01
11b. *Phyllocolpa* pustule morph	*S. lasiolepis*	Flagstaff, AZ	0.73	.01
12. *Phyllocolpa* sp.	*Populus tremuloides*	Flagstaff, AZ	0.70	.01
13. *Pontania* n. sp. near *pacifica*	*S. lasiolepis*	Flagstaff, AZ	0.90	.01
14. *Pontania pustulata*	*S. phylicifolia*	Joensuu, Finland	0.78	.01

Note: The variance in probability of attack accounted for by shoot length is given as a proportion (r²) and level of significance is a proportion of the time an estimated significant relationship may be in error (p).

Table 6
SPECIES OF INSECTS OTHER THAN TENTHREDINID SAWFLIES SHOWING A POSITIVE SIGNIFICANT RELATIONSHIP BETWEEN SHOOT LENGTH AND PROBABILITY OF ATTACK (SYMBOLS AS IN TABLE 4)

Insect taxa	Host plant species	Locality	r²	p
Cynipid wasps				
1. *Diplolepis fusiformans*	*Rosa arizonica*	Flagstaff, AZ	0.84	.01
2. *Diplolepis spinosa*[a]	*Rosa arizonica*	Flagstaff, AZ	0.92	.01
Gall midges				
3. *Rhabdophaga rosaria*	*Salix phylicifolia*	Joensuu, Finland	0.74	.01
Aphids				
4. Wax current aphid shoot galler	*Ribes cereum*	Flagstaff, AZ	0.64	.01
5. *Symydobius oblongus*	*Betula pubescens*	Kevo, Finland	0.76	.01
Moth				
6. *Amorpha* shoot galler	*Amorpha fruticosa*	Lincoln, NE	0.68	.01
Weevil				
7. *Rhynchites betulae*	*Betula pubescens*	Joensuu, Finland	0.75	.01

[a]   Correlation was between shoot diameter and probability of attack because gall is frequently terminal and prevents further shoot elongation.

high and first-instar survival is high compared with responses to plants of lower vigor. Many herbivores respond positively to the stronger-growing plants in a population (Table 7). And yet stressed plants have been frequently regarded as favorable for insect herbivores by White,[175-180] and this argument has entered into general theory on plant-herbivore interactions.[90,91,146-148,165] Dissension has been aired by Myers[118] based on

Table 7
**SOME EXAMPLES FROM THE LITERATURE IN WHICH ARTHROPOD
HERBIVORES ATTACK VIGOROUSLY GROWING TREES OR SHRUBS, OR
YOUNG PLANTS MORE THAN MATURE PLANTS, OR LARGE PLANT MODULES**

Insect taxon	Host plant taxon	O/F	W/P	Source
Coleoptera				
1. *White-pine weevil, *Pissodes strobi*	White pine, *Pinus strobus*	x	x	54
	Sitka spruce, *Picea sitchensis*			124
2. *Pales weevil, *Hylobius pales*	Pines, *Pinus*	x	x	5, 46
3. *Pine root-collar weevil, *Hylobius radicis*	Pine	x	x	47, 72
4. *Pine weevil, *Pissodes approximatus*	Pine	x	x	45
5. *Weevil, *Cylindrocopturus eatoni*	Pines, *Pinus ponderosa* and *P. jeffreyi*	x	x	41
6. *Lodgepole terminal weevil, *Pissodes terminalis*	Pines, *Pinus*	x	x	54
7. *Pine shoot beetle, *Tomicus piniperda*	Scots pine, *Pinus sylvestris*	(adult feeding)		75
8. *Poplar leaf beetle, *Chrysomela confluens*	Narrowleaf cottonwood, *Populus augustifolia*	x	—	71
9. Nutsedge leaf miner, *Taphrocerus schaefferi*	Yellow nutsedge, *Cyperus esculentus*	x	x	163
Lepidoptera				
10. *White-pine shoot borer, *Eucosma gloriola*	Jack pine, *Pinus banksiana*	x	x	5
11. *Jack-pine shoot moth, *Eucosma sonomana*	Jack pine, *Pinus banksiana*	x	x	5
	Ponderosa pine, *Pinus ponderosa*			160
12. *Cottonwood shoot borer, *Gypsonoma haimbachiana*	Eastern cottonwood, *Populus deltoides*	x	x	5
13. *Pitch nodule maker, *Petrova albicapitana*	Jack pine, *Pinus banksiana*	x	x	5
14. *European pine shoot moth, *Rhyacionia buoliana*	Pines, *Pinus*	x	x	5
15. *Nantucket pine tip moth, *Rhyacionia frustrana*	Pines, *Pinus*	x	x	5
16. *Aspen blotch miner, *Lithocolletis salicifoliella*	Aspen, *Populus tremuloides*	x	x	86
17. *Ponderosa twig moth, *Dioryctria ponderosae*	Ponderosa pine, *Pinus ponderosa*	x	x	54
18. *Pinyon pine cone and shoot borer, *Dioryctria albovitella*	Pinyon pine, *Pinus edulis*	x	x	106, 187
19. Hairstreak butterflies, *Satyrium* spp.	?	?	—	69
20. Oak leafminer, *Eriocrania subpurpurella*	Oak, *Quercus robur*	x	x	M. Crawley, pers. comm.
21. Oak leafminer, *Acrocercops brongniadella*	Oak, *Quercus robur*	x	x	M. Crawley, pers. comm.
22. Oak leafminer, *Phyllonorycter lautella*	Oak, *Quercus robur*	x	x	4
Hymenoptera				
23. Rose ball galler, *Diplolepis spinosa*	Arizona rose, *Rosa arizonica*	x	x	22
24. Rose shoot galler, *Diplolepis fusiformans*	Arizona rose, *Rosa arizonica*	x	x	22

Table 7 (continued)
SOME EXAMPLES FROM THE LITERATURE IN WHICH ARTHROPOD HERBIVORES ATTACK VIGOROUSLY GROWING TREES OR SHRUBS, OR YOUNG PLANTS MORE THAN MATURE PLANTS, OR LARGE PLANT MODULES

Insect taxon	Host plant taxon	O/F	W/P	Source
25. Cynipid oak galler, *Xantho-teras politum*	Oak, *Quercus stellata*	x	x	173
26. Cynipid gall wasp, *Disholcas-pis cinerosa*	Live oaks, *Quercus* spp.	x	x	50
27. Bud-galling sawfly, *Euura mu-cronata*	Willow, *Salix cinerea*	x	x	149
28. Shoot-galling sawfly, *Euura la-siolepis*	Arroyo willow, *Salix lasiolepis*	x	x	27
29. Shoot-galling sawfly, *Euura exiguae*	Coyote willow, *Salix exigua*	x	x	132
30. Oak galler, *Andricus kollari*	Oak, *Quercus robur*	x	x	M. Crawley, pers. comm.
31. Oak galler, *Andricus lignicola*	Oak, *Quercus robur*	x	x	M. Crawley, pers. comm.
Homoptera				
32. Gall-forming aphid, *Pemphi-gus betae*	Narrowleaf cottonwood, *Populus augustifolia*	N	x	181
33. Aphid, *Kallistaphidus betuli-cola*	Birch, *Betula pubescens*	N	—	49
34. *Balsam twig aphid, *Mindarus abietinus*	White fir, *Abies concolor*	N	—	69
Diptera				
35. "Pine-cone" gall midge, *Rhab-dophaga strobiloides*	Willow, *Salix chordata*	x	x	174
36. Willow gall midge, *Rhabdo-phaga terminalis*	Willows, *Salix alba, S. fragilis*	x	x	1
37. Gall needle midge, *Continaria* sp.	Douglas fir, *Pseudotsuga menziesii*	x	x	69
38. Alfalfa blotch leafminer, *Agro-myza frontella*	Alfalfa, *Medicago sativa*	x	x	40
Acarina				
39. *Spider mite, *Oligonychus subnudus*	Monterey pine, *Pinus radiata*	?	—	73
40. *Spruce spider mite, *Oligony-chius ununguis*	Spruce, *Abies*	?	—	82, 157
41. *Spider mite, *Oligonychus mil-leri*	Caribean pine, *Pinus caribeae*	?	—	70, 116
General				
42. Psyllids, leaf gallers, leaf min-ers	Birches, *Betula pubescens* and *B. pendula*			35
43. Mirids, Membracids, Cicadel-lids and Thrips	Creosote bush, *Larrea tridentata*			78

*   Species regarded as pests of trees

Note: The majority of species oviposit where larvae feed (x in the O/F column), and many species feed within plant parts (x in the W/P column). N in the O/F column signifies viviparous production of nymphs. — indicates that species does not fit the pattern. Many of the species regarded as pests of trees in the Coleoptera and Lepidoptera on conifers may become more abundant on stressed plants even though they attack young plants or relatively large plant modules. This is an interesting qualifier that needs to be developed in a finer classification of latent species.

some recent critical experimental work and other studies,[33,88,97,117] but the theory is widely, if not generally, recognized. Now we need to acknowledge another group of herbivores which are favored by unstressed plants. Many of the species listed in Table 7 may be latent species but others are regarded as pests by foresters as noted in the table. Whether any are truly eruptive species in this group we will examine later in this paper. Perhaps we have a dichotomy between latent species on vigorous plants and eruptive species on stressed plants? This raises the question of what the real differences are between latent and eruptive insect herbivores on woody plants. We attempt an answer in the next section.

IV. PHYLOGENETIC CONSTRAINTS, ADAPTIVE SYNDROMES, AND EMERGENT PROPERTIES: DIFFERENCES BETWEEN LATENT AND ERUPTIVE SPECIES

We use the perspective of evolutionary ecology and life history evolution to penetrate to the core characters which differentiate latent from eruptive species. Our approach is to identify the most constraining characters in the phylogeny of a group of insect herbivores — the **Phylogenetic Constraints** — and then to examine the suite of adaptive traits which circumvent or compensate for these constraints — the **Adaptive Syndrome** of the species. Once these are set for a species there are many inevitable properties of the population dynamics which result from, or are permitted by, the adaptive syndrome. These we call the **Emergent Properties.** We argue that this approach results in a new perspective on insect herbivore population dynamics which could alter the field considerably. A new paradigm could develop to replace the past emphasis on mortality factors and standard life table construction.

The kinds of phylogenetic constraints we emphasize are linked to the critical phases in the life history of an insect concerning female foraging and oviposition behavior, and establishment of first-instar larvae in favorable feeding sites. While we think the linkage of phylogenetic constraints, adaptive syndromes and emergent properties is new, there is a literature needing recognition which anticipates our development. An almost parallel view was developed by Gilbert[55] on *Heliconius* butterfly population and community interactions. His figure focused on individual behavior of adults which are behaviorally sophisticated because they are able to feed on pollen and so are long-lived, with a highly developed visual system and learning ability. This key set of characters helped to account for the plant-herbivore interaction, the pollination systems involved, and population and community ecology for the *Heliconius* butterflies, the *Passiflora* host plants, and the species of pollinated plants such as in the genus *Anguria*. In essence Gilbert identified an adaptive syndrome and its consequences, or emergent properties.

Root and Chaplin[151] were the first to define the term adaptive syndrome in the literature as far as we know, although Root[150] had used the term with the same meaning. "As organisms perfect a mode of life, their evolution is channeled so that a variety of adaptations are brought into harmony" (Reference 151, p. 139). Eckhardt[42] provided a more formal definition: "the coordinated set of characteristics associated with an adaptation or adaptations of overriding importance, e.g., the manner of resource utilization, predator defense, herbivore defense, etc." (p. 130). We use the term in precisely this way. Eckhardt[42] also recognized that species faced with similar problems in their evolutionary history are likely to evolve with similar adaptive syndromes so that guilds of species may possess similar adaptive syndromes. We argue the same case, not at the community level but for any insect herbivores with similar phylogenetic constraints. Thus, we focus on sets of species with similar constraints and syndromes which result in latent and eruptive species.

Nothnagle and Schultz (Reference 120, p. 71) also used the terms "phylogenetic constraints" and "suites of traits" and "syndromes" associated with eruptive forest pests. Their valuable analyses of the literature caused them to focus on the syndrome for pest species involved with female oviposition and feeding by young larvae. "Together with the loss of oviposition site selection by the female, massing offspring on individual hosts could make variable host quality (phenologically generated or otherwise) very influential." "One of the most influential environmental factors is likely to be variation in host plant quality for young larvae" (p. 74). Thus, Nothnagle and Schultz,[120] using a totally different perspective based on the pest insect literature, reached independently from our approach very similar conclusions on the key attributes in need of further understanding.

Emergent properties is not a new term (e.g., References 14, 121); only its coupling as we suggest is original.

We now focus on comparison of a latent species, *E. lasiolepis*, and a well-studied eruptive species, the eastern spruce Budworm, *Choristoneura fumiferana*. This is not an ideal comparison because one is a galler and the other is a bud and leaf feeder, and their phylogenetic backgrounds are so disparate. The advantage of this comparison is that both species are well studied and many of the relevant traits to be compared are known. Also, budworm is naturally eruptive in the absence of strong influences by man (cf. References 3, 21), so the example is uncomplicated by anthropogenic effects. A potentially stronger approach has been taken by Hanski and Otronen[60] and Hanski[59] who compare sympatric diprionid sawflies, all feeding on Scotts pine, some of which are eruptive and some latent. However, Hanski and Otronen[60] found some enigmatic properties, and cause and consequence were not clearly separable. Our approach separates cause and consequence as distinct in the adaptive syndrome-emergent properties scenario. Hanski[59] noted the link between gregarious larvae and high outbreak frequency and "the risk-prone reproductive strategy" (p. 327) of the gregarious species—the "risk concentrators" of Nothnagle and Schultz (Reference 120, p. 74). Our approach suggests that laying eggs in large clutches and/or gregariousness of larvae are emergent traits permitted by a particular adaptive syndrome, but not necessarily linked to the eruptive species' lifestyle. After all, although clusters of 10 or more eggs were characteristic of many eruptive species, 28% did not lay such large clusters, and 5 species (21%) laid single eggs, in the sample of eruptive species analyzed by Nothnagle and Schultz.[120] Therefore, although our comparison is not ideal we feel that it has provided us with a different perspective which should prove to be enlightening.

A. AN EXAMPLE OF A LATENT SPECIES

For the shoot-galling sawfly, *E. lasiolepis*, an important phylogenetic constraint is oviposition into living plant tissue, a trait common to all sawflies, hence their saw-like ovipositors. In addition, the gallers all oviposit into rapidly growing young plant parts because galling must be induced in undifferentiated meristematic tissues. The adaptive syndrome resulting from this constraint is that females evolve to effectively assess resource quality and lay eggs in high-quality sites where eggs and larvae survive best. These characters result in many emergent properties (Figure 4). (1) High-quality resources—rapidly growing shoots—are rare in a population of willow genets and within willow genets (Figure 1) so competition among females for optimal oviposition sites will be high. With spacing of eggs by females, larvae will not be likely to compete, resources are unlikely to be overexploited, host plants will not be killed, and insects are less likely to be serious pests. In addition, without serious depletion of food resources insect populations will remain steadier than in species which overexploit food resources, as in many of the eruptive species. Serious population crashes over several orders of

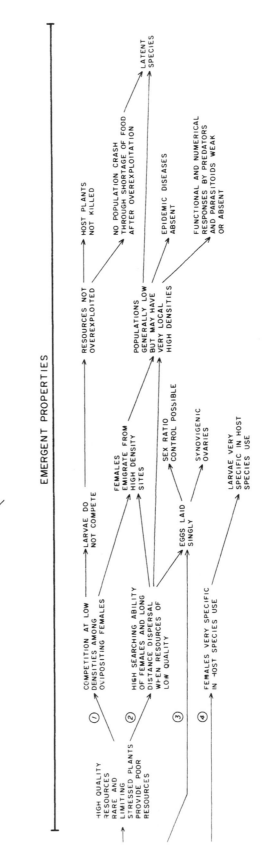

FIGURE 4. The phylogenetic constraint, adaptive syndrome, and emergent properties of the shoot-galling sawfly, *Euura lasiolepis*, a latent species. Numbers identify the lines of influence discussed in the test. The flow of influence is continuous from the adaptive syndrome to emergent properties, but the latter have been lowered to allow a larger image.

magnitude are unlikely. (2) Another pathway of effects is that females must be very efficient searchers for rare resources, and when these are seriously limiting females should disperse relatively long distances in search of patches of higher quality shoots. Together with competition among females the effect is that females emigrate from high sawfly density sites, a strongly stabilizing effect. This results in generally low densities of sawflies, but with high-density pockets on young vigorous willows in the rare wet sites in northern Arizona. Steady and generally low populations also have effects on diseases, parasitoids and predators. Epidemic disease is likely to be absent because contagious infection has a very low probability. When populations are steady we will not detect functional and numerical responses of predators and parasites through time and positive density-dependent mortality will be absent in time. (3) Another avenue of effects is that females must lay eggs singly or in very small groups of two or three at most. Being an arrhenotokous or haplodiploid species, the female can then adjust the sex of each egg for an optimal sex allocation in response to resource quality. Another result is that actively searching females, laying eggs slowly because searching time is high, are likely to have synovigenic ovaries, which produce a few mature eggs per unit time over the life of the insect. (4) The last line of effects is that natural selection for high-quality resource use will result in high specificity of females for host species use. This because special abilities are required: a) for detecting host species; b) for detecting high-quality resources within host species; c) for synchrony of emergence with rapidly developing plant tissues to enable oviposition in soft plant parts, and d) for positioning an egg accurately within those tissues. High female specificity will result in high larval specificity also because they are almost never exposed to selection for more generalized feeding ability. This scenario is summarized in Figure 4. The ramifications from the simple phylogenetic constraint and consequent adaptive syndrome are extensive, and impact practically every aspect of the insects' population dynamics. Life history evolution has dictated the population dynamics, not the other way around!

B. AN EXAMPLE OF AN ERUPTIVE SPECIES

The scenario for the eruptive eastern spruce budworm, *Choristoneura fumiferana*, is very different. Here the female has no plant-piercing ovipositor, as is typical for the Lepidoptera, and she lays eggs on foliage. She lays eggs on mature foliage in late July and early August in New Brunswick.[104] This trait is phylogenetically primitive in the genus *Choristoneura*, as all species with reported life histories lay eggs on mature foliage, and early larvae overwinter (cf. References 5, 54). This oviposition on old foliage we regard as the critical phylogenetic constraint on the budworm's life history. The larvae do not feed as first-instar larvae but spin a hibernaculum, moult to the second instar and overwinter. First-instar larvae may also drop on a silken thread and disperse to other plants.[104] Another phase of passive dispersal occurs in the spring when second-instar larvae emerge from their hibernacula. Feeding commences in the spring in opening staminate flowers, expanding vegetative buds or by mining into 1- and 2-year-old needles. Third- to sixth-instar larvae feed on newly developing foliage unless this is depleted and competition forces feeding on foliage developed in previous years.

Oviposition on old foliage, on which larvae generally do not feed, is a constraint because females cannot assess resource quality for larvae and therefore are unable to select high-quality sites (Figure 5). In addition larvae disperse passively on silken threads and again are unable to select on a large scale for high-quality foliage because they feed relatively close to the hibernaculum site, or where they land after drifting. Thus neither adult females nor larvae can evolve to be highly selective for resource quality. Natural selection will then favor larvae with a rather generalized capacity to deal with food of variable quality (e.g., Reference 94). In the budworm this means that on

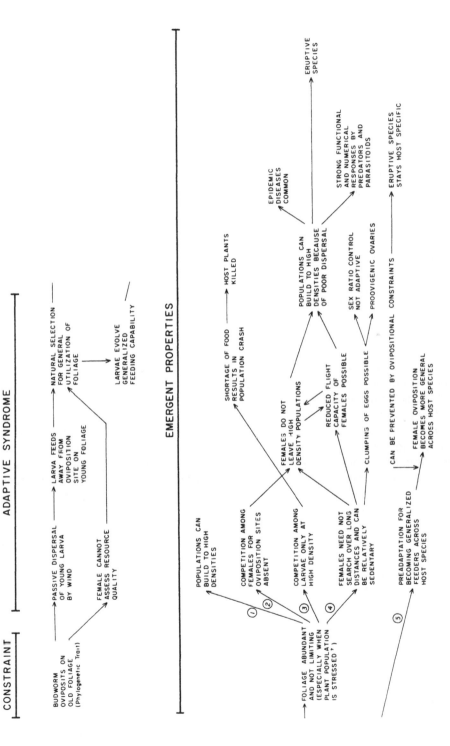

FIGURE 5. The phylogenetic constraint, adaptive syndrome, and emergent properties of the eastern spruce budworm, *Choristoneura fumiferana*, and eruptive species. Numbers identify the lines of influence discussed in the text. Emergent properties have been lowered as in Figure 4.

balsam fir almost any part of the tree's foliage can be utilized as food. This is shown by the capacity of larger larvae to feed on current shoots and those from earlier years if competition compels. Thus the adaptive syndrome for the budworm includes passive dispersal of larvae, which spreads the risk among progeny of being located at a very poor feeding site, if such exist at all, and their capacity as "generalists" on foliage quality within the host plant species. Budworm larvae fed on old fir foliage do survive although females are smaller with lower fecundity.[19,20,93,103]

Again, the emergent properties of this syndrome are very extensive. Resources are abundant for rather generalized feeders and are generally not limiting until insect densities become very high. This resource may be modified over large areas by water stress or nutrient stress. (1) This permits a population to increase to high densities under favorable conditions. (2) Oviposition sites are abundant so adult females do not compete, but lay eggs independently of others, or even aggregate on tall trees.[111] Therefore females are not likely to emigrate from dense populations, and high densities of larvae are likely to result. (3) Larvae are likely to compete but only at high densities, the shortage of food results in a population crash. Host plants are killed because larvae can feed very generally on foliage and most foliage can be consumed. (4) Females need not search intensively for oviposition sites because these sites are not critical to larval survival, and they are not limiting. Therefore females can become relatively sedentary, although not in the case of spruce budworm after half their eggs are laid.[93] They may even evolve with reduced flight capacity as in gypsy moth and winter moth. These traits reinforce the characteristics that females do not leave high-density locations leading to greater buildup of high densities. In high densities epidemic diseases can become common and significant in causing population decline. Also, strong functional and numerical responses in time by predators and parasitoids are likely and observable because such large ranges in density can be monitored. This line of effects may also permit clumping of eggs in large clutches because larvae are generalists and females cannot select high-quality sites. Laying large egg clutches probably selects for proovigenesis. If sex ratio were controlled, this would not be adaptive because sex ratio allocation could not be adjusted to high- or low-quality resources for larvae because the female cannot evaluate them. (5) Finally, larvae which can feed generally on plant resources within a host plant species are preadapted for becoming generalized feeders across host species. This would be facilitated by females which are rather unspecific in the sites they use for oviposition. Thus eruptive species may evolve to be more generalized feeders across host-plant species. However, ovipositional constraints may prevail, and prevent expansion of host-plant range. For example, if females oviposit into foliage like the diprionid sawflies, many of which are eruptive, this may well require enough specialized attributes in, for example, host recognition, phenological synchrony, ovipositor structure and oviposition behavior, that the species is constrained from broadening its host-plant range (note Table 8 for reasons why diprionid sawflies are likely to be eruptive). Thus the host range of an eruptive species may be broad or narrow and will not act as a diagnostic character for separating eruptive and latent species. In Nothnagle and Schultz's[120] sample of eruptive species larval diet breadth ranged from 1 host-plant species only to 60 host plant species, although many species were broadly polyphagous. Redfearn and Pimm[144] found no positive correlations between polyphagy and population variability, as one would predict if eruptive species were generally polyphagous.

Note that the adaptive syndrome is largely permissive, not deterministic, for the emergent properties. The syndrome allows populations to reach high densities; it does not drive them to high densities. Sooner or later in a permissive system the full range of possibilities is likely to be achieved and eruptions will occur. In this system we have

Table 8
EXAMPLES OF CHARACTERS ASSOCIATED WITH ERUPTIVE SPECIES[a]

Species	Characteristics
1. Spruce budworm, *Choristoneura fumiferana* (Tortricidae)	Oviposits on old foliage, larvae disperse passively, resource quality cannot be evaluated by female
2. Winter moth, *Operophtera brumata* (Geometridae)	Oviposition during winter when resource quality cannot be assessed; females flightless
3. Gypsy moth, *Porthetria dispar* (Lymantriidae)	Oviposition on bark, females do not fly, larvae disperse passively
4. Forest tent caterpillar, *Malacosoma disstria* (Lasiocampidae)	Oviposition on bark, larva feeds in the following spring
5. Jack pine sawfly, *Neodiprion swainei* (Diprionidae)	Female poor flyer, proovigenic with heavy load of eggs, oviposits on young needles, larvae feed on old needles (personal observations by PWP)
6. Fall cankerworm, *Alsophila pometaria* (Geometridae)	Female wingless, oviposition in winter, passive dispersal of larvae

[a] Based on information in References 5 and 54.

not attempted to isolate the conditions under which eruptions will occur as these are not yet clear and may be idiosyncratic for each species. Climatic release, host-plant stress or their interaction may well be important in many cases. Our scheme must also remain permissive because many insect species related to eruptive species are latent for reasons which are unknown. However, our contrast between the highly **restrictive system** for latent species, and the **permissive system** for eruptive species should help in focusing on the essential differences between related species which are latent and eruptive. The eruptive species' adaptive syndrome permits the evolution of the **risk concentrators,** to use Nothnagle and Schultz'[120] term, and the latent species' adaptive strategy restricts a species to be a **risk spreader.** Rhoades[148] recognized similar categories, which he called **opportunistic herbivores** and **stealthy herbivores,** respectively. Neither of the adaptive syndromes is so restrictive that a clear set of traits, which we regard as emergent properties, is always associated with each syndrome. For example both Rhoades[148] and Wallner[171] regard low fecundity and high fecundity as traits associated with latent and eruptive species, respectively, but this pattern is not supported by empirical data.[118,120]

Because of their fundamental differences we see very different dynamical properties between eruptive and latent species. Eruptive species show large amplitude fluctuations in density, and latent species show low amplitude fluctuations (Figure 6). Some of the eruptive-species cycles were summarized by McLeod:[100] black-headed budworm, eastern spruce budworm, and gray larch tortrix. He contrasted these with Swaine jack pine sawfly with relatively low amplitude changes compared to the 250 to 20,000 times differences in other eruptive species. However, McLeod[98-100] studied only epidemic populations, and not endemic populations, so we have modified McLeod's illustration for the Swaine jack pine sawfly to include the endemic situation. We do not think that periodicity is necessarily as illustrated, but wish only to emphasize that amplitude changes are probably much larger than the 68 times estimated by McLeod for epidemic populations.

These eruptive species contrast markedly with the one latent species in Figure 6, the arroyo willow shoot galler. Here we have taken the data in Figure 3 and translated it into densities per 1000 square feet (+1) for direct comparison with the eruptive species. We illustrate only a general impression of population change rather than the actual data, and have extrapolated it to the 32 years for comparison with the other

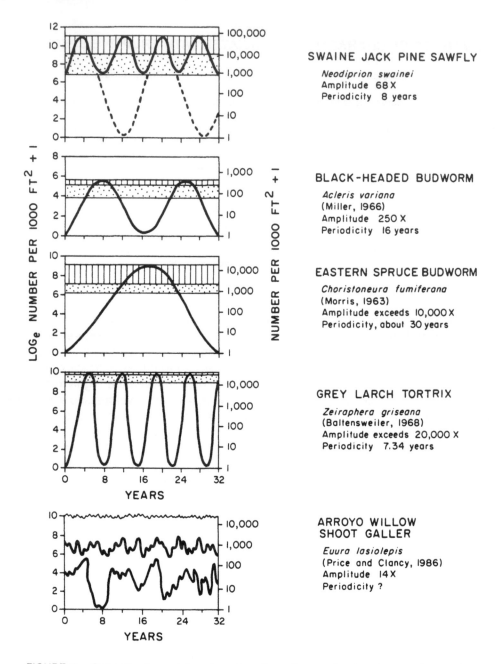

FIGURE 6. Generalized population dynamics of eruptive species (top four species) and a latent species (bottom species). For each species the amplitude of change and periodicity are given, and the ranges over which severe defoliation (vertical bars) and moderate defoliation (stippled) occur. For the arroyo willow shoot galler no defoliation occurs, and illustrations of typical densities are given for high water sites (highest densities), the mean for all sites (moderate densities), and for clones with the lowest densities observed, which have old ramets on dry sites. The eruptive species are based on McLeod[100] although the figure for the Swaine jack pine sawfly has been modified to include endemic populations not studied by McLeod (see text for details). *Euura lasiolepis* populations are based on Price and Clancy[133] and unpublished data.

species. With only 8 years data we cannot say with certainty that populations do not cycle, or that they do not fluctuate much more than we have observed. However, many lepidopteran species cycle within the range of 6- to 16-year periodicity,[118] so we should have captured much of the density range, if periodicity is not unusually long as in the eastern spruce budworm (cf. References 100, 118).

C. GENERAL CASES FOR LATENT AND ERUPTIVE SPECIES

Are these patterns for latent and eruptive species broadly general? The fundamental difference between latent and eruptive species according to our scenarios is that females of latent species can evaluate plant quality as a resource for larvae, and females of eruptive lepidopteran species cannot. Lack of evaluation by discriminating oviposition behavior by eruptive species females occurs because oviposition is away from larval feeding sites either in time or space and usually both. Some examples are given in Table 8. Common traits in the Lepidoptera are oviposition in winter, as in the winter moth, oviposition on twigs, as in the forest tent caterpillar, flightless females or poor fliers as in gypsy moth and fall cankerworm, and passive dispersal of larvae, as in spruce budworm and gypsy moth. Thus the scenario in Figure 4 fits the kinds of species listed in Table 8. The remarkable fact is that of the species of Macrolepidoptera listed by Nothnagle and Schultz[120] as eruptive species (both in their Table 2 and Appendix B), 97% of species fit the eruptive scenario, for species with enough life history information to enable a judgment (they list 41 species; 80% or 33 species have adequate information given in Baker[5] or Furniss and Carolin,[54] and 32 species fit the eruptive species scenario). For some clear reason relating to oviposition behavior and larval feeding, every species showed a reduced capacity of females to evaluate food quality for larvae, or this capacity was absent. The only species of questionable relationship to the eruptive species scenario was *Heterocampa manteo* for which there is a marginal amount of life history information. However, this species lays about 500 eggs in clusters of 30 to 300 on leaves,[189] suggesting that such high fecundity would prevent a female from making detailed decisions of preference on where to lay eggs. Hence, the species listed in Table 8 are representative of many others capable of outbreaks.

All the latent species we are most familiar with oviposit on or in living plant tissue at the site where larvae must establish and survive (see also Table 7). Females select oviposition sites based on resource quality. Some examples are given in Table 9, and we suspect that species listed in Tables 5 and 6 all fit the pattern, except of course that the aphids are viviparous but still reproduce during the summer on living plant tissues. Research by Whitham[181-185] on *Pemphigus betae*, which forms galls on narrowleaf cottonwood, shows patterns similar to the scenario for latent species (Figure 4). Females select the largest leaves on a tree for gall formation, they are territorial for the best oviposition sites where progeny production is highest, and populations are stable from year to year (T. G. Whitham, personal communication).

We have noted that many herbivores attack vigorously growing plants (Table 7) and many of them are regarded as pests of trees, for they are listed in Baker[5] or Furniss and Carolin.[54] Many of the shoot borers, for example, are highly selective of large vigorous shoots and oviposit preferentially on these. An example is the stem- and cone-boring moth, *Dioryctria albovitella*.[106,107,187] Females select vigorous young shoots which bear female cones, kill the shoots and thereby reduce female sex expression. Some sites have rather stable high populations and some low populations, and there is a genetic basis for plant resistance and susceptibility within sites.[107] These kinds of species behave with latent species characteristics.

Why are they pests? "We note that the literature suggests that many of the present-day forest pest species have become so only recently, since anthropogenic forest

Table 9
EXAMPLES OF CHARACTERS ASSOCIATED WITH LATENT SPECIES

Species	Characteristics
1. Arroyo willow shoot galler, *Euura lasiolepis* (Tenthredinidae)	Oviposits on young shoots where larva must survive, resource quality is evaluated by ovipositing female and the longest shoots are utilized preferentially[27,30,133]
2. Coyote willow shoot galler, *Euura exiguae* (Tenthredinidae)	Oviposits on young shoots where larva must survive, resource quality is evaluated by ovipositing female and the longest shoots are selected[132]
3. Willow bud galler, *Euura mucronata* (Tenthredinidae)	Oviposits into very young buds where larva must establish and survive, resource quality is evaluated by ovipositing female and buds on long shoots are selected[136,137]
4. Narrowleaf cottonwood gall aphid, *Pemphigus betae* (Aphidae)	Fundatrix (stem mother) selects gall site on very young leaf, resource quality is evaluated and the largest leaves are utilized[181-183]
5. Birch leafrolling weevil, *Rhynchites betulae* (Curculionidae)	Females roll large leaves in which larvae feed, resource quality is evaluated by female adult and large leaves on vigorous shoots are selected (personal observations by PWP)
6. Rose ball galler, *Diplolepis spinosa* (Cynipidae)	Oviposits on young shoots where larva must survive, resource quality is evaluated by adult female and vigorous shoots are selected[22]

habitation modification has become widespread" wrote Nothnagle and Schultz (Reference 120, p. 74). We believe this is particularly true for the kinds of species listed in Table 7. Young vigorous plants in a primeval forest would be much rarer than they are under current logging practices, which change dramatically the age structure of tree populations. Regeneration would be patchy due to local disturbance. Fires would result in larger areas of regeneration, but these would be intermittent in time and very episodic (cf. Reference 173). Hence resources for insect herbivore species requiring young vigorous growth would be generally rare and hard to find. Under these natural conditions, the herbivores would most likely be latent species and inconspicuous dwellers in the primaeval forest. With large-scale disturbance latent species may erupt if the colonizing populations have not been decimated during disturbance, as in the cynipid galler attacking young oak shoots after fire.[173]

An important general difference between latent and eruptive species appears to be in the form that competition appears. In latent species adult females are likely to compete for oviposition sites, even at low densities because the highest quality sites are rarest. In eruptive species larvae are likely to compete for food but only at high densities. This may account for the arguments made by Lawton and Strong[76] and Strong et al.[164] that competition among herbivorous insects is weak or absent. The eruptive species are uncommon, and larval competition, which is usually the focus of a study, is unlikely in latent species because females space out oviposition. However, adult female competition may be exceedingly important in latent species although it is less frequently studied. Thus we may have overlooked an important aspect of the life cycle and reached inaccurate conclusions. Certainly, territoriality among insect herbivores is common enough (e.g., Reference 129, 185), and is very well developed between females in such latent species as *Pemphigus betae*[182] and *E. lasiolepis*.[28,31] In the latter case it is actually the plant that produces a defensive chemical at the oviposition site, but the effect is interference competition, even though a female does not defend the site, as a strict use of the term territoriality would require. This emphasis on adult female competition should cause a reevaluation of the role of competition among herbivores which differs from the views expressed currently (e.g., Reference 76, 164). Competition even at low densities may be much more common than our present views admit (see especially References 185 and 186).

Although we have concentrated on galling insect herbivores as latent species we think the link to a more general picture of latent species can be made through the kinds of species listed in Table 7. These have many traits in common with the gallers. (1) They attack young, usually vigorously growing plants and young tissues so that female adults can evaluate resource quality for larvae. (2) Phenology of adult flight is tightly linked with host plant phenology. (3) Females are likely to compete for the best oviposition sites even at low densities. (4) There are steep gradients in resource quality within natural plant populations with constant directional selection on insect herbivores for utilization of rare high-quality resources. (5) Populations are relatively steady at any one site because resources are so limiting.

If we are to develop a strongly comparative understanding of the population dynamics of latent and eruptive species we need to focus on a method that enables the full range of characters in both types of species to be evaluated. Because so many of the dynamical properties of latent species are dependent on female behavior and plant quality, the present construction of life tables is not adequate. We feel the life table approach is still appropriate and the formality of life table construction is ideal for broad comparative work. Therefore, we turn to a reinterpretation of life table analysis and the changes needed for more comprehensive treatment of essential traits of insect herbivore populations.

V. A REINTERPRETATION OF LIFE TABLE STUDIES

A. AN EVOLUTIONARY PERSPECTIVE

Our major argument is that life table studies have concentrated on emergent properties and not on the ultimate and central forces influencing population dynamics— the phylogenetic constraints and the adaptive syndromes. Life table studies have concentrated on results, many of them inevitable, and not on causes. In fact the "tail" of emergent properties has wagged the "dog" of real effects hidden in the life history evolution of the insect species.

Recognition of the central issue as life history evolution, of course, could come only with the benefit of an evolutionary approach to ecology. This approach was emerging slowly only as the peak in insect population dynamics was being reached (cf. Reference 25). In fact, Collins[25] argued that the issues of density-independent and density-dependent regulation, hotly debated in the 1950s, provided a beginning to evolutionary ecology in North America when Orians[123] argued for recognition of both ecological causes and evolutionary causes—the proximate and ultimate factors clearly differentiated by Mayr.[96] This is precisely the dichotomy we argue has been missed in most studies on the population dynamics of insect herbivores. These are **ultimate factors** in the evolution of a species which force, dictate, or permit many of the **proximate factors** observed in ecological relationships. We cannot afford to disconnect the ultimate and proximate factors in insect population dynamics because they interact, as Hutchinson[65] emphasized in the title of his book, "The Ecological Theater and the Evolutionary Play." Bringing to light both the evolved traits of species and the ecological traits, and giving them equal emphasis illuminates insect herbivore population dynamics in a different and new way.

The life table is an ecological record, not an evolutionary one. So how can such a table unite the essential aspects of both the evolved traits and ecological traits of populations? Our scenarios in Figures 4 and 5 of latent and eruptive species provide the evolutionary perspective. This provides insight on the critical ecological interactions in need of detailed study which can be recorded in life table format. The life table should be designed to measure these critical interactions, rather than the emergent properties that have been studied traditionally. We will discuss now some of the issues we think are critical and how new studies may cause us to change the conventional wisdom.

B. FEMALE BEHAVIOR

At least for latent species we have argued that female behavior is of paramount importance in the population dynamics of the species we know best. By incorporating this behavior into life tables, using the cohort of eggs in searching females as the beginning of the life table this attribute of the population, so responsive to plant quality, is quantified well.[127] Competition for oviposition sites which reduced the egg cohort in the plants also needs to be evaluated, for competition between females is probably important in latent species.[31,182] In this way the most important differences between latent and eruptive species may be quantified.

Perhaps eruptive lepidopteran species have very little of their dynamical attributes tied to female behavior. Females are usually not involved in detailed selection of oviposition sites, and they usually lay many eggs rapidly. Therefore the appropriate start to the life table may well be the cohort of eggs on the plant populations, as in the original design of Morris and Miller.[113] This needs to be studied. For bark beetles, female behavior is crucial.

C. RESOURCE QUANTITY AND QUALITY

The role of plant growth and architecture and the resource quantity and quality provided for a herbivore population is critical for latent species, for this is what female behavior evaluates and responds to. Thus resource quantity and quality is reflected in the life table by the female behavior. However, to interpret the female behavior adequately we feel detailed studies on plant resources are needed. Where latent species respond positively to rapid growth, large leaves, long shoots, and so on, all probably closely correlated (e.g., see Reference 136 for correlated traits on *Salix cinerea*), it is probably more important to measure module size than chemical attributes. This can be done rapidly and easily and we think the growing awareness of plant architecture as an important ingredient of insect herbivore population dynamics will be a valuable expansion of our views (e.g., References 27, 108, 149).

Can we estimate what the carrying capacity per unit area is for insect populations dictated by plant resource quantity and quality? We can probably make predictions on potential population size per genet or plant population based on resource quantity and quality if these are influenced strongly by modular structure, such as shoot length. Because there is likely to be a hierarchical attack of high-quality modules before low-quality modules,[30] we can probably gauge quite accurately how close the population is to a carrying capacity. This capacity will be well below the level where damage to the plant approaches a lethal threshold or even becomes very conspicuous (cf. Reference 155).

For eruptive species, with more generalized larval feeding, the chemical attributes of the plant resources may well be more important than modular structure. If plant stress changes the chemical properties then such stress may play a positive role in permitting an eruption.[90,91] The carrying capacity of the plant population may well become all the foliage, as in spruce budworm[111] and Swaine jack pine sawfly.[98-100]

Resource quantity and quality cannot be included in the life table for the insect, but evaluation is essential in a mechanistic understanding of that life table. It is possible to erect life tables for plant modules when such plant parts are critical to the herbivore, as with many latent species. Such a life table would define accurately the change in carrying capacity in the plant population which can be considerable from year to year (e.g., Reference 133). Birth and death of high-quality plant parts, on resistant and susceptible plants in a population, may also be compared in a life table format.

D. WEATHER

We still do not understand the relative importance of direct effects of weather on insect populations and indirect effects via plant responses to weather. The climatic release hypothesis (e.g., References 56, 87, 162) has so many characteristics correlated with the plant stress hypothesis (e.g., References 90, 91) that separating them is difficult. However, in latent species which require vigorously growing plants precipitation may often be critical in supporting high-quality modules, plants, and plant populations. The indirect effects of weather may well be more important than the direct effects. For example, winter precipitation influenced the number of shoots initiated in the spring and the length of shoots on *S. lasiolepis* clones. The "resource index" based on number of shoots and mean shoot length was 2.5 times higher after a winter with high snowfall than after little snow on the same willow clones.[133]

In latent species that evaluate plant module quality, the effects of weather may be easily estimated by measuring module size. Shoot length would be an easy plant quality to measure, and would yield a direct estimate of the level of plant stress, so seldom used in insect herbivore population studies. For example White[175] proposed an estimate based on variation in precipitation that is exceedingly indirect, but quite typical of the literature.

Thus, we think that weather may work indirectly through plant quality in many latent species, as it does in bark beetles. Effects are likely to change plant resistance and the shape of the survivorship curve derived from the life table (see Reference 133). If eruptive species are commonly more exposed feeders then weather may well have more direct effects on development. This has been demonstrated elegantly for the fall webworm by Morris (Reference 112 and many other papers summarized in Reference 129).

Weather as a killer of insects in a population, usually emphasized in a life table analysis, may well become less important as we understand better the role of weather in changing plant quality. Weather may play very different roles in the dynamics of latent and eruptive species. Evaluating the role of weather in changing plant resource quantity and quality is therefore an essential ingredient for interpreting life tables.

E. COMPETITION

Larval competition and consequent starvation contribute to the crash of epidemic populations in many eruptive species, but competition among larvae in latent species may be rare or absent. The competition between ovipositing females in latent species has not been adequately evaluated and may prove to be a very important impediment to natality at high population levels, or where resources are very limiting. Thus the timing of competition in the life cycle of latent and eruptive species may be different. With typical life table construction this would be missed because adult competition is not evaluated.

F. NATURAL ENEMIES

In the galling sawflies we have shown how passive natural enemies are in the system. The windows of vulnerability of sawflies to parasitoids are short[32,134] and the plant interferes too much with access to insect hosts for parasitoids to become abundant or important in driving dynamics. They have become opportunists that attack when possible. They respond to dynamical trends in sawfly populations but do not cause them.

When the life table starts with eggs in searching females and much loss of this cohort results from female behavior and failure of larvae to establish in plant tissues, there is a relatively minor role for natural enemies to play. The egg cohort of *E. lasiolepis* was reduced to 14.9% in the low-water treatment experiment before parasitoids or predators could attack eggs or larvae (Table 4). We suspect that this will be found commonly in latent species.

We do not have a clear understanding of the role of predators and parasitoids or diseases even in eruptive species.[44,130] Much more experimental work is needed in evaluating their role. However, if females of eruptive species are often proovigenic and lay eggs rapidly there is much more possibility for effects of natural enemies to be important in dynamics. The scenario allows natural enemies to become abundant. Whether they follow host populations or drive them is in need of much more careful study. Evaluating these factors using life tables starting with ovipositing females will help place effects of natural enemies in a more comprehensive perspective.

G. NATALITY OR MORTALITY?

If eruptive species typically have females which lay their full complement of eggs rapidly, independently of plant quality, then the life table format developed by Morris and Miller[113] may well capture the essence of dynamical properties. Mortality factors will indeed be very important. Conversely, in latent species loss of potential eggs may be the primary factor in dynamics and natality needs more emphasis in life table studies. This would force more focus on plant qualities that dictate decisions to lay or withhold eggs (see also Reference 122). Even in eruptive species loss of egg potential may be critical as in the dynamics of bark beetles, with natural enemies only modifying, but not driving, dynamics (e.g., Reference 11). Therefore, the role of natality needs to be evaluated carefully in eruptive species also.

We hope that our new approach emphasizing phylogenetic constraints, adaptive syndromes and emergent properties, and the relevant changes needed in life table construction, will stimulate more critical evaluation of the differences between latent and eruptive species. We do not pretend to have answered the riddle. Many exceptions to our scenarios are likely to be identified by researchers more familiar with other taxa of insect herbivores. However, the exceptions may be better understood by using our approach and the concepts may then be broadened to accommodate them. Selecting two extremes in dynamical types obviously oversimplifies nature, but from two simple contrasts perhaps we can build a more complex and natural picture. Perhaps we need to include species that are eruptive in one location and more latent in others, for example, the gray larch tortrix, *Zeiraphera griseana*.[6,7] We need to recognize different kinds of eruptive species as Berryman and Stark[17] and Berryman[13] have done, and different kinds of latent species. However, much more work on latent species is needed, and many more directly comparative studies of sympatric latent and eruptive species with similar phylogenetic background are required before a comprehensive picture emerges. We are confident, however, that the study of latent species will greatly improve our understanding of eruptive species. The economic and heuristic incentives for more comparative work are evident.

VI. CONCLUSIONS

Identification of the most important **phylogenetic constraint** in the life cycle of an insect herbivore is the key to linking the evolutionary and ecological perspectives in population dynamics and permits identification of fundamental differences which result in radically divergent dynamical properties. Each species adapts in response to the phylogenetic constraint, and the resultant **adaptive syndrome** sets the limits and possibilities for population change. Dynamical behavior of a population is therefore constrained by the evolutionary ecology of the species and is manifested as **emergent properties** dictated or permitted by the phylogenetic constraint and adaptive syndrome.

The differences between latent and eruptive insect herbivore species on woody plants can be traced to major differences in phylogenetic constraints. Latent species oviposit into rapidly growing plant tissue where larvae must establish and survive. Females can evaluate plant quality. Ideal plant parts are rare and constrain the species to steady densities typical of latent species. Eruptive species oviposit away from the site where larvae establish and feed, or for other reasons cannot evaluate plant quality for larvae. Larvae then evolve the capacity to feed on a wide range of plant quality such that food is very abundant per unit area even though it may be suboptimal. The adaptive syndrome of these eruptive species permits a population to reach very high epidemic levels.

The emergent properties in population dynamics have typically been recorded in life table analysis, and have been the focus of attention in efforts to understand insect herbivore dynamics. However, these are results mediated by the phylogenetic constraints and adaptive syndromes, not causes. Thus, for eruptive species, it is almost inevitable that diseases, parasitoids and predators kill many insects and respond to high densities, although they may influence major dynamical properties very little. Conversely, latent species will seldom be subject to disease epidemics because contagion is not a property of such species. Positive density dependence will also be hard to detect because of the stability of populations through time. At least for latent species the forces affecting dynamics are the links between female behaviors in response to plant quality which dictate the number of potential eggs laid into high-quality plant modules where larvae establish and survive well. Female behavior may be much less critical in eruptive species because they usually do not select the feeding site for larvae. Thus the capacity of larvae to survive on a wide range of food quality may be of paramount importance.

Life table analysis has tended to minimize concern for the interaction between plant quality and female behavior, and maximize focus on mortality factors. But plant quality and female behavior dictate the natality of the population, and may be the most critical factors in the dynamics of many species. Life tables therefore need to include the results of female behavior involved with decisions during foraging for ovipostion sites and oviposition itself. When these factors are included the capacity for, say, competition or natural enemies to play a major role in population dynamics is diminished.

The generality of these arguments is in need of more examination, for we have not captured all the elements concerned with differences between latent and eruptive species. Some latent species have phylogenetic constraints identical to those of eruptive species, so factors other than those we emphasize must be involved. For example, not all members of genera listed by Nothnagle and Schultz[120] are eruptive species. There are more kinds of latent and eruptive species than we recognize, no doubt. We have clearly delineated the forces preventing latent species from erupting, but we have not identified the causes for epidemics of eruptive species. We simply emphasize that the adaptive syndrome of eruptive species is very permissive in allowing populations to increase, while that of latent species is very constraining. The degree to which our arguments will apply to herbivores on herbs and grasses needs exploration. Similar mechanisms to those we suggest should prevail, although much more study of latent species on herbs and grasses is needed before more general patterns to include them can be developed.

More comparative studies on latent and eruptive species in the same area on the same host plants will be rewarding. We believe that the new approaches we advocate using the progression from phylogenetic constraints, to adaptive syndromes, to emergent properties, coupled with new life table design to incorporate effects of female behavior, will aid in the development of comparative insect herbivore dynamics studies.

ACKNOWLEDGMENTS

We thank Alan A. Berryman and William J. Mattson for their important help with a former draft of this paper, without implying that they endorse all arguments we have developed. Their extensive reviews and knowledge, and the time invested are deeply appreciated. Financial support has been provided by the National Science Foundation for research on *Euura* species and, during the preparation of this paper, through grants DEB-8021754, BSR-8314594, and BSR-8705302. Peter Price thanks R. Frank Morris for stimulating interest in population dynamics while being an advisor on my Master's dissertation committee, and D. Russ Macdonald for employing me as a technician to study spruce budworm in New Brunswick 25 years ago.

REFERENCES

1. **Åman, I.,** Oviposition and larval performance of *Rhabdophaga terminalis* on *Salix* spp. with special consideration of bud size of host plants, *Entomol. Exp. Appl.,* 35, 129, 1984.

2. **Amman, G. D. and Cole, W. E.,** Mountain pine beetle dynamics in lodgepole pine forests. II. Population dynamics, USDA For. Serv. Gen. Tech. Rep. INT-145, 1983.

3. **Anderson, R. S., Davis, R. B., Miller, N. G., and Stuckenrath, R.,** History of late- and post-glacial vegetation and disturbance around Upper South Branch Pond, northern Maine, *Can. J. Bot.,* 64, 1977, 1986.

4. **Askew, R. R. and Shaw, M. R.,** Mortality factors affecting the leaf-mining stages of *Phyllonorycter* (Lepidoptera: Gracillariidae) on oak and birch. I. Analysis of the mortality factors, *Zool. J. Linn. Soc.,* 67, 31, 1979.

5. **Baker, W. L.,** Eastern forest insects, U. S. Dep. Agric. For. Serv. Misc. Pub. 1175, pp. 1-642, 1972.

6. **Baltensweiler, W.,** The cyclic population dynamics of the Grey Larch Tortrix, *Zeiraphera griseana* Hübner (= *Semasia diniana* Guenée) (Lepidoptera:Tortricidae), in *Insect Abundance,* Southwood, T. R. E., Ed., Blackwell Scientific, Oxford, 1968, 88.

7. **Baltensweiler, W.,** The relevance of changes in the composition of larch bud moth populations for the dynamics of its members, in *Dynamics of Populations,* den Boer, P. J. and Gradwell, G. R., Eds., Centre for Agricultural Publishing and Documentation, Wageningen, 1970, 208.

8. **Barbosa, P. and Schultz, J. C., Eds.,** *Insect Outbreaks,* Academic Press, New York, 1987.

9. **Beaver, R. A.,** The development and expression of population tables for the bark beetle *Scolytus scolytus* (F.), *J. Anim. Ecol.,* 35, 27, 1966.

10. **Berryman, A. A.,** Population dynamics of the fir engraver, *Scolytus ventralis* (Coleoptera: Scolytidae). I. Analysis of population behavior and survival from 1964 to 1971, *Can. Entomol.,* 105, 1465, 1973.

11. **Berryman, A. A.,** Population dynamics of bark beetles, in *Bark Beetles in North American Conifers,* Mitton, J. B. and Sturgeon, K. B., Eds., University of Texas Press, Austin, 1982, 264.

12. **Berryman, A. A.,** *Forest Insects: Principles and Practice of Population Management,* Plenum Press, New York, 1986.

13. **Berryman, A. A.,** The theory and classification of outbreaks, in *Insect Outbreaks,* Barbosa, P. and Schultz, J. C., Eds., Academic Press, New York, 1987, 3.

14. **Berryman, A. A.,** Equilibrium or nonequilibrium: Is that the question?, *Bull. Ecol. Soc. Am.,* 68, 500, 1987.

15. **Berryman, A. A., Ed.,** *Dynamics of Forest Insect Populations: Patterns, Causes, Implications,* Plenum Press, New York, 1988.

16. **Berryman, A. A.,** Towards a unified theory of plant defense, in *Mechanisms of Woody Plant Defenses Against Insects: Search for Pattern,* Mattson, W. J., Levieux, J., and Bernard-Dagan, C., Eds., Springer, New

York, 1988, 39.

17. **Berryman, A. A. and Stark, R. W.,** Assessing the risk of forest insect outbreaks, *Z. Ang. Entomol.,* 99, 199, 1985.

18. **Berryman, A. A., Stenseth, N. C., and Isaev, A. S.,** Natural regulation of herbivorous forest insect populations, *Oecologia,* 71, 174, 1987.

19. **Blais, J. R.,** The relationship of the spruce budworm *(Choristoneura fumiferana* (Clem.) to the flowering condition of balsam fir *(Abies balsamea* (L.) Mill.), *Can. J. Zool.,* 30, 1, 1952.

20. **Blais, J. R.,** Effects of the destruction of the current year's foliage of balsam fir on the fecundity and habits of flight of the spruce budworm, *Can. Entomol.,* 85, 446, 1953.

21. **Blais, J. R.,** Regional variation in susceptibility of eastern North American forests to budworm attack based on history of outbreaks, *For. Chron.,* 44, 7, 1968.

22. **Caouette, M. R. and Price, P. W.,** Growth of Arizona rose and attack and establishment of gall wasps *(Diplolepis fusiformans* (Ashmead) and *D. spinosa* (Ashmead) (Hymenoptera: Cynipidae), *Environ. Entomol.,* 18, 822, 1989.

23. **Chansler, J. F.,** Biology and life history of *Dendroctonus adjunctus* (Coleoptera: Scolytidae), *Ann. Entomol. Soc. Am.,* 60, 760, 1967.

24. **Clark, L. R., Geier, P. W., Hughes, R. D., and Morris, R. F.,** *The Ecology of Insect Populations in Theory and Practice,* Methuen, London, 1967.

25. **Collins, J. P.,** Evolutionary ecology and the use of natural selection in ecological theory, *J. Hist. Biol.,* 19, 257, 1986.

26. **Coulson, R. N.,** Population dynamics, in *The Southern Pine Beetle,* Thatcher, R. C., Searcy, J. L., Coster, J. E., and Hertel, G. D., Eds., USDA For. Serv. Sci. Educ. Admin. Tech. Bull. 1631, 71, 1980.

27. **Craig, T. P., Price, P. W., and Itami, J. K.,** Resource regulation by a stem-galling sawfly on the arroyo willow, *Ecology,* 67, 419, 1986.

28. **Craig, T. P., Itami, J. K., and Price, P. W.,** Plant wound compounds from oviposition scars used as oviposition deterrents by a stem-galling sawfly, *J. Insect Behav.,* 1, 343, 1988.

29. **Craig, T. P., Price, P. W., Waring, G., Clancy, K., and Sacchi, C.,** Forces preventing coevolution in the three-trophic-level system: willow, a gall-forming herbivore, and parasitoid, in *Chemical Mediation of Coevolution,* Spencer, K., Ed., Academic Press, New York, 1988, 57.

30. **Craig, T. P., Itami, J. K., and Price, P. W.,** A strong relationship between preference and performance in a shoot-galling sawfly, *Ecology,* 70, 1691, 1989.

31. **Craig, T. P., Itami, J. K., and Price, P. W.,** Intraspecific competition and facilitation by a shoot-galling sawfly, *J. Anim. Ecol.,* 59, 147, 1990.

32. **Craig, T. P., Itami, J. K., and Price, P. W.,** The window of vulnerability of a shoot-galling sawfly to attack by a parasitoid, *Ecology,* in press, 1990.

33. **Craig, T. P., Wagner, M. R., McCollough, D. G., and Frantz, D. P.,** Effects of experimentally altered plant moisture stress on the performance of *Neodiprion* sawflies, *For. Ecol. Manage.,* in press, 1990.

34. **Dahlsten, D. L.,** Preliminary life tables for pine sawflies in the *Neodiprion fulviceps* complex (Hymenoptera: Diprionidae), *Ecology,* 48, 275, 1967.

35. **Danell, K. and Huss-Danell, K.,** Feeding by insects and hares on birches earlier affected by moose browsing, *Oikos,* 44, 75, 1985.

36. **DeMars, C. J., Dahlsten, D. L., and Stark, R. W.,** Survivorship curves for eight generations of the western pine beetle in California, 1962-1965, and a preliminary life table, in *Studies on the Population Dynamics of the Western Pine Beetle Dendroctonus brevicomis* LeConte (Coleoptera: Scolytidae), Stark, R. W. and Dahlsten, D. L., Eds., University of California, Division of Agric. Sci., Berkeley, 1970, 134.

37. **Deevey, E. S.,** Life tables for natural populations of animals, *Q. Rev. Biol.,* 22, 283, 1947.

38. **den Boer, P. J. and Gradwell, G. R., Eds.,** *Dynamics of Populations,* Centre for Agricultural Publishing and Documentation, Wageningen, 1970.

39. **Denno, R. F. and McClure, M. S., Eds.,** *Variable Plants and Herbivores in Natural and Managed Systems,* Academic Press, New York, 1983.

40. **Drolet, J. and McNeil, J. N.,** Performance of the alfalfa blotch leaf miner, *Agromyza frontella* (Diptera: Agromyzidae), on four alfalfa varieties, *Can. Entomol.,* 116, 795, 1984.

41. **Eaton, C. B.,** Biology of the weevil, *Cylindrocopturus eatoni* Buchanan, injurious to ponderosa and Jeffrey pine reproduction, *J. Econ. Entomol.,* 35, 20, 1942.

42. **Eckhardt, R. C.,** The adaptive syndromes of two guilds of insectivorous birds in the Colorado Rocky Mountains, *Ecol. Monogr.,* 49, 129, 1979.

43. **Eberhardt, L. L.,** Correlation, regression, and density dependence, *Ecology,* 51, 306, 1970.

44. **Ewald, P. W.,** Pathogen-induced cycling of outbreak insect populations, in *Insect Outbreaks,* Barbosa, P. and Schultz, J. C., Eds., Academic Press, New York, 1987, 269.

45. **Finnegan, R. J.**, The pine weevil, *Pissodes approximatus* Hopk. in southern Ontario, *Can. Entomol.*, 90, 348, 1958.

46. **Finnegan, R. J.**, The pales weevil, *Hylobius pales* (Hbst.) in southern Ontario, *Can. Entomol.*, 91, 664, 1959.

47. **Finnegan, R. J.**, The pine root-collar weevil, *Hylobius radicis* Buch. in southern Ontario, *Can. Entomol.*, 94, 11, 1962.

48. **Fleischer, S. J. and Gaylor, M. J.**, *Lygus lineolaris* (Heteroptera: Miridae) population dynamics: nymphal development, life tables, and Leslie matrices on selected weeds and cotton, *Environ. Entomol.*, 17, 246, 1988.

49. **Fowler, S. V.**, Differences in insect species richness and faunal composition of birch seedlings, saplings and trees: the importance of plant architecture, *Ecol. Entomol.*, 10, 159, 1985.

50. **Frankie, G. W. and Morgan, D. L.**, Role of the host plant and parasites in regulating insect herbivore abundance, with emphasis on gall-inducing insects, in *A New Ecology: Novel Approaches to Interactive Systems*, Price, P. W., Slobodchikoff, C. N., and Gaud, W. S., Eds., John Wiley & Sons, New York, 1984, 101.

51. **Fritz, R. S., Gaud, W. S., Sacchi, C. F., and Price, P. W.**, Variation in herbivore density among host plants and its consequences for community structure: field studies on willow sawflies, *Oecologia*, 72, 577, 1987.

52. **Fritz, R. S., Gaud, W. S., Sacchi, C. F., and Price, P. W.**, Patterns of intra- and interspecific association of gall-forming sawflies in relation to shoot size on their willow host plant, *Oecologia*, 73, 159, 1987.

53. **Fritz, R. S. and Price, P. W.**, Genetic variation among plants and insect community structure: willows and sawflies, *Ecology*, 69, 845, 1988.

54. **Furniss, R. L. and Carolin, V. M.**, *Western Forest Insects*, U.S. Dep. Agric. For. Serv. Misc. Pub. 1339, 1977.

55. **Gilbert, L. E.**, Ecological consequences of a coevolved mutualism between butterflies and plants, in *Coevolution of Animals and Plants*, Gilbert, L. E. and Raven, P. H., Eds., University of Texas Press, Austin, 1975, 210.

56. **Greenbank, D. O.**, The role of climate and dispersal in the initiation of outbreaks of the spruce budworm in New Brunswick. I. The role of climate, *Can. J. Zool.*, 34, 453, 1956.

57. **Haack, R. A., Wilkinson, R. C., Foltz, J. L., and Corneil, J. A.**, Spatial attack pattern, reproduction, and brood development of *Ips calligraphus* (Coleoptera: Scolytidae) in relation to slash pine phloem thickness: a field study, *Environ. Entomol.*, 16, 428, 1987.

58. **Haack, R. A., Wilkinson, R. C., and Foltz,**

J. L., Plasticity in life-history traits of the bark beetle *Ips calligraphus* as influenced by phloem thickness, *Oecologia*, 72, 32, 1987.

59. **Hanski, I.**, Pine sawfly population dynamics: patterns, processes, problems, *Oikos*, 50, 327, 1987.

60. **Hanski, I. and Otronen, M.**, Food quality induced variance in larval performance: comparison between rare and common pine-feeding sawflies (Diprionidae), *Oikos*, 44, 165, 1985.

61. **Harcourt, D. G., Guppy, J. C., Drolet, J., and McNeil, J. N.**, Population dynamics of alfalfa blotch leafminer, *Agromyza frontella* (Diptera: Agromyzidae), in eastern Ontario: analysis of numerical change during the colonization phase, *Environ. Entomol.*, 16, 145, 1987.

62. **Hassell, M. P. and Huffaker, C. B.**, The appraisal of delayed and direct density-dependence, *Can. Entomol.*, 101, 353, 1969.

63. **Hines, G. S., Taha, H. A., and Stephen, F. M.**, Model for predicting southern pine beetle population growth and tree mortality, in *Modelling Southern Pine Beetle Populations: Symposium proceedings*, Stephen, F. M., Searcy, J. L., and Hertel, G. D., Eds., USDA For. Serv. Tech. Bull. 1630, 1980.

64. **Houseweart, M. W. and Kulman, H. M.**, Life tables of the yellowheaded spruce sawfly, *Pikonema alaskensis* (Rohwer) (Hymenoptera: Tenthredinidae) in Minnesota, *Environ. Entomol.*, 5, 859, 1976.

65. **Hutchinson, G. E.**, *The Ecological Theater and the Evolutionary Play*, Yale University Press, New Haven, 1965.

66. **Isaev, A. S., Khlebopros, R. G., Nedorezov, L. V., Kondakov, Y. P., and Kiselev, V. V.**, Population dynamics of forest insects, *Nauka*, Novosibirsk (in Russian), 1984.

67. **Ives, W. G. H.**, Population and mortality assessment during the egg and larval stages of the larch sawfly, *Pristiphora erichsonii* (Htg.), *Can. Entomol.*, 94, 256, 1962.

68. **Ives, W. G. H. and Turnock, W. J.**, Estimation of cocoon populations of the larch sawfly, *Pristiphora erichsonii* (Htg.), *Can. Entomol.*, 91, 650, 1959.

69. **Karban, R.**, Herbivory dependent on plant age: a hypothesis based on acquired resistance, *Oikos*, 48, 336, 1987.

70. **Kazimi, S. K.**, Control of spider mite on Carib pine plantations, U.N. Development Programme, F.A.O., FO:DP/JAM/73/007 Terminal Rep., 1975.

71. **Kearsley, M. C. and Whitham, T. G.**, Developmental changes in resistance to herbivory: implications for individuals and populations, *Ecology*, 70, 422, 1989.

72. **Kennedy, P. C. and Wilson, L. F.**, Pine root collar weevil damage to red pine plantations in Michigan related to host age, tempera-

ture, and stand location, *Can. Entomol.*, 103, 1685, 1971.

73. **Landwehr, V. R. and Allen, W. W.**, Life history of *Oligonychus submidus* and *O. milleri* (Acari: Tetranychidae) and influence of temperature on development, survival and oviposition, *Ann. Entomol. Soc. Am.*, 75, 340, 1982.

74. **Langor, D. W. and Raske, A. G.**, Mortality factors and life tables of the eastern larch beetle, *Dendroctonus simplex* (Coleoptera: Scolytidae), in Newfoundland, *Environ. Entomol.*, 17, 959, 1988.

75. **Långström, B.**, Distribution of pine shoot beetle attacks within the crown of Scots pine, *Stadia Foretalia Suecica*, 154, 1, 1980.

76. **Lawton, J. H. and Strong, D. R.**, Community patterns and competition in folivorous insects, *Am. Nat.*, 118, 317, 1981.

77. **LeRoux, E. J., Paradis, R. O., and Hudon, M.**, Major mortality factors in the population dynamics of the eye-spotted bud moth, the pistol casebearer, the fruit-tree leafroller, and the European corn borer, *Mem. Entomol. Soc. Can.*, 32, 67, 1963.

78. **Lightfoot, D. C. and Whitford, W. G.**, Variation in insect densities on desert creosotebush: is nitrogen a factor?, *Ecology*, 68, 547, 1987.

79. **Louda, S. M.**, Distribution ecology: variation in plant recruitment over a gradient in relation to insect seed predation, *Ecol. Monogr.*, 52, 23, 1982.

80. **Louda, S. M.**, Seed predation and seedling mortality in the recruitment of a shrub, *Haplopappus venetus* (Astoraccae), along a climatic gradient, *Ecology*, 64, 511, 1983.

81. **Louda, S. M., Farris, M. A., and Blua, M. J.**, Variation in methylglucosinolate and insect damage to *Cleome serrulata* (Capparaceae) along a natural soil moisture gradient, *J. Chem. Ecol.*, 13, 569, 1987.

82. **Loyttyniemi, K.**, Zur Biologie der Nadelholzspinnmilbe *(Oligonychus ununguis* (Jacobi), Acarina: Tetranychidae) in Finnland, *Acta Entomol. Fenn.*, 27, 1, 1970.

83. **Luck, R. F.**, An appraisal of two methods of analyzing insect life tables, *Can. Entomol.*, 103, 1261, 1971.

84. **Luck, R. F. and Dahlsten, D. L.**, Natural decline of a pine needle scale *(Chionaspis pinifoliae* [Fitch]), outbreak at south Lake Tahoe, California, following cessation of adult mosquito control with malathion, *Ecology*, 56, 893, 1975.

85. **Maelzer, D. A.**, The regression of Log N_{n+1} or Log N_n as a test of density dependence: an exercise with computer-constructed density-independent populations, *Ecology*, 51, 810, 1970.

86. **Martin, J. L.**, The bionomics of the aspen blotch miner, *Lithocolletis salicifoliella* Cham.

(Lepidoptera: Gracillariidae), *Can. Entomol.*, 88, 155, 1956.

87. **Martinat, P. J.**, The role of climatic variation and weather in forest insect outbreaks, in *Insect Outbreaks*, Barbosa, P. and Schultz, J. C., Eds., Academic Press, New York, 1987, 241.

88. **Martinat, P. J. and Allen, D. C.**, Relationship between outbreaks of saddled prominent *Heterocampa guttivitta* (Lepidoptera: Notodontidae) and drought, *Environ. Entomol.*, 16, 246, 1987.

89. **Mason, R. R.**, Nonoutbreak species of forest Lepidoptera, in *Insect Outbreaks*, Barbosa, P. and Schultz, J. C., Eds., Academic Press, New York, 1987, 31.

90. **Mattson, W. J. and Haack, R. A.**, The role of drought stress in provoking outbreaks of phytophagous insects, in *Insect Outbreaks*, Barbosa, P. and Schultz, J. C., Eds., Academic Press, New York, 1987, 365.

91. **Mattson, W. J. and Haack, R. A.**, The role of drought in outbreaks of plant-eating insects, *BioScience*, 37, 110, 1987.

92. **Mattson, W. J. Levieux, J., and Bernard-Dagan, C., Eds.**, *Mechanisms of Woody Plant Defenses against Insects: Search for Pattern*, Springer-Verlag, New York, 1988.

93. **Mattson, W. J., Simmons, G. A., and Witter, J. A.**, The spruce budworm in eastern North America, in *Dynamics of Forest Insect Populations*, Berryman, A. A., Ed., Plenum Press, New York, 1988, 309.

94. **Mattson, W. J., Slocum, S. S., and Koller, C. N.**, Spruce budworm *(Choristoneura fumiferana)* performance in relation to foliar chemistry of its host plants, in *Forest Defoliator-Host Interactions: A Comparison Between Gypsy Moth and Spruce Budworms*, Talerico, R. L. and Montgomery, M., Eds., USDA For. Serv. Gen. Tech. Rep. NE-85, 1983, 55.

95. **May, R. M.**, *Stability and Complexity in Model Ecosystems*, Princeton University Press, Princeton, 1973.

96. **Mayr, E.**, Cause and effect in biology, *Science*, 134, 1501, 1961.

97. **McCullough, D. G. and Wagner, M. R.**, Influence of watering and trenching ponderosa pino on a pine sawfly, *Oecologia*, 71, 382, 1987.

98. **McLeod, J. M.**, Results of an aerial spraying operation against the Swaine jack-pine sawfly, *Neodiprion swainei* Middleton, in Quebec utilizing the insectide phosphamidon, *For. Chron.*, 44(5), 14, 1968.

99. **McLeod, J. M.**, The epidemiology of the Swaine jack-pine sawfly, *Neodiprion swainei* Midd., *For. Chron.*, 46(2), 126, 1970.

100. **McLeod, J. M.**, The Swaine jack pine sawfly, *Neodiprion swainei* life system: evaluating the long-term effects of insecticide appli-

cations in Quebec, *Environ. Entomol.*, 1, 371, 1972.

101. **McNeill, S.**, The dynamics of a population of *Leptopterna dolabrata* (Heteroptera:Miridae) in relation to its food resources, *J. Anim. Ecol.*, 42, 495, 1973.

102. **Mellors, W. K. and Helgesen, R. G.**, Life table analysis for the alfalfa blotch miner, *Agromyza frontella* in central New York, *Environ. Entomol.*, 9, 738, 1980.

103. **Miller, C. A.**, A technique for estimating the fecundity of natural populations of the spruce budworm, *Can. J. Zool.*, 35, 1, 1957.

104. **Miller, C. A.**, The spruce budworm, in *The Dynamics of Epidemic Spruce Budworm Populations*, Morris, R. F., Ed., *Mem. Entomol. Soc. Can.*, 31, 12, 1963.

105. **Miller, C. A.**, The black-headed budworm in eastern Canada, *Can. Entomol.*, 98, 592, 1966.

106. **Mopper, S. and Whitham, T.**, Natural bonsai of Sunset Crater, *Nat. Hist.*, 95, 42, 1986.

107. **Mopper, S., Mitton, J., Whitham, T. G., Cobb, N., and Christensen, K.**, Heterozygote advantage in pinyon pine in response to stress and herbivory, *Evolution*, in press, 1990.

108. **Mopper, S., Maschinski, J., Cobb, N., and Whitham, T. G.**, A new approach to plant structure: interactions between herbivores and the architecture of plants, in *Habitat Structure: The Physical Arrangement of Objects in Space*, Bell, S. S., McCoy, E. D., and Mushinsky, H. R., Eds., Chapman and Hall, New York, in press, 1990.

109. **Morris, R. F.**, The development of sampling techniques for forest insect defoliators, with particular reference to the spruce budworm, *Can. J. Zool.*, 33, 225, 1955.

110. **Morris, R. F.**, Single-factor analysis in population dynamics, *Ecology*, 40, 580, 1959.

111. **Morris, R. F., Ed.**, The dynamics of epidemic spruce budworm populations, *Mem. Entomol. Soc. Can.*, 31, 1, 1963.

112. **Morris, R. F.**, Approaches to the study of population dynamics, in *Forest Insect Population Dynamics*, Waters, W. E., Ed., U.S. For. Serv. Res. Paper NE-125, 1969, 9.

113. **Morris, R. F. and Miller, C. A.**, The development of life tables for the spruce budworm, *Can. J. Zool.*, 32, 283, 1954.

114. **Morris, R. F. and Royama, T.**, Logarithmic regression as an index of responses to population density, *Can. Entomol.*, 101, 361, 1969.

115. **Mott, D. G.**, The analysis of determination in population systems, in *Systems Analysis in Ecology*, Watt, K. E. F., Ed., Academic Press, New York, 1966, 179.

116. **Muma, M. H. and Apeji, S. A.**, *Oligonychus milleri* on *Pinus caribaea* in Jamaica. *Fla. Entomol.*, 53, 241, 1970.

117. **Myers, J. H.**, Interactions between western tent caterpillar and wild rose: a test of some general plant herbivore hypotheses, *J. Anim. Ecol.*, 50, 11, 1981.

118. **Myers, J. H.**, Can a general hypothesis explain population cycles of forest Lepidoptera? *Adv. Ecol. Res.*, 18, 179, 1988.

119. **Neilson, M. M. and Morris, R. F.**, The regulation of European spruce sawfly numbers in the Maritime Provinces of Canada from 1937 to 1963, *Can. Entomol.*, 96, 773, 1964.

120. **Nothnagle, P. J. and Schultz, J. C.**, What is a forest pest? in *Insect Outbreaks*, Barbosa, P. and Schultz, J. C., Eds., Academic Press, New York, 1987, 59.

121. **Odum, E. P.**, *Ecology and Our Endangered Life-Support Systems*, Sinauer, Sunderland, MA, 1989.

122. **Ohgushi, T. and Sawada, H.**, Population equilibrium with respect to available food resource and its behavioural basis in an herbivorous lady beetle, *Henosepilachna niponica*, *J. Anim. Ecol.*, 54, 781, 1985.

123. **Orians, G. H.**, Natural selection and ecological theory, *Am. Nat.*, 96, 257, 1962.

124. **Overhulser, D., Gara, R. I., and Johnsey, R.**, Emergence of *Pissodes strobi* (Coleoptera: Curculionidae) from previously attacked Sitka spruce, *Ann. Entomol. Soc. Am.*, 65, 1423, 1972.

125. **Podoler, H. and Rogers, D.**, A new method for the identification of key factors from life-table data, *J. Anim. Ecol.*, 44, 85, 1975.

126. **Pottinger, R. P. and LeRoux, E. J.**, The biology and dynamics of *Lithocolletis blancardella* (Lepidoptera: Gracillariidae) and apple in Quebec, *Mem. Entomol. Soc. Can.*, 77, 1, 1971.

127. **Preszler, R. W. and Price, P. W.**, Host quality and sawfly populations: a new approach to life table analysis, *Ecology*, 69, 2012, 1988.

128. **Price, P. W.**, Toward a holistic approach to insect population studies, *Ann. Entomol. Soc. Am.*, 64, 1399, 1971.

129. **Price, P. W.**, *Insect Ecology*, 2nd ed., John Wiley & Sons, New York, 1984.

130. **Price, P. W.**, The role of natural enemies in insect populations, in *Insect Outbreaks*, Barbosa, P. and Schultz, J. C., Eds., Academic Press, New York, 1987, 287.

131. **Price, P. W.**, Inversely density-dependent parasitism: the role of plant refuges for hosts, *J. Anim. Ecol.*, 57, 89, 1988.

132. **Price, P. W.**, Clonal development of coyote willow, *Salix exigua* (Salicaceae), and attack by the shoot-galling sawfly, *Euura exilguae* (Hymenoptera: Tenthredinidae). *Environ. Entomol.*, 18, 61, 1989.

133. **Price, P. W. and Clancy, K. M.**, Multiple effects of precipitation on *Salix lasiolepis* and populations of the stem-galling sawfly, *Euura lasiolepis*, *Ecol. Res.*, 1, 1, 1986.

134. **Price, P. W. and Clancy, K. M.,** Interactions among three trophic levels: gall size and parasitoid attack, *Ecology,* 67, 1593, 1986.

135. **Price, P. W. and Craig, T. P.,** Life history, phenology, and survivorship of a stem-galling sawfly, *Euura lasiolepis* (Hymenoptera:Tenthredinidae), on the arroyo willow, *Salix lasiolepis,* in northern Arizona, *Ann. Entomol. Soc. Am.,* 77, 712, 1984.

136. **Price, P. W., Roininen, H., and Tahvanainen, J.,** Plant age and attack by the bud galler, *Euura mucronata, Oecologia,* 73, 334, 1987.

137. **Price, P. W., Roininen, H., and Tahvanainen, J.,** Why does the bud-galling sawfly, *Euura mucronata,* attack long shoots? *Oecologia,* 74, 1, 1987.

138. **Price, P. W., Waring, G. L., Julkunen-Tiito, R., Tahvanainen, J., Mooney, H. A., and Craig, T. P.,** The carbon/nutrient balance hypothesis in within-species phytochemical variation of *Salix lasiolepis, J. Chem. Ecol.,* 15, 1117, 1989.

139. **Prokopy, R. J.,** Epideictic pheromones that influence spacing patterns of phytophaghous insects, in *Semiochemicals: Their Role in Pest Control,* Nordlund, D. A., Jones, R. L., and Lewis, W. J., Eds., Wiley, New York, 1981, 181.

140. **Raffa, K. F. and Berryman, A. A.,** Physiological differences between lodgepole pines resistant and susceptible to the mountain pine beetle and associated microorganisms, *Environ. Entomol.,* 11, 486, 1982.

141. **Raffa, K. F. and Berryman, A. A.,** The role of host plant resistance in the colonization behavior and ecology of bark beetles (Coleoptera:Scolytidae), *Ecol. Monogr.,* 53, 27, 1983.

142. **Raffa, K. F. and Berryman, A. A.,** A mechanistic computer model of mountain pine beetle populations interacting with lodgepole pine stands and its implications for forest managers, *For. Sci.,* 32, 789, 1986.

143. **Raffa, K. F. and Berryman, A. A.,** Interacting selective pressures in conifer-bark beetle systems: a basis for reciprocal adaptations? *Am. Nat.,* 129, 234, 1987.

144. **Redfearn, A. and Pimm, S. L.,** Population variability and polyphagy in herbivorous insect communities, *Ecol. Monogr.,* 58, 39, 1988.

145. **Redfern, M. and Cameron, R. A. D.,** Population dynamics of a yew gall midge *Taxomyia taxi* (Inchbald) (Diptera: Cecidomyiidae), *Ecol. Entomol.,* 3, 251, 1978.

146. **Rhoades, D. F.,** Evolution of plant chemical defense against herbivores, in *Herbivores: Their Interaction With Secondary Plant Metabolites,* Rosenthal, G. A. and Janzen, D. H., Eds., Academic Press, New York, 1979, 3.

147. **Rhoades, D. F.,** Herbivore population dynamics and plant chemistry, in *Variable Plants and Herbivores in Natural and Managed Systems,* Denno, R. F. and McClure, M. S., Eds., Academic Press, New York, 1983, 155.

148. **Rhoades, D. F.,** Offensive-defensive interactions between herbivores and plants: their relevance in herbivore population dynamics and ecological theory, *Am. Nat.,* 125, 205, 1985.

149. **Roininen, H., Price, P. W., and Tahvanainen, J.,** Field test of resource regulation by the bud-galling sawfly, *Euura mucronata,* on *Salix cinerea, Holarctic Ecol.,* 11, 136, 1988.

150. **Root, R. B.,** Some consequences of ecosystem texture, in *Ecosystem Analyses and Prediction,* Levin, S. A., Ed., Soc. Indust. Appl. Math., Philadelphia, 1975, 83.

151. **Root, R. B. and Chaplin, S. J.,** The lifestyles of tropical milkweed bugs, *Oncopeltus* (Hemiptera:Lygaeidae) utilizing the same hosts, *Ecology,* 57, 132, 1976.

152. **Royama, T.,** Population persistence and density dependence, *Ecol. Monogr.,* 47, 1, 1977.

153. **Royama, T.,** Fundamental concepts and methodology for the analysis of animal population dynamics, with particular reference to univoltine species, *Ecol. Monogr.,* 51, 473, 1981.

154. **Royama, T.,** Evaluation of mortality factors in insect life table analysis, *Ecol. Monogr.,* 51, 495, 1981.

155. **Sacchi, C. F., Price, P. W., Craig, T. P., and Itami, J. K.,** Impact of shoot galler attack on sexual reproduction in the arroyo willow, *Ecology,* 69, 2021, 1988.

156. **Samarasinghe, S. and LeRoux, E. J.,** The biology and dynamics of the oystershell scale, *Lepidosaphes ulmi* (L.) (Homoptera: Coccidae) on apple in Quebec, *Ann. Entomol. Soc. Quebec,* 11, 206, 1966.

157. **von Scheller, H. D.,** Zur Biologie und Schadwirkung der Nadelholzspinnmilber *Oligonychus ununguis* Jacobi (Acar. Tetr.) und der Fichtenrohrenlaus *Liosomaphis abietina* Walker (Hom. Aphid.) 1: *Oligonychus ununguis* J., *Z. Angew Entomol.,* 51, 69, 1962.

158. **Snedecor, G. W.,** *Statistical Methods,* 4th ed., Iowa State College Press, Ames, 1946.

159. **Southwood, T. R. E., Ed.,** *Insect Abundance,* Blackwell Scientific, Oxford, 1968.

160. **Sower, L. L. and Mitchell, R. G.,** Host-tree selection by western pine shoot borer (Lepidoptera: Olethreutidae) in ponderosa pine plantations, *Environ. Entomol.,* 16, 1145, 1987.

161. **St. Amant, J. L. S.,** The detection of regulation in animal populations, *Ecology,* 51, 823, 1970.

162. **Stark, R. W.,** Population dynamics of the lodgepole needle miner, *Recurvaria starki* Freeman, in Canadian Rocky Mountain parks, *Can. J. Zool.,* 37, 917, 1959.

163. **Story, R. N., Robinson, W. H., Pienkowski, R. L., and Kok, L. T.,** The biology and immature stages of *Taphrocerus schaefferi,* a leaf-miner of yellow nutsedge, *Ann. Entomol. Soc. Am.,* 72, 93, 1979.

164. **Strong, D. R., Lawton, J. H., and Southwood, R.,** *Insects on Plants: Community Patterns and Mechanisms,* Harvard University Press, Cambridge, 1984.

165. **Tuomi, J., Niemela, P., Haukioja, E., and Siren, S.,** Nutrient stress: an explanation for plant anti-herbivore responses to defoliation, *Oecologia,* 57, 298, 1984.

166. **Turnbow, R. H. and Franklin, R. T.,** Bionomics of *Brachys tessellatus* in coastal plain scrub oak communities, *Ann. Entomol. Soc. Am.,* 74, 351, 1981.

167. **Varley, G. C. and Gradwell, G. R.,** Key factors in population studies, *J. Anim. Ecol.,* 29, 399, 1960.

168. **Varley, G. C. and Gradwell, G. R.,** Population models for the winter moth, in *Insect Abundance,* Southwood, T. R. E., Ed., Blackwell Scientific, Oxford, 1968, 132.

169. **Varley, G. C., Gradwell, G. R., and Hassell, M. P.,** *Insect Population Ecology: An Analytical Approach,* Blackwell Scientific, Oxford, 1973.

170. **Walgenbach, D. D., Elliott, N. C., and Kieckhefer, R. W.,** Constant and fluctuating temperature effects on developmental rates and life table statistics of the greenbug (Homoptera: Aphididae), *J. Econ. Entomol.,* 81, 501, 1988.

171. **Wallner, W. E.,** Factors affecting insect population dynamics: differences between outbreak and non-outbreak species, *Annu. Rev. Entomol.,* 32, 317, 1987.

172. **Waring, G. L. and Price, P. W.,** Consequences of host plant chemical and physical variability to an associated herbivore, *Ecol. Res.,* 3, 205, 1988.

173. **Washburn, J. O. and Cornell, H. V.,** Parasitoids, patches, and phenology: their possible role in the local extinction of a cynipid gall wasp population, *Ecology,* 62, 1597, 1981.

174. **Weis, A. E. and Kapelinski, A.,** Manipulation of host plant development by the gall-midge *Rhabdophaga strobiloides, Ecol. Entomol.,* 9, 457, 1984.

175. **White, T. C. R.,** An index to measure weather-induced stress of trees associated with outbreaks of psyllids in Australia, *Ecology,* 50, 905, 1969.

176. **White, T. C. R.,** The nymphal stage of *Cardiaspina densitexta* (Homoptera: Psyllidae) on leaves of *Eucalyptus fasciculosa, Aust. J. Zool.,* 18, 273, 1970.

177. **White, T. C. R.,** A hypothesis to explain outbreaks of looper caterpillars, with special reference to populations of *Selidosema suavis* in a plantation of *Pinus radiata* in New Zealand, *Oecologia,* 16, 279, 1974.

178. **White, T. C. R.,** Weather, food and plagues of locusts, *Oecologia,* 22, 119, 1976.

179. **White, T. C. R.,** The importance of a relative shortage of food in animal ecology, *Oecologia,* 33, 71, 1978.

180. **White, T. C. R.,** The abundance of invertebrate herbivores in relation to the availability of nitrogen in stressed food plants, *Oecologia,* 63, 90, 1984.

181. **Whitham, T. G.,** Habitat selection by *Pemphigus* aphids in response to resource limitation and competition, *Ecology,* 59, 1164, 1978.

182. **Whitham, T. G.,** Territorial behavior of *Pemphigus* gall aphids, *Nature,* 279, 324, 1979.

183. **Whitham, T. G.,** The theory of habitat selection: examined and extended using *Pemphigus* aphids, *Am. Nat.,* 115, 449, 1980.

184. **Whitham, T. G.,** Costs and benefits of territoriality: behavioral and reproductive release by competing aphids, *Ecology,* 67, 139, 1986.

185. **Whitham, T. G.,** Evolution of territorial behavior by herbivores in response to plant defenses, *Am. Zool.,* 27, 359, 1987.

186. **Whitham, T. G.,** Host resistance and competition at low densities by *Pemphigus* gall aphids, *Am. Nat.,* in press, 1990.

187. **Whitham, T. G. and Mopper, S.,** Chronic herbivory: impacts on architecture and sex expression of pinyon pine, *Science,* 228, 1089, 1985.

188. **Whittaker, R. H.,** A study of summer foliage insect communities in the Great Smoky Mountains, *Ecol. Monogr.,* 22, 1, 1952.

189. **Wilson, L. F. and Surgeoner, G. A.,** Variable oakleaf caterpillar, *For. Ins. Dis. Leaflet,* 67, 1, 1979.

2

The Role of Experience in Host Plant Choice by Phytophagous Insects

Árpád Szentesi and Tibor Jermy
Plant Protection Institute
Hungarian Academy of Sciences
Budapest, Hungary

TABLE OF CONTENTS

I. Introduction...40
II. Types of Learning 40
    A. Habituation..41
    B. Sensitization and Pseudoconditioning....................................41
    C. Imprinting...42
    D. Associative Learning 43
III. Neural Mechanisms Underlying Experience-Induced Changes of
    Behavior..43
IV. Categories of Plants as Related to Host Selection 45
V. Experience-Induced Changes in Feeding Behavior 46
    A. Habituation to Food-Related Stimuli....................................46
    B. Food Imprinting (Induction of Food Preference, IFP)....................47
        1. Incidence of Food Imprinting.......................................48
        2. Degree of Food Imprinting 48
        3. Persistence of Food Imprinting 50
        4. Phytochemicals Involved in Food Imprinting.......................51
        5. Food Imprinting as Related to Host Range 51
    C. Food-Aversion Learning...52
    D. Dietary Self-Selection...54
    E. Learning in Food Finding..55
VI. Experience-Induced Changes in Oviposition Behavior 55
VII. Pre- and Early-Imaginal Experience 59
VIII. Adaptive Significance of Experience-Induced Behavioral Changes 60
IX. Behavioral Changes and Speciation...64
X. Conclusions 65
Acknowledgments 66
References 66

I. INTRODUCTION

Insect behavior was generally thought to be determined largely by closed genetic programs, leaving little room for learning in behavioral development. This picture has changed over the last decades, especially since the highly developed learning ability of social Hymenoptera has been demonstrated. The number of publications dealing with learning in phytophagous species is also increasing rapidly. As a consequence, there is now a growing danger of overemphasizing the role of learning in insects, because most authors stress abilities without pondering the limits and imperfections of processes involved, and this may easily lead to false conclusions. In addition, in some insect species learning could not be demonstrated.[53] A further problem is that the experiments, especially with phytophagous insects, are carried out mostly in the laboratory and it is difficult to estimate how the learning ability demonstrated under such conditions influences the survival and fitness of insects in nature.

For students of insect-plant relationships it is of considerable importance to know how plastic the phytophagous insects' nervous system is in "solving" tasks in natural circumstances. If the insects' food-related behavior can be optimal[120,121] (see criticism on optimal foraging[163]), it is learning that might contribute to optimal "decisions" (see more in Section VIII).

With these constraints in mind, we try to give an objective survey on the state-of-the-art of investigations into the experience-induced behavioral changes in phytophagous insects as related to food selection and host finding. The majority of mostly theoretical papers[93,141,142,171,220,221] are not considered although these are of exceptional value in defining the criteria and circumstances for learning studies, or approximating the ultimate mechanisms of learning by the use of mutant strains[61,88] or various sensory modalities.[18,71,197] The studies of accompanying events, for instance central excitatory state (CES)[222] and experience-dependent behavioral elements,[85] are equally important and briefly dealt with.

II. TYPES OF LEARNING

In the following, a list of learning phenomena similar to the one presented in a recent review[58] and thought to be involved in the host selection behavior of phytophagous insects is prepared with definitions based on works of several authors.[83,96,102,140,141,160] Learning will be divided into nonassociative (habituation, sensitization, and pseudoconditioning) and associative (classical and instrumental conditioning) types (see definitions under separate headings). Imprinting is discussed separately as it is usually considered a special type of learning,[229] although not all authors agree on this.[211]

The generally accepted broad definition of learning that will be used also by the authors is modification of behavior through experience. With learning, a central nervous system process is generally implied, in contrast to peripheral events, such as (sensitivity) changes in receptor function which may also result in behavioral changes (see Section III).

Learning types in insects can range from simple habituation to complex associative processes. The proper conditions set for the definition of different types of associative learning are quite specific (see below), but involve a clear demonstration of association between the unconditioned stimulus (US) which stimulates the behavioral response prior to learning, and the conditioned stimulus (CS) which becomes a signal for the response as result of learning.

The term "conditioning" is generally reserved for associative (classical and instrumental) learning. Therefore, it is misleading to use it in its "loose sense"[161] and to equate induced preference with conditioning as "conditioning of host-plant preference"[154] because induced preference does not meet all the criteria set for associative learning.[57]

In the literature there are other aspects of confusion in the use of the terminology of learning, and especially in the way in which changes in food preferences are discussed. For example, Blaney and Simmonds[21] refer to short-term changes in food selection behavior as learning and long-term changes by contrast as "induction" although they do not define the time scale involved. However, we believe that long-lasting experiences in phytophagous insects can involve both associative and nonassociative learning[138] and that the well known phenomenon of induced preference for the rearing food[115] could well involve learning (see below).

A. HABITUATION

Habituation is the waning of response to a repeatedly presented stimulus over time.[83,96,140,207] Kandel and Spencer[123] (cited by Eisestein[63]) listed criteria that define habituation. A response decrement characteristic of the habituation process may occur for a number of reasons and several different mechanisms may underlie it. It is usually thought of as the most primitive, simplest type of learning.

There is also some controversy on certain aspects of habituation. For instance, its retention depends on the rate of stimulation. Thompson and Spencer[207] predicted a faster response decrement to more frequent stimulation, while Thon[208] found the opposite with *Calliphora vomitoria*. Weak and frequently presented stimuli are more generally likely to elicit habituation than strong ones. Though certain stimuli (e.g., a predator's silhouette for a young bird) will never produce a habituation response, the importance of different types of stimuli is generally small.

Habituation can be elicited by both stimulatory and inhibitory stimuli. Dethier[50] (cited by Duerr and Quinn[62]) observed a response decrement upon presentation of the phagostimulant sucrose to *Phormia*; the change was mediated centrally. Szentesi and Bernays[203] demonstrated that habituation of *Schistocerca gregaria* to a deterrent was also a central nervous phenomenon. (See Section V.A for further examples of habituation.) All sensory modalities can be involved in eliciting a habituation response, but the habituation is specific to the particular stimuli participating in its induction, and as a consequence underlying mechanisms may also differ.[125]

Finally, Thon and Pauzie[209] (p. 119) concluded from a study conducted on cardiac and motor response habituation (i.e., on two "response systems") in *C. vomitoria* the "habituation processes develop independently in different response systems, even if they have the same sensory input."

B. SENSITIZATION AND PSEUDOCONDITIONING

Sensitization may be defined as a state of excitation or increased responsiveness of the organism to biologically significant stimuli. However, applying the more precise learning paradigm terminology[141] (p. 481), "we speak of sensitization if a change in response occurs after repeated presentation of a conditioned stimulus (CS) that has not been paired with an unconditioned stimulus (US)." By pairing US and CS we expect an enhanced response to CS, although increased responses can also appear from presenting the US alone.

Authors generally agree that sensitization is a nonassociative type of learning[83,140,141] although sensitization can be considered "as a necessary precursor of associative learning or conditioning . . . "[140] (p. 339). The dual-process theory[83] presumes that the two contrasting processes (habituation and sensitization) have evolved and occur independently, but interact. As it has been pointed out " . . . the strength of the behavioral response elicited by a repeated stimulus is the net outcome of the two independent processes of habituation and sensitization" (pp. 441-442).[83] This can be illustrated by the following. Städler and Hanson[199] showed that wheat germ extract was deterrent for larvae of *Manduca sexta* reared on host plant leaves, but the response to the deterrent was significantly reduced in larvae reared on wheat germ diet. On the other hand, diet-reared larvae preferred diet to host plant leaves. It may be that ingestion of a certain food leads, through sensitization, to a habituation to food-specific deterrents.

In connection with sensitization the phenomenon of the central excitatory state (CES) has to be mentioned. It was first described by Dethier et al.[59] who found that immediately after the presentation of a drop of sucrose solution to the labellar hair of a water-satiated *Phormia regina*, the fly would extend its proboscis immediately afterwards when water only is offered. This increased reactivity was attributed by the authors to an excitatory state of the central nervous system (CNS). The duration of the CES varies significantly: e.g., 15 s in *Leptinotarsa decemlineata*;[110] up to 120 s in *P. regina*;[59,219] (cited in Reference 222) at least 10 min in *Drosophila*.[222] We agree with Duerr and Quinn's opinion[62] who came to the conclusion, when measuring sensitization after sucrose stimulation in *Drosophila*, that this phenomenon is basically identical with the CES described by Dethier et al.[59] We also propose that sensitization in this broader sense is relevant to the explanation of induced feeding preference and early imaginal (oviposition) experiences in phytophagous insects (see Sections V.B and VII).

Upon repeated presentation only of an unconditioned stimulus (US) without pairing it with a conditioned one (CS), the increased response to the CS shown by the animal is termed *pseudoconditioning*.[141] For our further discussion the involvement of US has a special importance, as it refers to food or other vitally important environmental features. There is usually a CS (e.g., shape of food) present, although much separated in time from the reinforcing US (e.g., chemical features). It is the time separation feature which is used to separate it from associative learning. Thus pseudoconditioning is used as a proper control for associative learning.[150,171] Krasne[125] equates sensitization with pseudoconditioning, emphasizing the role of the US only, while McGuire[141] and others separate the two phenomena, as we do, by altered responses to either CS or US.

C. IMPRINTING

It has frequently been observed that animals possess one or more especially sensitive periods in ontogenesis to environmental stimuli, such as those representing parent, mate, habitat, or food. As a result they tend to persistently respond to stimuli corresponding to the first experiences. The learning process through which the perceptual change is acquired was termed imprinting.[102,140]

Imprinting is a quick learning process, resulting in long-term memory.[229] Some students of animal behavior[91,211] do not consider it as a separate type of learning, while others (e.g., Wallace[229]) argue that it is a special learning type that can be characterized by its fast formation. It is similar to instrumental conditioning, except that usually only one experience is sufficient and no further reinforcement is required. The persistence of imprinted experience may be life long.

Its concept has developed considerably since the original description.[134] Immelmann[102] discussed in detail the cases of imprinting known so far and came to the conclusion that there are only two important criteria: (1) the existence of a sensitive period, and (2) the subsequent stability (rigidity) of response to experience gained during that period. Imprinting-like behavior has mostly been demonstrated in vertebrates, and with the exception of parasitic insects (e.g., *Nemeritis canescens*[212] and *Asobara* species[223]), the number of instances with insects are few. Apart from involvement in various social contexts, "ecological" imprinting[102] involves food preference, habitat or host selection.

The notion of ecological imprinting provides the framework within which we interpret induction of food preference (see Section V.B).

D. ASSOCIATIVE LEARNING

Associative learning is by far the most complex type of learning, embracing at least two main types: classical and instrumental conditioning. A behavioral change is generally termed conditioning when a strict timely coincidence or pairing of a conditioned stimulus, the CS (which is quite ineffective in this context in itself), and an unconditioned stimulus, the US, occurs, and after only one or several experiences the CS serves to elicit a response previously elicited by the US, but not the CS (classical conditioning). While in classical conditioning a close temporal relationship of stimulus and reinforcer is provided, in instrumental (operant) conditioning a connection of response and reinforcer is required,[140,229] which may also be the essential process in trial-and-error learning.[211] Careful experimentation must exclude pseudoconditioning, CES or sensitization, and habituation.

One of the early reports on conditioning with arthropods (locusts) is that of Horridge.[97] Subsequently Nelson[150] showed convincingly that the blowfly, *Phormia regina* is able to acquire a conditional response. Fukushi[66] used houseflies (*Musca domestica*) to demonstrate olfactory conditioning. The increase of CS-US interval, i.e., "weakening" of the contingency of the US upon the CS decreased the probability of a conditioned response. Fukushi[67] conditioned houseflies also to perform a characteristic food searching behavior, "dance"[151] for sucrose droplets containing any of the test substances as CS. The information about the odor of the food source was stored for at least 3 h without loss of performance. A fast discrimination learning was demonstrated with intact *Periplaneta americana*.[8] Operant conditioning has also been demonstrated in a number of cases with insects.[28]

The above and similar reports convincingly demonstrate that many insect species are capable of performing learning tasks and have the capacity to store important information about the environment, at least temporarily.

It is perhaps the importance of testing a given insect species in circumstances resembling its natural environment that explains some more successful associative learning studies. Some early work on learning, applying naturalistic conditions, demonstrates the importance of relevant conditions (see Fukushi[67] p. 248, for references). The carabid beetle, *Pterostichus melanarius*, was unable to learn a maze, but successful learning was demonstrated in the open field, using an apparatus in which the insects could contact the wall.[165] Water-deprived beetles showed a marked decrease in time necessary to locate a centrally placed water source after trials on 4 consecutive days, but the information was not retained over 24 h.

Focusing here on phytophagous insects, we consider associative learning mainly in food-aversion learning as well as food and oviposition site finding (see Sections V and VI).

III. NEURAL MECHANISMS UNDERLYING EXPERIENCE-INDUCED CHANGES OF BEHAVIOR

The decision whether to eat a plant or not depends on the sensory pattern provided primarily by the chemoreceptors as well as on the processing and retention of the incoming information by the CNS.

Several experimental studies have shown that feeding experience may change receptor characteristics. Schoonhoven[184] exposed *Manduca sexta* larvae for 2.5 days to an artificial diet containing the feeding inhibitor salicin that stimulated the deterrent receptor. After exposure, the sensitivity of that receptor was considerably lower than in unexperienced larvae. As the effect developed gradually, sensory adaptation of receptors could be excluded. The same author later[185] adopted the view that long-term sensitivity changes occurred in the chemosensory system. The characteristics of the receptor sites remained constant but the spike generating process was changed. The role of the CNS in the process of changes in response to deterrents was not raised.

When allelochemicals such as azadirachtin, nicotine, or sinigrin were added to the artificial diet on which larvae of *Spodoptera* spp. were reared, the sensitivity of the chemoreceptors to these compounds was strongly reduced with experience.[20,194] Schoonhoven[183] found that in *M. sexta* larvae that were fed on an artificial diet, the maxillary sensilla responded differently to saps of various plants compared with those of plant-fed larvae. When caterpillars were grown on two different host plant species, their responses to the saps of these plants also differed.[198] The responsiveness of the deterrent receptor to strychnine in *Pieris rapae* larvae was somewhat lowered relative to control larvae reared on untreated leaves[143] (cited by Blaney et al.[19]) when the larvae were reared on cabbage leaves sprayed with this chemical. The possible mechanisms regulating receptor responsiveness have been discussed in detail[19] but very few experimental data are available for phytophagous insects.

The central mechanisms regulating receptor sensitivity may be neural and/or hormonal, and it is possible that chemicals form the hemolymph act on the development and function of receptors (Bernays, personal communication). It is a general belief that there are no efferent pathways from higher nervous centers actually modifying receptor responses in insects.[19] However, axons of unknown origin have been found close to the receptor neurons in the A1 sensilla of the clypeo-labrum in *Locusta migratoria*.[36] As for the possibility of hormonal regulation, it was found that in *L. migratoria*, midgut distension stimulates release of a hormone from the corpora cardiaca that activates the closing mechanism of the palp-tip sensilla, and by this change their responsiveness.[13,14] Several authors have demonstrated the endocrine control of receptor sensitivity in adults of Diptera (see Blaney et al.[19] for references). It has been found that specific chemical stimuli are able to change the responsiveness of a single receptor type, and this argues against hormonal control. Therefore, it is most probable that the receptor cells are actively involved in controlling their own sensitivity,[19] although the mechanism by which this is achieved is not understood.

No changes in peripheral receptors were found by Ma[135] in the larvae of *Pieris brassicae* however, when they were reared on an artificial diet instead of on the host plant. Therefore, he proposed that in this case changes in preference behavior may be attributed entirely to changes in the CNS.

Memory is here defined as the timely persistence of experience, not referring to the neuroanatomical locus where it is formed. Therefore, it also applies to habituation, though its development and maintenance, partly at least, take place at early synaptic levels.[29] In the literature the persistence of information is usually classified as short- and long-term memory.[3] This applies also to arthropods, where the examples of non-associative and associative learning infer various "kinds" and duration of memory, the proximate processes of which are almost unknown. Longer lasting memory may often cover associative learning while shorter term memory may at least sometimes correspond with certain kinds of nonassociative learning (e.g., pseudoconditioning).

From behavioral studies it is well established that certain lifestyles, especially eusociality, is accompanied by the ability to retain information gained from experience, and the memory of previous experience has been best demonstrated with Hymenoptera.[7] Experience may be retained in many insect species in spite of the dramatic changes taking place during metamorphosis.[4,76] Short-term memory contributed to more efficient foraging in *Colias* and *Pieris* butterflies.[132,200] Visual spatial and landmark memory are supposed to operate with bees,[5,77] and in the ant *Cataglyphis*.[234] A "novelty-effect" was described with *Musca domestica* induced by odor stimuli that lasted for ca. 30 min, and it was assumed that this resulted from memory.[240] Similarly, the information gained by females of the moth *Acrolepiopsis assectella* about host-plant allelochemicals was retained for ca. 1 day.[206]

A semblance of memory can be given by the "chemical legacy" effects.[38] This hypothesis states that the effect of larval chemosensory experience on adult responsiveness may depend on traces of chemical cues remaining inside or outside the insect's body after molting and influences adult behavior as a result of experience at eclosion (see also Sections VII and VIII).

With regard to the neural mechanisms underlying the four types of learning discussed in Section II, the following can be said:

- Habituation is believed to result from depression of transmitter release at synaptic junctions,[29,125] or from collateral inhibition.[177] Duerr and Quinn[62] directly demonstrated the central mediation of habituation with *Drosophila*. Sucrose stimulation on the right prothoracic tarsus resulted in a decreased responsiveness by the contralateral leg to the same subsequent stimulation.
- The mechanism(s) underlying sensitization are not known,[125] although it is supposed that it occurs in the CNS regions that set the level of responsiveness of the animal, and not along the stimulus-response pathways.[83]
- Food imprinting may arise partly through changes in chemoreceptor sensitivity resulting from feeding experience. Though direct experimental evidence is lacking, the role of the CNS cannot be excluded either.[199]
- As for associative learning, both elementary physiological processes and cellular level correlates of associative learning are discussed in recent papers.[28,125] In case of invertebrates, the study of *Drosophila* mutants has proved to be especially useful in understanding such mechanisms (e.g., References 61, 88, and 171).

IV. CATEGORIES OF PLANTS AS RELATED TO HOST SELECTION

In the literature on host plant selection by phytophagous insects the plant species are classified mostly as host plants and non-host plants, respectively. The former are represented by plants that are readily acceptable by and are suitable food for a certain insect species, while the latter are not accepted. However, the border between these two categories is often indistinct. For example, in case of the solanaceous feeder *Leptinotarsa decemlineata*, the plant species tested can be arranged in a "triangle of food preferences" based on the intensity of acceptable and unacceptable attributes (Figure 1).[109] The plant species located on the upper part, left of the dotted line, are the primary and secondary solanaceous host plants that are also attacked in nature, while further down on the left side are nonsolanaceous species that are more or less readily consumed (also in nature) by water- and food-deprived beetles. Plant species to the right of the dotted line are rejected even in forced-feeding situations. Based on this kind of distinction de Boer and Hanson[47] classified the plant species with respect to their acceptability by the oligophagous *Manduca sexta* as (1) host plants, (2) acceptable non-host plants that are not fed upon in nature but can act as food plants in the laboratory, and (3) unacceptable non-host plants, emphasizing that the hierarchy shows a graded continuum of acceptability extending across host and non-host species. In the following discussions these three categories are used.

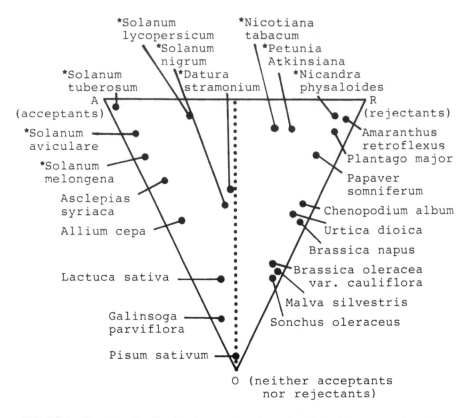

FIGURE 1.   The "triangle of food preferences" as determined by leaf disc tests and sandwich tests for *Leptinotarsa decemlineata*. The effect of acceptants and rejectants decreases with the distance from A and R, respectively. Asterisks indicate solanaceous species. (From Jermy, T., *Acta Zool. Acad. Sci. Hung.*, 7, 119, 1961. With permission.)

It can be assumed that this categorization is valid for most oligophagous and poly-phagous insects. To which category a certain plant species belongs in relation to an insect species presumably varies within limits depending on varietal differences and growing conditions that determine plant quality.[112] On the other hand, host-plant hier-archies may also change with experience gained by the insect[47,135] and may differ in relation to different populations of the same insect species.[100]

V. EXPERIENCE-INDUCED CHANGES IN FEEDING BEHAVIOR

A. HABITUATION TO FOOD-RELATED STIMULI

Besides theoretical aspects, studies on habituation have been initiated by consid-erations of the practical use of feeding inhibitory substances, since it has been observed that by the repeated contact of feeding insects with a deterrent-containing diet will increase its acceptance over time.[75,184,201] Schoonhoven and Jermy,[186] on the basis of field experiments conducted by Murbach and Corbaz[149] and Murbach[148] on the efficiency of Bordeaux mixture [a suspension of $CaSO_4 \cdot Cu(OH)_2 \cdot 3Ca(OH)_2$], against *Leptinotarsa decemlineata*, and of their own behavioral observations on the same species and on *Pieris brassicae*, came to the conclusion that habituation at least to Cu-ions is not probable. On the other hand, Chapman[31] expressed the opinion that habituation to deterrent substances was a possibility. Bernays[12] did report data of reduced effective-ness of a neem extract sprayed against *Zonocerus variegatus* when tested after 12 days of treatment, that was attributed to habituation and not to inactivation of active component(s). Subsequently, an increased consumption of deterrent-containing food over time was demonstrated in laboratory with the polyphagous *Mamestra brassicae* and *Schistocerca gregaria*, and the oligophagous *Pieris brassicae* and *Locusta migratoria.*[114,116,203]

In an examination of habituation to a feeding deterrent (nicotine hydrogen tartrate, NHT) with last instar nymphs of *S. gregaria*,[203] small pieces of nylon tubing were attached to both maxillary palpi of the nymphs. For the experimental group, the tubing was filled with NHT solution for given periods of time daily. The control group received distilled water in the tubes. On the test day, no chemical was applied to the capillaries, but feeding response was measured by the quantity of plant material consumed when treated with the same chemical. The insects that had been repeatedly exposed to the chemical consumed more of the treated leaves compared with the control group, despite the fact that direct perception of the chemical by the sensilla on the maxillary palpi was prevented by the nylon tubing still in place. The experiment proved that by the time of final test habituation had occurred and was mediated centrally.

The study with *S. gregaria*[203] demonstrated that weak stimuli can be more successful in eliciting habituation than stronger ones.[96] Habituation to NHT was shown by 1% concentration of the chemical, but not by 5 or 20%.

Jermy et al.[114] observed differences with species differing slightly in their patterns of habituation to different chemicals. For example, the pattern of development of the responses by *M. brassicae* and *P. brassicae* to strychnine was different, and the same is true of *P. brassicae* tested on strychnine or quinine. Furthermore, the method of presentation of the stimulus affected the outcome considerably. If NHT was presented to either *S. gregaria* or *L. migratoria* on living plant tissue, habituation to the chemical occurred. However, if the same substance was offered on a glass fiber disk, the insects developed an aversion (see Section V.C for further discussion).

From the few cases known so far it seems that polyphagous species habituate to single chemicals in the laboratory more readily than oligophagous species.[114]

B. FOOD IMPRINTING (INDUCTION OF FOOD PREFERENCE, IFP)

Entomologists have long known that lepidopterous larvae, after having fed on one host-plant species, would only reluctantly accept other host species. Darwin[44] (P. 293) mentions the observation made by Michely:[145] "The caterpillars of the *Bombyx hesperus* feed in a state of nature on the leaves of the *Café diable*, but, after having been reared on the *Ailanthus*, they would not touch the *Café diable*, and actually died of hunger."

The term IFP was given to this phenomenon by Jermy et al.[115] who thoroughly investigated it. Dethier[57] emphasized that this neutral term should be preserved because the nature of the phenomenon is largely unknown. The latter opinion has been shared by many others until recently,[112] although the possible relationship of IFP with learning has been noted. Bernays and Wrubel[17] wrote in connection with food selection learning (see more details in Section V.E): "This process, known as induction, could involve elements of learning" (p. 359).

We assume that IFP is a type of learning and it meets two main criteria listed for imprinting:[102] the existence of a sensitive period and the subsequent stability (rigidity, "irreversibility") of the resulting behavior. As for the first criterion: in some lepidopterous species early food experience with certain plant species definitively determines food preference for the larvae[65,135,241] indicating that the young instars are very sensitive in this respect. Although induction proved to be possible also in later instars,[115] so that there is not such a clearly definable sensitive period, in nature it may be of decisive importance that the first instar lepidopterous larvae become "imprinted" on the plant on which they emerge from the eggs.[236]

The criterion of stability is better met; for example, food experience of early instar *Pieris brassicae* larvae may last till the end of larval development.[135] It can be formed after brief (e.g., 4 h long[135]) experience with the food, and can be so rigid that the insect may rather starve to death than consume another host plant.[86,135,191] The degree of rigidity of induction is a relative measure. This is demonstrated by the induction index[47] (see Section V.B.2). In several cases induced food preference was found to be changeable,[30,135] but similar lability has also been observed in cases of typical imprinting,[102] as well as individual variability of imprintability.[102]

Based on the above we argue that IFP basically meets the criteria of imprinting on food, and should, therefore, be referred to as "food imprinting". Moreover, the term "food imprinting" more explicitly refers to a learning process. (It would be tempting to use the term "host-plant imprinting"; however, lepidopterous larvae can be imprinted also on artificial diets.[199])

1. Incidence of Food Imprinting

Several authors have dealt with this phenomenon in various groups of insects, especially over the last two decades. As can be seen from Table 1, food imprinting may occur in all the main groups of phytophagous insects.

Very few data have been published on the absence of food imprinting. This may partly be due to the fact that negative results are less often published than positive ones. Table 2 shows the insect species with which experience-induced changes of feeding behavior have been looked for and not found.

A comparison of Tables 1 and 2 indicates that with the same insect species (*Acyrtosiphon pisum, Phratora vitellinae, Leptinotarsa decemlineata, Hyphantria cunea, Pieris rapae*) some authors did and some did not find food imprinting. This may be due to (1) different experimental procedures, (2) different ontogenetic stages of the insects, (3) different plants used in the experiments, since it has been found repeatedly that host plants may differ considerably in the degree of food imprinting they are able to induce (see Section V.B.2), and (4) intraspecific differences among populations, e.g., two geographically distant populations of *Ph. vitellinae* differed significantly in the extent to which food imprinting of the same plant species occurred.[178]

2. Degree of Food Imprinting

The degree of food imprinting may be expressed as an "induction index" calculated as follows[47] (p. 179): "The difference in feeding preferences for two plant species is expressed in a choice index (range, -100 to $+100$) which measures the mean consumption of plant A minus that of plant B. The degree to which preferences are induced is expressed as an induction index (range, 0 to 200) which is the absolute difference between the choice indices for plant pair A,B obtained for two groups of larvae reared on either plant A or plant B."

Surveying the data on lepidopterous larvae there was an inverse correlation between the degree of food imprinting and taxonomic relatedness of plant species paired in the above sense.[47] This was supported by the data on *Pieris brassicae*,[135] *Papilio machaon*,[236] *Callosamia promethea, Polygonia interrogationis, Antheraea polyphemus*,[86] *Lymantria dispar*,[9,233] *Manduca sexta, Limenitis astyanax, L. hybrid rubidus, L. archippus*, and *Heliothis zea*.[47] Presumably this inverse correlation is related to similarities in available gustatory stimuli from closely related plants. In this connection it has to be emphasized, however, that taxonomic relatedness does not always reflect similarity in the plants' chemical "Gestalt" perceived by the insects. The discrepancy between taxonomy and chemical "Gestalt" may explain that, e.g., only weak or no food imprinting was found with *Lymantria dispar* when plant pairs belonging to the same family (Fa-

Table 1
LIST OF INSECT SPECIES IN WHICH EXPERIENCE WAS
FOUND TO CHANGE FEEDING PREFERENCES

Species	Stage tested	Ref.
Orthopetera		
Schistocerca gregaria	L	225
Phasmatodea		
Carausius morosus	L, A	30
Bacillus rossius	L, A	166
Heteroptera		
Dysdercus koenigi	L	180
Homoptera		
Acyrthosiphon pisum	A	101
Schizaphis graminum	A	188
Coleoptera		
Epilachna pustulosa	A	104
Subcoccinella 24-punctata	L, A	2
Haltica lythri	L, A	162
Galerucella lineola	L, A	124
Phratora vitellinae (Swiss population)	A	178
Leptinotarsa decemlineata	A	238
Lepidoptera		
Noctuidae		
Heliothis armigera	L	1
H. zea	L	115, 239
Spodoptera eridania	L	191
Lymantriidae		
Euproctis chrysorrhoea	L	73, 128
Lymantria dispar	L	9, 73, 233
Arctiidae		
Hyphantria cunea	L	81
Sphingidae		
Manduca sexta	L	47, 65, 87, 115, 182, 241
Saturniidae		
Antheraea pernyi	L	73, 196
A. polyphemus	L	86
Callosamia promethea	L	86
Hyalophora cecropia	L	79
Limenitis archippus	L	86
L. astyanax	L	86
L. hybrid rubidus	L	86
Pieridae		
Pieris brassicae	L	118, 135
P. rapae	L	99
Papilionidae		
Papilio aegeus	L	202
P. glauca	L	47
P. machaon	L	236
Nymphalidae		
Chlosyne lacinia	L	214
Polygonia interrogationis	L	86
Pyraustidae		
Loxostege sticticalis	L	43

Partial data from Jermy, T., *Perspectives in Chemoreception*, Chapman, R. F., Bernays, E. A., and Stoffolano, J. G., Jr., Eds., Springer-Verlag, New York, 1987, chap. 9.

Table 2
LIST OF INSECT SPECIES WITH
WHICH EXPERIENCE WAS FOUND
NOT TO CHANGE FEEDING
PREFERENCES

Species	Stage tested	Ref.
Coleoptera		
Leptinotarsa decemli-neata	A	25
Tribolium castaneum	A	11
Phratora vitellinae (French population)	A	178
Homoptera		
Acyrthosiphon pisum	A	147
Lepidoptera		
Noctuidae		
Mamestra brassicae	L	112
Arctiidae		
Hyphantria cunea	L	112
Pieridae		
Pieris rapae	L	33
Pieris napi macdun-noughii	L	33

gaceae) were tested,[233] but a significant effect could be achieved with *Papilio machaon* when plant pairs chosen also from one family (Umbelliferae) were used.[236]

Since oligophagous insects *ab ovo* can be tested only with closely related plant species, it seems obvious that polyphagous species should be more prone to food imprinting than oligophagous ones. This has been proposed mainly as a result of electrophysiological investigations.[194] However, no correlation could be shown between the degree of polyphagy and the degree of food imprinting when the relevant literature was surveyed.[47] The variability of food imprinting depends more probably on the taxonomic and/or phenetic similarity of plants tested than on the degree of polyphagy.[233]

The degree of food imprinting also shows significant individual variation within insect populations.[30,115,135,162] At present nothing is known about the possible physiological and/or genetic basis of this variability.

3. Persistence of Food Imprinting

Although only a few insects have been studied in this respect, it seems that persistence does vary greatly both with insect and plant species. Jermy et al.[115] have shown with *Heliothis zea* larvae that imprinting on a plant species may persist through two molts and a whole instar between them, when that instar was fed an artificial diet not containing host-plant constituents. In *Carausius morosus* imprinting could be reversed, although this flexibility strongly decreased with ontogeny.[30]

In some cases striking rigidity of food imprinting has been observed. Ma[135] demonstrated that when larvae of *Pieris brassicae* were reared on *Brassica oleracea* and the young fifth instar larvae were transferred to *Tropaeolum majus*, all larvae died of starvation although the latter plant supported normal larval development from the first instar on. A similar observation was made with different plants using larvae of *Callosamia promethea*[86] and *Spodoptera eridania*.[191] This phenomenon was called the "starving-to-death-at-Lucullian-banquets" phenomenon.[112]

Grabstein and Scriber[78] drew attention to the fact that strong and permanent (rigid) food imprinting may strongly hinder discrimination between behavioral (preingestive) and physiological (post-ingestive) effects of different foods on insects. Imprinting has to be taken into consideration also when phytophagous insect species, introduced for biological control of weeds, have to be checked for their possible host plant range in the region of introduction.

4. Phytochemicals Involved in Food Imprinting

It is not generally clear which phytochemicals are involved in food imprinting. The ability of an insect to discriminate among plant species, however, does not guarantee that imprinting will occur,[47] as other factors may limit induction.

Manduca sexta larvae may become imprinted on artificial diets and nutrients, especially lipid components, were found to be responsible for both plant discrimination and food imprinting, although aqueous fractions were also involved.[199] Imprinting for corn oil or linoleic acid alone was also demonstrated. It was concluded[199] that the artificial diet used for rearing the larvae of *M. sexta* was not neutral relative to causing changes in feeding behavior, as had been implicitly suggested by earlier studies.[115,183,241]

Recent investigations[48] have indicated that solanaceous alkaloids are most probably not part of the "chemosensory profile"[47] ("biochemical profile"[111]) perceived by *M. sexta* larvae and thus do not provide chemical information for discrimination and/or food imprinting.

M. sexta larvae reared on two diets containing liquidized leaves of one of two host-plant species, became imprinted on the diet, i.e., they preferred the one on which they were reared, but they did not discriminate between the two plant species used in preparing the diets. Similarly, larvae reared on one of the two plant species did not discriminate between the two diets. This indicates that the larvae were imprinted by a composite taste quality of the leaf diet which was different from that of the leaves, and vice versa.[199]

Behavioral tests carried out with caterpillars that were fed on deterrent-containing diets indicated reduced sensitivity to the compounds used. For example, linolenic acid was deterrent for *M. sexta* larvae in behavioral choice tests but when the larvae were reared on a diet containing this compound, it became less deterrent. Wheat germ diet extract was much less deterrent to larvae reared on wheat germ diet than to plant-reared ones.[199] This implies that food imprinting may result from decreased sensitivity to deterrents. However, increased responses to stimulating extracts with *M. sexta* larvae that were accordingly food-imprinted have also been shown,[199] although the possibility that this was due to decreased responsiveness to deterrent components of the extracts was not excluded.

With regard to the role of volatile substances in food imprinting, it was shown[181,182] that in *M. sexta* larvae close range orientational response to a certain food increased as a result of previous experience. Since *Malacosoma americanum* larvae were found[54] to rely heavily on specific volatile substances for their initial discrimination of plants, and *Leptinotarsa decemlineata* adults were attracted to their host plants by a specific blend of general green leaf volatiles,[136,224] it is possible that olfactory experience of specific plant volatiles may be involved in food imprinting more generally than thought so far.

5. Food Imprinting as Related to Host Range

It was suggested[241] that first-instar larvae of the solanaceous feeder, *Manduca sexta*, were polyphagous because they accepted also nonsolanaceous plants when force-fed. Furthermore, *Verbascum thapsus* (Scrophulariaceae) supported larval development to the fourth instar and *Vigna sinensis* (Fabaceae) was suitable for complete larval development. the conclusion was that oligophagy was the consequence of food imprinting in this species.[65,241] Also Ma[135] has shown that the crucifer-feeding *Pieris brassicae* could be reared normally on *Tropaeolum majus* (Tropaeolaceae) from the first instar on, while *Brassica*-fed larvae did not accept *Tropaeolum*. Similarly, it was found that the first instar larvae of *Papilio machaon* had a genetically determined "spectrum of potential host plants" that was wider than that of the older instars.[236]

On the other hand, de Boer and Hanson[47] found with *M. sexta* that the relative preference for host over acceptable non-host plant species was maintained at the same level by rearing on the former plant species and was reduced by rearing on the latter. Therefore, they concluded that oligophagy in this species is inherited and not the result of experience.

The common feature of all such experiments was that the first instar larvae rejected many non-host plant species, i.e., "unacceptable non-host plants" (see Section IV). It is merely a question of semantics whether the somewhat broader potential food plant spectrum of the newly hatched larvae is regarded as a sign of polyphagy or as of less restricted oligophagy. Nevertheless, these findings clearly demonstrate the slightly greater potential of newly hatched larvae with respect to host-plant range as compared with older larvae. It also suggests and important role of food imprinting in forming the actual host plant spectrum of natural insect populations which may be narrower than the inherited potential spectrum. These results also suggest that care must be taken in experimentally determining the potential host plant ranges even in case of phytophagous insect species known as narrow food specialists.

C. FOOD-AVERSION LEARNING

It is well known that vertebrate generalist feeders (e.g., rats) sample a novel food by taking a small quantity when they first encounter it. After taking a mere mouthful they wait for visceral consequences, and if symptoms of sickness develop, they subsequently, following recovery, avoid that food. The process by which post-ingestional effects are associated with the taste or smell of the sampled food, to induce an altered behavior, is termed food-avoidance or food-aversion learning. Food aversion learning differs from many associative learning processes because of the long interval between the presentation of the CS (in this case the sensory cues of the food plant), and the US (post-ingestive malaise). (See general descriptions in References 10 and 140.)

While the phenomenon is well established with some vertebrates, there are only relatively few examples known in invertebrate animals. However, a similar phenomenon does occur and is best known with molluscs.[49,69,235]

Dethier and Yost[60] tested an oligophagous insect, *Manduca sexta*, for food-aversion learning. They conducted two types of experiments: one using *Atropa belladonna* or *Nerium oleander* leaves sprayed with insecticides, and the other using only *Petunia* sp. without treatment but known to have toxic effects on the insect. On both types of leaves the larvae began feeding and in varying time intervals they became "ill". Only larvae that had recovered from illness were tested again on the same but unsprayed foliage. Surprisingly, larvae that had recovered from the illness did not show a reduction in relative preference for the same plants subsequently.

However, two polyphagous species (*Diacrisia virginica* and *Estigmene congrua*) did show food-aversion learning.[55] Larvae fed *Petunia* for 24 h subsequently showed symptoms of illness. After recovery, the consumption of *Petunia* was compared with two other plant species (*Gaultheria* and *Unifolium*) in preference tests; *Petunia* was significantly less preferred by these insects compared with naive larvae. Bernays and Lee[16] suggest that the altered behavior of the caterpillars in this case may have "resulted from processes other than aversion learning. The alternative food may have become more acceptable...." Nevertheless, the above results are strongly suggestive of the occurrence of food-aversion learning. It is difficult to say how much and how serious symptoms need to be before rejection occurs at the next encounter. It may be that no conspicuous consequence of consumption on a toxic plant can be detected. In any case, *physiological* feedback is normally considered a prerequisite of food-aversion learning, but when no symptoms are measurable, this concept becomes entirely theoretical. Another approach would be to consider it as an associative process directed toward the avoidance of "negative" influences of which timely discovery can happen at sensory levels prior to swallowing or ingestion. For example, two reports with oligophagous species demonstrate another way in which rejection of food can be enhanced by experience[22,23] involving oligophagous species. In the first, *Locusta migratoria* on unpalatable plant species, like *Senecio* and *Brassica*, initially palpated and bit before rejecting the plants, and on a subsequent occasion rejection occurred following palpation without biting. That is, insects apparently learned to associate unpleasant taste stimuli at biting with other characters (e.g., olfactorily acting substances, or materials dissolved in a waxy layer or adhering to the surface of the plant), stored the information and next time on encountering the same species, palpation proved to be enough to cause rejection. Similar results and more detailed analyses of the rejection behavior are given on *Locusta*[23] and *Spodoptera exempta*.[21] The dynamics of rejection showed that on subsequent occasions the frequency of biting decreased and rejection at palpation became dominating.

These findings may possibly be explained by sensitization only, as strong and possibly noxious stimuli might produce a high level CES which would be strengthened by each further biting trial. Sensitization was not controlled for in these experiments, and by definition[141] it can occur when a CS (here specific substances of the plant leading to rejection) *is not paired* with an US (the noxious character of the plant species), though less rigorous definitions also can be found.[83] Perhaps also the notion of aversion learning should be widened or differentiated according to the degree of food specialization shown by the organism under consideration.

It is probable that poly- and oligophagous herbivorous insects possess the same level of chemosensory sophistication, but the processing of information in the CNS is different.[56] As a consequence, polyphagous insects may be less able to differentiate among plant species so substantial consumption of a plant having toxic constituents is more likely. Depending on the quantity and/or noxiousness of such substances, polyphagous insect herbivores may either detoxify or show food-aversion learning at next encounter. An oligophagous species being perhaps less able to tolerate alien plant substances internally will show rejection at the sensory level on the basis of information gained from tiny bites and learn from them. So we should perhaps term both types of reaction as food-aversion learning; this is contrary to the view that "...polyphagous species should be expected to show food-avoidance learning but oligophagous species would not"[70] (p. 202). It may be that one should look for more subtle effects with oligophages.

Several examples of food-aversion learning, and especially those on the polyphagous *S. gregaria*, do not enable us to make sharp distinctions on the basis of food specialization. When *S. gregaria* was fed with NHT-treated *Sorghum, habituation* was found, while to the same chemical the nymphs developed an *aversion*-type response if it was presented on glass fiber disk + sucrose. No sign of sickness was reported, however.[114] It is possible that not only the "typical" physiological response (illness) was unnoticeable or missing, but the outcome (habituation or food-aversion-type learning) depended on the stimulus situation provided (but see also Section V.D).

Bernays and Lee[16] studied the food-aversion learning of *Schistocerca americana*, a polyphagous feeder, on acceptable plants: *Brassica oleracea* and *Spinacea oleracea*, artificially associating a meal with subsequent injection of NHT solution into the body. Continuous observations were made using the following parameters: initiation of each meal, length of each meal, interfeed time. Immediately after a meal, either on *Brassica* or *Spinacea*, a dose of 2% aqueous NHT solution was injected into the abdomen of each insect which was then returned to its place where the next meal could be on either one of the two plant species. The results showed that on *Spinacea,* which was generally the less acceptable of the two plants, NHT injection was associated with the plant and caused a significant decrease in the next meal-size. The size of this meal was also significantly less than the meals on different controls. However, this was not the case if the next meal was on *Brassica* or if injection was followed by a meal on it. The experiment, then, demonstrated food-aversion learning, induced by coupling an artificial negative feedback with feeding on a particular plant. Generally, a single meal has been found to be adequate to establish food-aversion (e.g., References 49 and 243).

Another point of theoretical interest is the range of plant species on which food-aversion learning can be seen with phytophagous insects. Bernays and Lee[16] did not find aversion learning with *Schistocerca americana* on a fairly acceptable host, broccoli "A" (*Brassica*). At least for rats, it is known that on less preferred food, aversion may be acquired more easily[64] (cited by Bernays and Lee[16]), but the possibility of inducing food-aversion learning on the most preferred foods cannot be excluded. The use of toxic compounds on highly acceptable plant species to deter further feeding is, in principle, similar to the experimental circumstances provided for aversion learning. It would be interesting to discover in what percentage of cases decreases in consumption reportedly due to antifeedant activity actually involve food-aversion learning.

Experiments with rats[153] indicated another possibility—that of second order conditioning of odor with post-ingestive malaise. The odor alone was not a sufficient cue to associate with a poisonous food; the odor, in turn, had to be associated with taste before it resulted in refusal of the food. Second order conditioning has not yet been demonstrated with phytophagous insects.

D. DIETARY SELF-SELECTION

It has been found that some vertebrates will eat several kinds of food in proportions yielding an optimal nutrient balance (see Waldbauer and Feiedman [228] for references). Such dietary self-selection has been demonstrated also in the polyphagous species *Tribolium confusum*[226] and *Heliothis zea*.[35,227] It can be assumed that self-selection occurs in other phytophagous insects.[228]

The most probable behavioral mechanisms underlying self-selection is proposed by the "malaise hypothesis". Cohen et al.[34] supposed that feeding on a food lacking or deficient in some nutrients causes metabolic disturbance (malaise) that, through a feedback mechanism, results in exploratory behavior until the insect comes to another kind of food that contains the missing nutrient. Feeding on the latter the insect associates the sensory stimuli from that food with the alleviation of the malaise. In this way the insect will switch back and forth between nutritionally incomplete but complementary diets.

It can be added that the first step of the self-selection behavior, i.e., the abandoning of the deficient diet that causes malaise, may involve a learning process. Namely, the chemical stimuli of that diet are associated with the malaise, therefore, the diet becomes sensorially unacceptable and will be avoided. This process is analogous to food-aversion learning, although the aversive behavior may be much less persistent and is easily deleted by feeding on another diet. The second step, i.e., the feeding on a complementary diet, is analogous to food imprinting. Thus, self-selection requires delayed learning mechanisms that have been found in several animals.[176,192]

E. LEARNING IN FOOD FINDING

There are no data on how widespread learning is in the food finding processes of phytophagous insects. Contrary to the abundant examples of oviposition site finding and acceptance behavior in these insects (see Section VI.A), there are, to our knowledge, only a few studies demonstrating learning in food finding.[17,132,181,182]

Bernays and Wrubel[17] attempted to associate visual cues (color and light intensity) with food in nymphs of the grasshopper *Melanoplus sanguinipes*. As a general experimental design, they used a cage in which two 3-sided cardboard boxes painted green or yellow were placed. Experimental insects within the cage stayed on a warm wire roost between feeding bouts and there were three stages of the experiment: (1) pretraining, where they could visit the empty colored boxes without food, (2) training, where they could find food either in the green or the yellow box and were allowed to feed freely for ca. 18 h, and (3) post-training, when both boxes were replaced by new empty boxes. The nymphs were tested both for positional and color learning.

The experiment proved that *M. sanguinipes* nymphs could associate visual stimuli with food rewards. While in the case of the yellow box loaded with food the median length of stay post-training did not increase over the length of stay in the pretraining period, the value for green significantly increased. The association was developed without a positional effect by the boxes. With untrained insects there was a bias toward yellow, but a significant enhancement of the preference for it, when the yellow box contained food, was not demonstrated perhaps because it was already so high. The learning process was quick, as a single trial was enough to locate food faster subsequently.

Saxena and Schoonhoven[181,182] drew attention to the fact that food-imprinting in lepidopterous larvae changes not only feeding behavior but also olfactory orientation that may be of importance in food finding. As a result, changes in host plant preferences due to experience are more obvious in choice tests where olfactory discrimination of plants tested is also involved as compared to no-choice tests where this possibility is excluded.

In laboratory experiments, Lewis[132] demonstrated that in *Pieris rapae* butterflies the time required to find nectar in flowers decreased during subsequent encounters, i.e., they actually learned to discover nectar more efficiently. For switching to a new source then back again to the previous one, first they had to learn then successively relearn the way of "handling" a flower in order to improve performance. (Also see remark in Section VIII.)

VI. EXPERIENCE-INDUCED CHANGES IN OVIPOSITION BEHAVIOR

In this section emphasis is placed on experimental evidence of experience-induced changes in oviposition. However, development of a searching image is also discussed (1) as it might be involved in oviposition behavior, and (2) since there are some shared elements in the learning mechanism. Experience-induced changes of oviposition preferences have been demonstrated only in a few phytophagous insect species so far (see Table 3).

Studies on learned oviposition preference (induced oviposition preference, IOP) fall into the category either of sensitization or of associative learning (both classical and instrumental conditioning, as well as trial-and-error learning). In most cases the adult

Table 3
LIST OF INSECT SPECIES IN WHICH
EXPERIENCE WAS FOUND TO
CHANGE OVIPOSITON PREFERENCE

Species	Ref.
Caleoptera	
Callosobruchus maculatus	137, 232
Lepidoptera	
Pieridae	
Pieris rapae	216, 217, 218
Papilionidae	
Papilio machaon	237
Battus philenor	74, 154, 155, 156, 172
Heliconius sp.	74
Diptera	
Tephritidae	
Rhagoletis pomonella	167
Ceratitis capitata	37

must "evaluate" and associate visual (spatial, light intensity and reflectance, shape), tactile, and chemical (contact chemical, olfactory) information. As learning would be expected in all these it may seem more appropriate to replace the term IOP by "learned oviposition preference".

In addressing the term "induction" as commonly used in the literature on oviposition, Dethier[57] noticed that "conditioning" (that is to say associative learning) generally had to be excluded as a possible mechanism because of the lack of rigorous criteria set for learning studies. However, we believe that although in earlier studies of induction of oviposition preference the conditions were not designed to meet the criteria for associative learning, nevertheless, elements can be recognized. The more recent studies of associative learning in *oviposition* behavior fall into this group.

Traynier[215] demonstrated a long-term change in the oviposition behavior of the cabbage butterfly, *Pieris rapae*, induced by contact with plants. Gravid females were placed in cages and were allowed contact either with cabbage or lettuce disks for 30 min, then deprived of any leaf disks for 24 and 72 h, and finally retested on cabbage or lettuce. "Cabbage-experienced" butterflies landed on cabbage and laid eggs, whereas on lettuce they frequently moved from disk to disk but rarely laid eggs. The females "experienced" on cabbage landed more frequently on lettuce than those becoming "experienced" on lettuce, but few eggs were laid. Those that had experienced lettuce on each occasion showed a low level of responsiveness in both tests, because the tendency to land decreased after the first two minutes. It could be that the females gained oviposition site "imprinting" or induction and had learned to respond to green disks but it is also possible, as suggested by Traynier,[216] to explain it by sensitization. Likewise, a state of sensitization may account for the results obtained with ovipositing leek moths (*Acrolepiopsis assectella*). The information obtained about specific host substances caused a response corresponding roughly to the duration of exposure and was retained over time.[206]

Traynier[216,217,218] has repeatedly and successfully demonstrated associative learning in ovipositing *Pieris rapae*. Sinigrin stimulates oviposition in this species. "Training" was given by exposing the gravid females for 30 min to green paper disks treated with 10 ppm aqueous sinigrin solution. The oviposition response of females was tested after 1.5 and 21.5 h, on water-only disks. There was a substantial difference between the number of eggs laid by "sinigrin-experienced" and by those butterflies not having contact with sinigrin. It appears that the butterflies developed an association between green color and chemical suitability of disks as oviposition sites.[216] Since the butterflies showed an intrinsic preference for a green color, the experiment was balanced for such a bias, but it did not control for sensitization and pseudoconditioning.

A subsequent paper[217] on oviposition site learning put more emphasis on visual components. *Pieris rapae* females oviposit readily on paper disks treated with an aqueous sinigrin solution. However, certain colors (violet, red) or black are inherently not preferred, and on these even sinigrin could not release an egglaying response and oviposition occurred only on water-treated disks of acceptable colors. Associative learning in egglaying site selection has been proved to be possible and important also in other phytophagous insect species, such as *Rhagoletis pomonella*,[167-169] *Ceratitis capitata*,[37] and *Battus philenor*.[154-157]

Prokopy et al.[167] trained naive apple maggot flies (four successive oviposition bouts separated by 2-min intervals) on apple. After the last training period the flies were offered a fruit of *Crataegus mollis* which they rejected; conversely, flies trained on *C. mollis* did not accept apple. The experiment showed that flies could learn to accept or reject an oviposition substrate, and that the possible mechanism was associative learning. The fruit flies must have associated several (?) attributes of one fruit with its suitability as oviposition site.

Subsequently, however, this study has received considerable criticism from students of learning and behavioral geneticists. Holliday and Hirsch[93] rejected the learning interpretation for the experiment,[141] objecting to the experimental design. In a reply by McGuire,[142] it was stated (p. 469) that "the study of Prokopy et al. (1982) was not designed to be a learning study and therefore lacks some learning controls. Nevertheless, he regarded their study as an example of instrumental conditioning. In a similar study,[168] naive flies without training showed a great deal of variation in response to an oviposition substrate other than that which served as larval food. In contrast, in field assays, among females that had just finished ovipositing into apple or *Crataegus,* most preferred to continue ovipositing in the same type of fruit. The results could be explained as genetically determined food preference and/or sensitization. However, in the field test, and in a separate case when females were trained to a given fruit type in the laboratory, the females consistently rejected a subsequently offered different test fruit. This behavior can best be explained by associative learning. In the next study[169] the "true nature" of the learned performance was addressed. The question was posed, whether: (a) experienced females acquired a greater propensity to accept the fruit-type on which they were trained as compared to naive females, or (b) experienced females had a greater inclination to reject a novel fruit type. The results suggested that the second alternative was more probable. With a similar experimental set-up learning was demonstrated in *Ceratitis capitata*.[37] An interesting addition is that this species is a highly polyphagous one. In general accordance with the selective advantages of learning to an oligophagous species[169] we expect that polyphagous feeders have an even higher propensity to perform different types of learning. Variability has been found in the level of conditioning with a given fruit type in a wild population, which may reflect the difference in the "quantity" of conditioning stimuli provided by the substrate.[37] We think that the length of exposure, that is, the "quantity of experience" obtained previously on that or a similar type of fruit may also be important.

What characters of the host are associated with the development of the response? Papaj and Prokopy[158,159] studied this and concluded that with *Rhagoletis pomonella* it was fruit size and the surface chemistry which counted. Papaj,[154,155] pursuing Traynier's line, successfully demonstrated with *Battus philenor* association of visual cues with relevant characteristics of the host plant. The butterfly is able to discriminate between the two host *Aristolochia* species visually by leaf shape.[172,174] (See "Searching image" below for details.) The butterflies demonstrated instrumental conditioning in this case. Pseudoconditioning and sensitization were excluded as possible mechanisms. It has been supposed that leaf buds may improve host-seeking behavior in the same way as leaf shape did.[155]

With regard to *Coleoptera*, very few examples[137,232] of learned oviposition preference have been found, all with a bruchid species, *Callosobruchus maculatus*. The plant species tested, however, differed: *Cajanus cajan* and *Phaseolus radiatus* in the earlier study, and *Vigna unguiculata* and *Cicer arietinum* in the later one. A learning process is possible in these studies, since in addition to the effect of inherently preferred host seeds, other specific stimuli such as the seed's surface curvature,[6] chemical constituents of the seed coat, oviposition marking substances, and seed size can all serve as associating reinforcers and therefore influence the response.[146,152]

Searching image. This term was coined following Tinbergen's[213] and Royama's[179] classical works on bird food seeking and food selection behavior. They observed that, in spite of the availability of an alternative food, the birds "insisted" on foraging only a certain type, and ignored others. As an explanation, it was supposed that the foraging animal, being inexperienced with all available food types in a given environment, would discover a cryptic food item and then would look for similar ones, while others went unnoticed, at least until another was suddenly recognized again.

It is assumed that searching image is accompanied by a perceptual change in the predator that enhances its ability to discover cryptic prey. Searching image presupposes discriminating ability and selective attention.[140] It results in a preference[103] for a given food by learning. During foraging, switch-over to another type of cryptic food may occur and new experience (with reinforcement) leads to the formation of another searching image. It is argued[126] that the proper expression for the mechanism was learning to see, because the process was actually breaking the camouflage of a cryptic prey which became suddenly visible, and recognizable for the forager. From the moment of recognizing something edible the process is a quick, one-trial associative learning process reinforced by rewards from the now suddenly visible and similar-looking food items.

Although there are relatively few experimental examples of searching image, a recent paper[84] questions even the by now "classical" results,[41,45,46] stating that they proved only a change in "search rate" and were better explained by this than by the formation of a searching image. (Search rate furnishes the predator with an ability to *vary* the time required for scanning a given part of the environment while foraging.) Furthermore, Krebs[126] collected and listed all those situations where a predator learns those behaviors, sites, etc. that led to more efficient foraging, and found no support for development of a searching image.

Courtney[39] predicts by means of a model that a female insect's searching for the most preferred host by employing a searching image may result in a higher overall host discovery rate as compared to "searching for all host-plant species" (p. 317). (See also Section VIII).

However, in phytophagous insects there are no unambiguously demonstrated instances of searching image formation. It is not surprising that all supposed cases concern species with well developed visual acuity, although nothing would exclude the formation of an olfactory searching image. Formation of a searching image in ovipositing females occurs mostly in connection with certain plant parts, and a similar behavior can be expected in flower visiting. Though the phenomenon shows a superficial resemblance to imprinting, they are probably different, as the latter lasts for a relatively long time, even a lifetime, while foraging requires frequent switching in searching image throughout a season. *Heliconius* species are known to differentiate between leaf shapes, as well as small details of leaves (egg mimics) while seeking for an egg-laying site.[74] Albeit no searching image formation was shown with *Heliconius*, we assume it is likely to be occurring, especially since Gilbert explicitly states that an associative learning process takes place not only for leaf shape but also for spatial position of particular *Passiflora* vines. Wiklund[237] describing *Papilio machaon* females after laying an egg "on the same plant species a number of times...evidently acquire a 'searching image' of this plant, as it is not uncommon to see swallowtails making initial oviposition approaches towards non-host plants which bear a general superficial resemblance to the dominant host plant of the locality" (p. 194). Rausher[172] discovered by observation in nature that searching image formation took place in *Battus philenor* female butterflies when they were looking for oviposition sites. The two host-plant species, *Aristolochia serpentaria* and *A. reticulata* differ in leaf shape and once a female discovered a host plant, it tended to look for similar ones.

Lawrence and Allen[129] argued that a searching image interpretation is not appropriate for the oviposition site searching behavior in *B. philenor,* because Rausher[172] overlooked alternative learning possibilities[45,46,126] which must first be excluded. The particular alternative is the "preference or avoidance of a prey over others that is independent of the predator's ability to see the different types"[129] (p. 313). It is indeed an inherent preference that is shown by *B. philenor* females for both wide and narrow leaf shapes of *Aristolochia* species, the food plant. Unlike a generalist predator, where a wide spectrum of prey species may be available, there are only two alternatives for *B. philenor* butterflies: to search in one or in the other leaf-shape mode. Females adopt mostly one type of search mode at a time, so although searching image might be involved, this interpretation is not necessary, because instrumental learning may also be involved.

Lawrence and Allen[129] also criticize Rausher[172] for not demonstrating *Aristolochia* to be "cryptic", though it probably was. Among similar looking leaf shapes of other plant species *B. philenor* females searched only in one mode and they were not able to differentiate perfectly between hosts and non-hosts solely by visual cues. It was the chemical information that gave the clue for identification. During this process mistakes of identification did occur;[155] that is also an accompaniment of searching image development. Once again, a critical point that is common in all such studies is the lack of demonstration whether the increased ability to discover food items by the forager comes from changing the search rate[84] (e.g., Figure 2, p. 200 in Reference 172).

Another study with *B. philenor*[154] experimentally proved that the alighting response was reinforced by chemoreceptive experience of the host plant and this in turn aided the recognition of further hosts.

Summing up, it may be stated that, though formation of a searching image has not been proven in insects and such studies certainly face exceptional difficulties, it is still possible that such a learning process does occur and plays a role in the foraging of insects. The formation of a searching image may be restricted to visually searching phytophagous insects.

VII. PRE- AND EARLY-IMAGINAL EXPERIENCE

The discussion of pre- and early-imaginal experience under a separate heading is justified only by the special character and historical significance of the phenomenon in insect-plant interactions. As for its placement into a learning category, opinions differ widely.[141]

For more than a century entomologists suggested that larval experience gained on a specific food would increase the adult's preference for that food.[230,231] This phenomenon was called the "Hopkins Host Selection Principle" (HHSP) because Hopkins[95] was among the first to draw attention to it. Several authors claimed to have proven HHSP in phytophagous insects,[40,99,128,161] in *Drosophila*[90,210] and in parasitoids.[195,212]

Experiments to prove the validity of HHSP should meet the following conditions:

1.   The possibility of selection of genetically based oviposition preference should be excluded. Therefore, all experiments that have been carried out using adult field populations for oviposition preference tests,[128] or insects that were reared for several generations on a certain plant,[99,161] are not convincing, because differences in behavior might be determined genetically.
2.   The contact of emerging adults with the larval food or its remainders must be excluded since the cases of learned oviposition preference (see Section VI) clearly proved the influence of early adult experience on adult preference behavior. In several cases the above precaution was not taken or at least the authors do not report on it.[90,99,161,195,212] This is especially true for experiments with parasitoid insects where contact of the emerging adults with the remainders of the host (pupal exoskeleton, cocoon, meconium) was presumably not prevented.

Unfortunately, no attempts have been made so far to find out, whether larval (nymphal) feeding experience is transferable to adults in hemimetabolous insects, where the last molt is not accompanied by such basic morphological and physiological changes as in holometabolous groups. Based on experimental evidence several authors explicitly denied the validity of the HHSP.[51,98,106,124,137,162,164,169,204,205,242]

Jaenike[106] (p. 324) proposed a "neo-Hopkins host selection principle" suggesting that the "exposure of adult insects to a particular type of host will often, though not always, increase the subsequent acceptability of that host as an oviposition site" (see also Section VI).

The role of early adult experience has already been demonstrated;[210] washing the fully developed larvae or newly formed puparia of *Drosophila*, i.e., freeing them from the remainders of the rearing media, reduced the effect of the larval food on adult behavior, although it did not eliminate it totally. This and similar cases led to Corbet's[38] "chemical legacy" hypothesis. She claimed that even if the puparia were washed, some chemical cues, originating from the larval food, might persist within the puparial exoskeleton (see Corbet[38] for references), and thus might be contacted by the newly emerging adult.

VIII. ADAPTIVE SIGNIFICANCE OF EXPERIENCE-INDUCED BEHAVIORAL CHANGES

Studies on changes due to experience in feeding and oviposition behavior of phytophagous insects naturally also include discussions on the presumed adaptive advantages of the changes. In this connection one often finds Panglossian types of assumptions that regard the observed learning modalities as perfect solutions for performance of tasks faced by insects.

The function of a certain behavior is often regarded as its adaptive value[127] which implies that the behavior originates from and is maintained by natural selection, and is genetically determined.

Jamieson,[108] however, convincingly proved by concrete examples that a purely functional approach to the causality of the origin, diversity and persistence of behavior patterns may lead to unwarranted and unsupported selectionist explanations, because it ignores extragenetic factors evoking and maintaining new behaviors. For example, it has been shown that in some species generalist or specialist types of behavior depend on the environment where the individuals were raised.[80] Thus behavioral evolution may arise also through adaptation of an unaltered genotype to a persistent environmental change[119] (cited by Jamieson[108]). In most cases of behavioral plasticity learning processes are probably involved.[92] Life cycle studies have also led to the conclusion that phenotypic plasticity can be genetically determined and is subject to selection, but the position of an individual within the range of phenotypic variation is not determined.[26,27]

Thus, if a given intraspecific difference in a behavior proves to be of epigenetic origin, there is no point in measuring costs and benefits of that behavior on the population's overall fitness, because in such cases selection does not influence the average genotype.

Experience may modify individual adult lifetime parameters and "fitness". For instance, males of *Drosophila melanogaster* obtain experience from fertilized females that subsequently modifies their courtship behavior.[193] Studies with *Drosophila* and *Phormia* species,[82] and papers cited therein, demonstrate that learning ability is an individual characteristic, and that roughly 30% of a population would respond as "good learners". Selection experiments and work with mutant strains corroborate that a level of genetic determination is involved (see Section II.D). It is unfortunate that, to the best of our knowledge, there are no similar studies on phytophagous insects. Thus nothing is known about the genetic background of either the behavioral changes due to experience, or the ability of the insects to change. In the following, only the presumed adaptive significance of the behavioral changes themselves will be discussed.

Habituation to feeding deterrents. This enables the insect to feed normally on plant species that belong to the potential host plant spectrum but are in some degree deterrent to "naive", unexperienced individuals. This might be of adaptive value where there is not a strong correlation between deterrence and toxicity of plant phytochemicals. Habituation to food-related stimuli is a type of phenotypic plasticity, the range of which is determined genetically. How widespread the ability to habituate to plants with deterrent stimuli might be is not known, although generally no habituation was found in *Mamestra brassicae* larvae to several acceptable but slightly deterrent non-host plants.[116]

Whether or not the great individual variation in the capacity for habituation to deterrents observed in several phytophagous insect species[114,116,203] is genetically determined is not known, so that we cannot yet comment on whether or not it could be subject to selection.

Food-aversion learning. This prevents the ingestion of deleterious quantities of a poisonous food if there is not a genetically based deterrent response to the food. On the other hand, one might argue that making complex decisions about the food whether learned or unlearned is highly developed in polyphages (E. A. Bernays, personal communication). It is, therefore, not surprising that the capacity for aversion learning should be demonstrated mostly in polyphagous animals,[55,72,114] although it may also have adaptive significance for oligophagous species. Significant individual variability has been observed also with this behavior.[116] If such variability is of epigenetic nature, selection would not alter the average genotype of the population.

Food imprinting. As for the selective advantage of the capacity for food imprinting several authors have assumed that it reduces the probability that in a mixed plant stand an insect would frequently change food plants which, in some cases, has been found to reduce the efficiency of food utilization.[78,79,187,189-191,202] However, in other cases feeding on a sequence of different host-plant species enhanced development and fecundity (see also Section V.D) or had no effect at all.[42,73,79,144]

It is conceivable that in mobile insects foods imprinting enhances the fast refinding of the food at the end of intervals between two meals.[79,181] This might be of special importance, e.g., in some noctuid moth species whose larvae spend the day in the soil and thus have to refind the food plant every evening. Nevertheless, it is difficult to envisage what adaptive advantage may be attributed to cases when larvae become imprinted for (acceptable) non-host plants[47] that do not permit normal development. Furthermore, rigid food imprinting may be clearly a disadvantage in situations when, e.g., a host plant becomes defoliated by lepidopterous larvae before they complete development and no other plants of the same species occur in the vicinity. With our present knowledge, it is hard to conceive that rigid food imprinting, leading to the "starving-to-death-at-Lucullian-banquets" phenomenon,[112] is adaptively advantageous, although host fidelity could have ecological advantages.[15]

Jermy[112] proposed that food imprinting may simply reflect the limited flexibility of the insect's nervous system, i.e., the restricted capability of recognizing different chemical "Gestalts" as food plants. A specific chemical "Gestalt" becomes "imprinted" in the sensory system, i.e., the insect is not able to "abstract" a general "notion" (chemical "Gestalt" image) of "host plant" that would incorporate the chemical "Gestalts" of all suitable host plant species and would exclude all non-host plant species. When a newly hatched larva begins to feed on one plant species belonging to the potential host plant spectrum, which may contain plant species of quite different chemical composition (see Section V.B.5), it becomes tuned to that plant, but also loses the capability of recognizing another possible host-plant species, the chemical "Gestalt" of which more or less strongly differs from the experienced one. The occurrence of such a "chemical tunnel vision" is also supported by the findings that food imprinting is stronger (high induction index) the more distantly two plant species are related botanically, i.e., the more their chemistry differs.[47]

In conclusion, food imprinting may result from the insects' incomplete plant recognition and relearning capabilities. In some cases it may serve simply for fast refinding of the food by mobile insects; in other cases it might be more or less irrelevant with respect to the fitness of the individual, while in some ecological situations it might be disadvantageous. Since food imprinting is an epigenetic phenomenon, even the occasional disadvantageous consequences do not affect the average genotype of the population in ecological time. The genotype determines the function of the nervous system which, in some insect species, leads to food imprinting. Whether or not there has been selection for or against the capacity for imprinting cannot be determined.

Host finding. Learning is considered to be important in the foraging behavior of higher animals.[122] It is assumed that behaviors incur costs and/or benefits, and as such, measures of "profitability" of different behaviors are often undertaken. The scientific area that has grown on such conceptual grounds is optimality theory. Most examples came from the domain of foraging theory, and specifically on prey-predator relationships (see, e.g., Pyke[170] for a review). However, aspects like mate finding and choice (e.g., Reference 24) and others have also been developed. Optimality theory has received considerable criticism (e.g., Reference 133). Recently Pierce and Ollason,[163] collecting convincing arguments against various aspects of the theory, concluded that it may not be possible at all to test whether an animal behaves optimally.

Optimality studies are generally scanty with phytophagous insects. The following examples demonstrate both the scarcity of knowledge in this domain, and the controversies in their explanation.

Following Holling[94] and Charnov,[32] Futuyma[68] has shown by a mathematical model that under certain conditions a specialized type of host-seeking behavior, like the use of searching image in some *Lepidoptera*, may increase fitness and thus may be selected for. Courtney[39] came to the same conclusion. A model has been constructed on optimal oviposition behavior in phytophagous insects suggesting that imprinting is useful.[105] By another model[131] it was demonstrated that if an egg-laying female insect has only limited capacity for distinguishing between suitable and toxic plants, then it may reach highest fitness by ovipositing only on one identifiable palatable plant species.

Prokopy et al.[169] emphasized that the adaptive significance of learning to discriminate among fruit species by *Rhagoletis pomonella* females is unknown. Nevertheless, they suggested that (p. 105): "learning to reject a novel fruit may reduce the likelihood that an experienced female entering a patch of the rare host type would stay in that patch and continue to search" and by this may enhance the probability of finding abundant conspecific host fruits. This is supported by the findings[175] that the female flies left a tree more quickly after encountering an unsuitable fruit.

It was suggested that learning increases the searching efficiency in females of *Colias* spp. because the host plants are rather cryptic in their natural surroundings.[200] Papaj,[156] discussing the results of investigations into the host plant searching behavior in *Battus philenor*, concludes that selection favors rapid learning by naive females and conservativism in switching from one host to another by experienced ones, both enhancing efficiency of host finding. But this searching behavior is not very accurate, because (p. 306) "even after considerable experience...females land on many non-host leaves whose shape is similar to that of the preferred host.[155]

As regards food finding, grasshoppers were found to associate visual cues (colored boxes in the laboratory) with the presence of food.[17] There is also evidence that they may be able to associate non-nutritive volatile chemical cues (mint) with food, so that they move upwind when exposed to it.[130] Both types of learning may enhance food finding, but evidence is lacking on their importance under natural conditions.

It has been shown that experienced bumblebees may remain constant to a flower type even after it deteriorates as a resource because relearning in foraging is costly in time and energy.[89] In this connection, however, one may also argue that conservatism in switching behavior is not necessarily adaptive, but may originate from the inability of the insects to relearn, as in the case of food imprinting (see above). This is more likely in the case of short-lived butterflies. The wealth of publications on learned oviposition, although they deal with only a few species, tend to suggest that egglaying is, in general, fairly accurate. For example, *Battus philenor* females discriminate "between hosts and non-hosts, among host species, and among individuals within a host species. The facultative adaptation of that preference enables the females to select the most suitable individual plants from a complex non-host background, even as suitability changes over phenological time"[155] (p. 306-307). The other side of the picture, namely, the inability or the restricted ability to learn as well as the imperfections in finding and recognizing the host, were seldom studied although the latter are important for the understanding of insect population dynamics.[52] As an excellent example, Dethier[52] has clearly demonstrated that females of the butterfly *Euchaetias egle* fail to find many scattered host plants in a meadow and the number of eggs laid on individual plants is assessed inadequately in relation to the quantity of available food. This often results in defoliation of some plants before the larvae have completed development, while other individual plants remain unattacked. In such cases the larvae may die of starvation or be killed by predators because they are unable to find more distant host plants. In another example, *Leptinotarsa decemlineata* females often lay eggs on non-host plants, probably because the leaf surface structure of the latter is more suitable for depositing the eggs than that of the host plant. As a result, a considerable number of eggs may be situated several decimeters from the host, though the chance of finding the host by the newly hatched larvae rapidly decreases within a few centimeters.[113] These examples indicate that the use and value of learning in relation to oviposition may vary considerably from species to species.

Significant individual variation has been observed in the host searching behavior of butterflies. Assuming that the variation is genetically determined, Rausher[173] developed a model of the modification of egg-laying behavior by selection. However, if the variation is only of epigenetic origin, which cannot be excluded, the model becomes questionable.

The common feature of both food imprinting and learned oviposition preference is that the insects involved show narrower host-plant specialization than the "naive" ones, the latter presumably indicating the inherited host plant spectrum. It seems, therefore, logical to ask: What is the adaptive advantage of narrowing down the host range during ontogenesis? This leads to the never-ending debate on the adaptive advantage or disadvantage of specialist vs. generalist behaviors. Since it would fall outside the scope of this paper even to outline the essence of this debate, we only refer to two recent publications on the evolution of host-plant specialization in phytophagous insects.[15,117] However, if the behavioral narrowing down of host-plant range results simply from "tunnel vision" caused by experience, and is explainable by the limits of neural mechanisms, then the problem of adaptive advantage or disadvantage becomes dubious.

As can be seen from the foregoing, learning ability is extremely variable in phytophagous insects both in inter- and intraspecific contexts. This does not necessarily reflect the advantages and disadvantages of learning ability. It may simply indicate that in some species the evolution of the nervous system reached a level which enabled the selection for complex learning processes.

IX. BEHAVIORAL CHANGES AND SPECIATION

Mayr[139] expressed the view that changes in behavior may represent the initial step in the subdivision of a species, i.e., in speciation. Naturally this occurs only if the behavioral differences also result in at least some reproductive isolation between the populations involved.[102] With this in mind, the following can be said about the probability of reproductive isolation provided by experience-induced changes of feeding and oviposition behavior.

Food imprinting of larvae itself obviously cannot result in reproductive isolation with endopterygote insects, since it is not transferable from the larval to the adult stage. However, a seeming transfer may result from early-imaginal experience (see Section VII). In such cases the perpetuation of a specific host preference through generations is conceivable, and may be enhanced by selective mating that takes place on the host plant. For example, this may happen in *Haltica* species.[162]

The gene flow between populations imprinted to different host species may be impeded if selective mating also occurs. Although no suitable data are available on food imprinting with Hemimetabola, the transfer of imprinting between larva and adult would seem to be most likely with this group.

Food imprinting in herbivorous adults, if it is connected with oviposition on the same plant species, theoretically may result in isolation of populations in the same way as has been discussed above in case of early-imaginal experience. Without genetic changes, however, the emergence of such "behavioral host races" would not persist in nature. Therefore, when a food-imprinted insect population is found in nature, detailed and long-term investigations are needed before it can be revealed whether the specific behavior persists over subsequent generations or is restricted only to the one in which it was first found.

Learned oviposition preference has been found so far only in *Lepidoptera* and *Diptera*. Since there is no feedback of information from the larval to the adult stage, specific preference does not persist through generations; thus, it cannot result in isolation. However, early-adult experience deserves attention also in these insects. For example, since tephritid flies living in fruit trees pupate in the soil under the tree, the newly emerged adults will probably first encounter that tree on which they developed as larvae. By this the learned specific oviposition behavior may become a seemingly "inherited" trait and may temporarily isolate populations. However, such a process is rather improbable in sympatry without simultaneous and accidental genetic changes promoting reproductive isolation. Gene flow by emigration/immigration could significantly reduce the degree of isolation, especially in highly mobile species.

Jaenike,[107] studying the genetics of host selection behavior in egg-laying *Drosophila tripunctata*, concluded that sequential phases of oviposition behavior are regulated by independently segregating loci. Therefore, the probability of resource-based host race formation and subsequent speciation is unlikely in this species.

The "chemical legacy" hypothesis[38] stresses in general that even minute chemical traces of the host that are transferred from one ontogenetic stage to the next could have significant behavioral effects if they are present in the immediate environment of an insect's sensitive developmental stage. Corbet assumes that cases of host specificity that were hitherto interpreted in genetic terms could also be accounted for by this hypothesis. One of its predictions is that host shifts in phytophagous insects might result from "accidental contacts of a sensitive stage with a novel host"[38] (p. 150). Modern methods of chemical analysis are sensitive enough to prove or disprove the above hypothesis. However, considering the results of behavioral experiments and observations carried out so far by many authors, it is very unlikely that accidental contacts would cause shifts to novel host plants outside the inherited host spectrum.

Thus experimental and observational data available so far do not allow us to draw firm conclusions about the possible role of experience-induced behavioral changes of feeding and oviposition in the processes of speciation.

X. CONCLUSIONS

In recent years, increased attention has been focused on experience-induced changes in the host-selection behavior of phytophagous insects. The studies have provided a wealth of information on a series of behavioral phenomena, but their interpretation, especially in the context of modern learning theories, is still controversial. Such studies often lack planned control experiments which sometimes make them uninterpretable or at least ambiguous. This is especially true for cases where fine distinctions have to be made, e.g., to decide whether a given response is only sensitization or whether associative learning played a role. On the other hand, the set of criteria that defines a certain type of learning is difficult to meet experimentally; therefore, the evidence gained often allows only an *a posteriori* and frequently inadequate explanation. Further studies could answer the questions of general importance: How fixed is the host selection behavior in various groups of phytophagous insects (e.g., euryphagous vs. stenophagous species) and what are the limits of behavioral changes, i.e., how far can learning change behavior? Such investigations may provide important contributions to the knowledge of learning capabilities of insects.

Optimality approach views behavior as a series of processes in which cost/benefit measurement can be done. Taking into consideration the arguments and debates surrounding the topic, optimality models should be predictive enough to allow understanding elementary rules in animal decision making. Nevertheless, once learning is incorporated as a possible behavior modifying factor, models may become too complex to handle or understand; the better learning abilities an animal has the more alternative behavioral outcomes are possible. Naturally, not only the learning abilities, but also the limits of learning, of a given animal species should be known.

Another large group of open questions relates to the neural mechanisms underlying behavioral changes. The interesting results of electrophysiological experiments obtained so far indicate that further studies may shed new light on the function of the insect peripheral nervous system and its relation to learning.

The overwhelming majority of experiments and observations have shown that considerable individual and interpopulational variations of behavioral phenomena exist. How much is due to genetic or to epigenetic factors is totally unknown. The behavioral genetic approach should urgently be extended to those phytophagous insect species whose behavior has already been studied in detail.

A further serious insufficiency of the studies carried out so far is the almost total lack of experiments and detailed observations in nature. Clearly, such investigations meet with enormous methodological difficulties. Without such studies, however, it is very dangerous to extrapolate the results of laboratory experiments to natural populations, and trying to estimate the "raison d'être" of any experience-induced behavioral change may provide very vague conclusions. At the present time the genetic or epigenetic nature of the behavioral phenomena is an open question. If only epigenetic causes are involved, the learned behavior is not affected by selection, because it has no effect on the average genotype of the population. Thus, at present most speculations about the evolutionary significance of experience-induced changes in host selection behavior of phytophagous insects are built on loose ground.

ACKNOWLEDGMENTS

The authors are deeply indebted to Professors Elizabeth A. Bernays, Reginald Chapman, Endre Grastyán, Dr. György Kemenes, and to three anonymous reviewers for critical comments, advice, and corrections. Special thanks are due to Mrs. Judit Horváth for practical assistance in many different ways.

REFERENCES

1. **Aboul-Nasr, A., Mansour, M. H., and Salem, N. Y.,** The phenomenon of induction of preference in the cotton bollworm *Heliothis armigera, Z. Pflanzenkr. Pflanzenschutz,* 88, 116, 1981.

2. **Ali, M.,** Studies on the induction of food preference in alfalfa ladybird, *Subcoccinella* 24-*punctata* L., (Coleoptera, Coccinellidae), *Symp. Biol. Hung.,* 16, 23, 1976.

3. **Alloway, T. M.,** Learning and memory in insects, *Annu. Rev. Entomol.,* 17, 43, 1972.

4. **Alloway, T. M.,** Retention of learning through metamorphosis in the grain beetle *(Tenebrio molitor), Am. Zool.,* 12, 471, 1972.

5. **Anderson, A. M.,** A model for landmark learning in the honey bee, *J. Comp. Physiol.,* 114, 335, 1977.

6. **Avidov, Z., Berlinger, M. J., and Applebaum, S. W.,** Physiological aspects of host specificity in the Bruchidae. III. Effect of curvature and surface area on oviposition of *Callosobruchus chinensis* L., *Anim. Behav.,* 13, 178, 1965.

7. **Baerends, G. P.,** Fortpflanzungsverhalten und Orientierung der Grabwespe, *Ammophila campestris* Jur., *Tijdschr. Entomol.,* 84, 68, 1941.

8. **Balderrama, N.,** One trial learning in the American cockroach, *Periplaneta americana, J. Insect Physiol.,* 26, 499, 1980.

9. **Barbosa, P., Greenblatt, J., Withers, W., Cranshaw, W., and Harrington, E. A.,** Host-plant preferences and their induction in larvae of the gypsy moth, *Lymantria dispar, Entomol. Exp. Appl.,* 26, 180, 1979.

10. **Barker, L. M., Best, M. R., and Domjan, M., Eds.,** *Learning Mechanisms in Food Selection,* Baylor University Press, Waco, Texas, 1977.

11. **Bergson, O. and Wool, D.,** Attraction of flour beetles, *Tribolium castaneum* (Herbst) (Coleoptera, Tenebrionidae) to wheat flour: heritable character or "conditioning"?, *J. Appl. Entomol.,* 104, 179, 1987.

12. **Bernays, E. A.,** Antifeedants in crop pest management, in *Natural Products for In-*

novative *Pest Management,* Whitehead, D. L. and Bowers, W. S., Eds., Pergamon Press, Oxford, 1983, 586.

13. **Bernays, E. A., Blaney, W. M., and Chapman, R. F.,** Changes in chemoreceptor sensilla on the maxillary palps of *Locusta migratoria* in relation to feeding, *J. Exp. Biol.,* 57, 745, 1972.

14. **Bernays, E. A. and Chapman, R. F.,** The control of changes in the peripheral sensilla associated with feeding in *Locusta migratoria* (L.), *J. Exp. Biol.,* 57, 755, 1972.

15. **Bernays, E. A. and Graham, M.,** On the evolution of host range in phytophagous arthropods, *Ecology,* 69, 886, 1988.

16. **Bernays, E. A. and Lee, J. C.,** Food aversion learning in the polyphagous grasshopper *Schistocerca americana, Physiol. Entomol.,* 13, 131, 1989.

17. **Bernays, E. A. and Wrubel, R. P.,** Learning by grasshoppers: association of colour/light intensity with food, *Physiol. Entomol.,* 10, 359, 1985.

18. **Beugnon, G.,** Learned orientation in landward swimming in the cricket *Pteronemobius lineolatus, Behav. Process.,* 12, 215, 1986.

19. **Blaney, W. M., Schoonhoven, L. M., and Simmonds, M. S. J.,** Sensitivity variations in insect chemoreceptors; a review, *Experientia,* 42, 13, 1986.

20. **Blaney, W. M. and Simmonds, M. S. J.,** Electrophysiological activity in insects in response to antifeedants, *Trop. Dev. Res. Inst. Misc. Pub. London.,* 219, 1983.

21. **Blaney, W. M. and Simmonds, M. S. J.,** Experience: a modifier of neural and behavioural sensitivity, in *Insects-Plants, Proc. 6th Int. Symp. on Insect-Plant Relationships,* Pau, Labeyrie, V., Fabres, G., and Lachaise, D., Eds., W. Junk, Dodrecht, 1987, 237.

22. **Blaney, W. M. and Winstanley, C.,** Food selection behaviour in *Locusta migratoria,* in *Proc. 5th Int. Symp. Insect-Plant Relationships,* Wageningen, Visser, J. H., and Minks, A. K., Eds., PUDOC, Wageningen, 1982, 365.

23. **Blaney, W. M., Winstanely, C., and Simmonds, M. S. J.,** Food selection by locusts: an analysis of rejection behaviour, *Entomol. Exp. Appl.,* 38, 35, 1985.

24. **Blum, M. S. and Blum, N. A.,** Eds., *Sexual Selection and Reproductive Competition in Insects,* Academic Press, New York, 1979, 463.

25. **Bongers, W.,** External factors in the host plant selection of the Colorado beetle *Leptinotarsa decemlineata* Say, *Meded. Landbouwhogesch. Opzoekingsstn. Staat Gent,* 30, 1516, 1965.

26. **Bonner, J. T.,** *Size and Cycle. An Essay on the Structure of Biology,* Princeton University Press, Princeton, 1965.

27. **Bradshaw, W. E.,** Pervasive themes in insect life cycle strategies, in *the Evolution of Insect Life Cycles,* Taylor, F. and Karban, R., Eds., Springer Verlag, New York, 1986, 287.

28. **Byrne, J. H.,** Cellular analysis of associative learning, *Physiol. Rev.,* 67, 329, 1987.

29. **Callec, J. -J.,** Synaptic transmission in the central nervous system of insects, in *Insect Neurobiology,* Treherne, J. E., Ed., North-Holland, Amsterdam, 1974, 450.

30. **Cassidy, M. D.,** Development of an induced food plant preference in the Indian stick insect, *Carausius morosus, Entomol. Exp. Appl.,* 24, 287, 1978.

31. **Chapman, R. F.,** The chemical inhibition of feeding by phytophagous insects, a review, *Bull. Entomol. Res.,* 64, 339, 1974.

32. **Charnov, E. L.,** Optimal foraging: attack strategy of a mantid, *Am. Nat.,* 110, 141, 1976.

33. **Chew, F. S.,** Foodplant preferences of *Pieris* caterpillars (Lepidoptera), *Oecologia (Berlin),* 46, 1980.

34. **Cohen, R. W., Heydon, S. L., Waldbauer, G. P., and Friedman, S.,** Nutrient self-selection by the omnivorous cockroach *Supella longipalpa, J. Insect Physiol.,* 33, 77, 1987.

35. **Cohen, R. W., Waldbauer, G. P., Friedman, S., and Schiff, N. M.,** Nutrient self-selection by *Heliothis zea* larvae: a time-lapse film study, *Entomol. Exp. Appl.,* 44, 65, 1987.

36. **Cook, A. G.,** The ultrastructure of the A1 sensilla on the posterior surface of the clypeo-labrum of *Locusta migratoria migratorioides* (R & F), *Z. Zellforsch. Mikrosk. Anat.,* 134, 539, 1972.

37. **Cooley, S. S., Prokopy, R. J., McDonald, P. T., and Wong, T. T. Y.,** Learning in oviposition site selection by *Ceratitis capitata* flies, *Entomol. Exp. Appl.,* 40, 47, 1986.

38. **Corbet, S. A.,** Insect chemosensory responses: a chemical legacy hypothesis, *Ecol. Entomol.,* 10, 143, 1985.

39. **Courtney, S. P.,** Models of host plant location by butterflies: the effect of search images and search efficiency, *Oecologia (Berlin),* 59, 317, 1983.

40. **Craighead, F.,** Hopkins host-selection principle as related to certain cerambycid-beetles, *J. Agric. Res.,* 22, 189, 1921.

41. **Croze, H.,** Searching image in carrion crows, *Z. Tierpsychol.,* 5, 1, 1970.

42. **Dadd, R. H.,** Feeding behaviour and nutrition in grasshoppers and locusts, *Adv. Insect Physiol.,* 1, 47, 1963.

43. **Danilevsky, A. S.,** Rol' pytayushtshikh rastenii v biologii lugovogo motyl'ka, *Entomol.*

Obozr., 26, 91, 1935.

44. **Darwin, C.,** *The Variation of Animals and Plants Under Domestication,* 2nd ed., Methuen, London, 1875, 495.

45. **Dawkins, M.,** Perceptual changes in chicks: another look at the "search image" concept, *Anim. Behav.,* 19, 566, 1971.

46. **Dawkins, M.,** Shifts of "attention" in chicks during feeding, *Anim. Behav.,* 19, 575, 1971.

47. **de Boer, G. and Hanson, F. E.,** Food plant selection and induction of feeding preference among host and non-host plants in larvae of the tobacco hornworm *Manduca sexta, Entomol. Exp. Appl.,* 35, 177, 1984.

48. **de Boer, G. and Hanson, F. E.,** Feeding responses to solanaceous allelochemicals by larvae of the tobacco hornworm, *Manduca sexta, Entomol. Exp. Appl.,* 45, 123, 1987.

49. **Delaney, K. and Gelperin, A.,** Post-ingestive food-aversion learning to amino acid deficient diets by the terrestrial slug *Limax maximus, J. Comp. Physiol.,* 159, 281, 1986.

50. **Dethier, V. G.,** Adaptation to chemical stimulation of the tarsal receptors of the blowfly, *Biol. Bull.,* Woods Hole, Mass., 103, 1952.

51. **Dethier, V. G.,** Evolution of feeding preferences in phytophagous insects, *Evolution,* 8, 33, 1954.

52. **Dethier, V. G.,** Egg-laying habits of Lepidoptera in relation to available food, *Can. Entomol.,* 91, 554, 1959.

53. **Dethier, V. G.,** *The Hungry Fly. A Physiological Study of the Behavior Associated with Feeding,* Harvard University Press, Cambridge, 1976, 489.

54. **Dethier, V. G.,** Responses of some olfactory receptors of the eastern tent caterpillar *(Malacosoma americanum)* to leaves, *J. Chem. Ecol.,* 6, 213, 1980.

55. **Dethier, V. G.,** Food-aversion learning in two polyphagous caterpillars, *Diacrisia virginica* and *Estigmene congrua, Physiol. Entomol.,* 5, 321, 1980.

56. **Dethier, V. G.,** Evolution of receptor sensitivity to secondary plant substances with special reference to deterrents, *Am. Nat.,* 115, 45, 1980.

57. **Dethier, V. G.,** Mechanism of host plant recognition, *Entomol. Exp. Appl.,* 31, 49, 1982.

58. **Dethier, V. G.,** Analyzing proximate causes of behavior, in *Evolutionary Genetics of Invertebrate Behavior,* Huettel, M.D., Ed., Plenum Press, New York, 1987, 319.

59. **Dethier, V. G., Solomon, R. L., and Turner, L. H.,** Sensory input and central excitation and inhibition in the blowfly, *J. Comp. Physiol. Psychol.,* 60, 303, 1965.

60. **Dethier, V. G. and Yost, M. T.,** Oligophagy and absence of food-aversion learning in tobacco hornworms, *Manduca sexta, Physiol. Entomol.,* 4, 125, 1979.

61. **Dudai, Y. and Bicker, G.,** Comparison of visual and olfactory learning in *Drosophila, Naturwissenschaften,* 65, 495, 1978.

62. **Duerr, J. S. and Wuinn, W. G.,** Three *Drosophila* mutations that block associative learning also affect habituation and sensitization, *Proc. Nat. Acad. Sci. U.S.A.,* 79, 3646, 1982.

63. **Eisestein, E. M.,** Learning and memory in isolated insect ganglia, *Adv. Insect Physiol.,* 9, 111, 1972.

64. **Etscorn, F.,** Effects of a preferred vs a nonpreferred CS in the establishment of a taste aversion, *Physiol. Psychol.,* 1, 5, 1973.

65. **Flowers, R. W. and Yamamoto, R. T.,** Feeding on non-host plants by the tobacco hornworm, *Manduca sexta,* (Lepidoptera, Sphingidae), *Fla. Entomol.,* 65, 523, 1982.

66. **Fukushi, T.,** Properties of olfactory conditioning in the housefly, *Musca domestica, J. Insect Physiol.,* 25, 155, 1979.

67. **Fukushi, T.,** The role of learning on the finding of food in the searching behavior of the housefly, *Musca domestica* (Diptera: Muscidae), *Entomol. Gener.,* 8, 241, 1983.

68. **Futuyma, D. J.,** Selective factors in the evolution of host choice by phytophagous insects, in *Herbivorous Insects, Host-Seeking Behavior and Mechanisms,* Ahmad, S., Ed., Academic Press, New York, 1983, chap. 8.

69. **Gelperin, A.,** Rapid food-aversion learning by a terrestrial mollusk, *Science,* 189, 567, 1975.

70. **Gelperin, A.,** Neuroethological studies of associative learning in feeding control systems, in *Neuroethology and Behavioral Physiology,* Huber, F. and Markl, H., Eds., Springer-Verlag, Berlin, 1983, 189.

71. **Gelperin, A.,** Plasticity in control systems for insect feeding behavior, in *Perspectives in Chemoreception and Behavior,* Chapman, R. F., Bernays, E. A., and Stoffolano, J. G., Jr., Eds., Springer-Verlag, New York, 1987, chap. 3.

72. **Gelperin, A. and Forsythe, D.,** Neuroethological studies of learning in mollusks, in *Simpler Networks and Behavior,* Fentress, J. C., Ed., Sinauer Associates, Sunderland, 1975, 239.

73. **Getzova, A. B. and Lozina-Lozinskii, L. K.,** Rol' povedeniya nasekomykh v processe prisposobleniya ikh k rastitel'noy pishtshe, *Zool. Zh.,* 34, 1066, 1955.

74. **Gilbert, L. E.,** Ecological consequences of a coevolved mutualism between butterflies and plants, in *Coevolution of Animals and plants,* Gilbert, L. E. and Raven, P. H., Eds., University of Texas Press, Austin, 1975, 246.

75. **Gill, J. S.,** Studies on Insect Feeding Deterrents with Special Reference to the Fruit Extracts of the Neem Tree, *Azadirachta indica* A. Juss, Ph.D. thesis, University of Lon-

don, 1972.

76. **Goldsmith, C. M., Hepburn, H. R., and Mitchell, D.,** Retention of an associative learning task after metamorphosis in *Locusta migratoria migratorioides, J. Insect Physiol.,* 24, 737, 1978.

77. **Gould, J. L.,** The locale map of honey bees: do insects have cognitive maps?, *Science,* 232, 861, 1986.

78. **Grabstein, E. M. and Scriber, J. M.,** The relationship between restriction of host plant consumption and postingestive utilization of biomass and nitrogen in *Hyalophora cecropia, Entomol. Exp. Appl.,* 31, 202, 1982.

79. **Grabstein, E. M. and Scriber, J. M.,** Host-plant utilization by *Hyalophora cecropia* as affected by prior feeding experience, *Entomol. Exp. Appl.,* 32, 262, 1982.

80. **Gray, L.,** Genetic and experiential differences affecting foraging behavior, in *Foraging Behavior. Ecological, Ethological and Psychological Approaches,* Kamil, A. C. and Sargent, T. D., Eds., Garland STPM Press, New York, 1981, 455.

81. **Greenblatt, J. A., Calvert, W. H., and Barbosa, P.,** Larval feeding preferences and inducibility in the fall webworm, *Hyphantria cunea, Ann. Entomol. Soc. Am.,* 71, 605, 1978.

82. **Grossfield, J.,** Non-sexual behavior of *Drosophila,* in *The Genetics and Biology of Drosophila,* Vol. 2b, Ashburner, M., and Wright, T. R. F., Eds., Academic Press, 1978, chap. 10.

83. **Groves, P. M. and Thompson, R. F.,** Habituation: a dual-process theory, *Psychol. Rev.,* 77, 419, 1970.

84. **Guilford, T. and Dawkins, M. S.,** Search images not proven: a reappraisal of recent evidence, *Anim. Behav.,* 35, 1838, 1987.

85. **Hall, J. C.,** Learning and rhythms in courting, mutant *Drosophila, Trends Neuro. Sci.,* 9, 414, 1986.

86. **Hanson, F. E.,** Comparative studies on induction of food choice preferences in lepidopterous larvae, *Symp. Biol. Hung.,* 16, 71, 1976.

87. **Hanson, F. E. and Dethier, V. G.,** Role of gustation and olfaction in food plant discrimination in the tobacco hornworm, *Manduca sexta, J. Insect Physiol.,* 19, 1019, 1973.

88. **Hay, D. A.,** Strain differences in maze-learning ability of *Drosophila melanogaster, Nature (London),* 257, 44, 1975.

89. **Heinrich, B.,** "Majoring" and "minoring" by foraging bumblebees, *Bombus vagans:* an experimental analysis, *Ecology,* 60, 245, 1979.

90. **Hershberger, W. A. and Smith, M. P.,** Conditioning in *Drosophila melanogaster, Anim. Behav.,* 15, 259, 1967.

91. **Hinde, R. A.,** *Animal Behaviour. A Synthe-*

sis of Ethology and Comparative Psychology, McGraw-Hill, London, 1966, 534.

92. **Hinde, R. A. and Tinbergen, N.,** The comparative study of species-specific behavior, in *Behavior and Evolution,* Roe, A., and Simpson, G. S., Eds., Yale University Press, New Haven, 1958, 251.

93. **Holliday, M. and Hirsch, J.,** A coment on the evidence for learning in Diptera, *Behav. Genet.,* 16, 439, 1986.

94. **Holling, C. S.,** Some characteristics of simple types of predation and parasitism, *Can. Entomol.,* 91, 385, 1959.

95. **Hopkins, A. D.,** A discussion of C. G. Hewitt's paper on "Insect Behavior", *J. Econ. Entomol.,* 10, 92, 1917.

96. **Horn, G. and Hinde, R. A., Eds.,** *Short-term Changes in Neural Activity and Behaviour,* Cambridge University Press, Cambridge, 1970, 628.

97. **Horridge, G. A.,** Learning of leg positions by headless insects, *Nature (London),* 193, 697, 1962.

98. **Hough-Goldstein, J. A.,** Oviposition site selection by seedcorn maggot flies (Diptera: Anthomyiidae) unaffected by prior experience of larval host, *Environ. Entomol.,* 14, 289, 1985.

99. **Hovanitz, W. and Chang, V. C. S.,** Ovipositional preference tests with *Pieris, J. Res. Lipid.,* 2, 185, 1963.

100. **Hsiao, T. H.,** Host plant adaptations among geographic populations of the Colorado potato beetle, *Entomol. Exp. Appl.,* 24, 437, 1978.

101. **Hubert-Dahl, M. L.,** Aenderung des Wirtswahlverhaltens dreier Biotypen von *Acyrthosiphon pisum* Harris nach Anzucht auf verschiedenen Wirtspflanzen, *Beitr. Entomol.,* 25, 77, 1975.

102. **Immelmann, K.,** Ecological significance of imprinting and early learning, *Annu. Rev. Ecol. Syst.,* 6, 15, 1975.

103. **Irwin, F. W.,** An analysis of the concepts of discrimination and preference, *Am. J. Psychol.,* 58, 152, 1958.

104. **Iwao, S.,** Some experiments on the host-plant preference in a phytophagous lady beetle, *Epilachna pustulosa* Kono, with special reference to its individual variation, *Insect Ecol.,* 8, 10, 1959.

105. **Jaenike, J.,** On optimal oviposition behavior in phytophagous insects, *Theor. Popul. Biol.,* 14, 350, 1978.

106. **Jaenike, J.,** Induction of host preference in *Drosophila melanogaster, Oecologia (Berlin),* 58, 320, 1983.

107. **Jaenike, J.,** Genetic complexity of host-selection behavior in *Drosophila, Proc. Natl. Acad. Sci. U.S.A.,* 83, 2148, 1986.

108. **Jamieson, I. G.,** The functional approach to behavior: is it useful?, *Am. Nat.,* 127, 195,

1986.

109. **Jermy, T.,** On the nature of the oligophagy in *Leptinotarsa decemlineata* Say (Coleoptera, Chrysomelidae), *Acta Zool. Acad. Sci. Hung.,* 7, 119, 1961.

110. **Jermy, T.,** Biological background and outlook of the antifeedant approach to insect control, *Acta Phytopathol. Acad. Sci. Hung.,* 6, 253, 1971.

111. **Jermy, T.,** Multiplicity of insect antifeedants in plants, in *Current Themes in Tropical Science,* Odhiambo, T. R., Ed., Vol. 2, *Natural Products for Innovative Pest Management,* Whitehead, D. L. and Bowers, W. S., Eds., Pergamon Press, Oxford, 1983, chap. 12.

112. **Jermy, T.,** The role of experience in the host selection of phytophagous insects, in *Perspectives in Chemoreception,* Chapman, R. F., Bernays, E. A., and Stoffolano, J. G., Jr., Eds., Springer-Verlag, New York, 1987, chap. 9.

113. **Jermy, T.,** unpublished.

114. **Jermy, T., Bernays, E. A., and Szentesi, A.,** The effect of repeated exposure to feeding deterrents on their acceptability to phytophagous insects, in *Proc. 5th Int. Symp. Insect-Plant Relationships,* Wageningen, Visser, J. H. and Minks, A. K., Eds., PUDOC, Wageningen, 1982, 25.

115. **Jermy, T., Hanson, F. E., and Dethier, V. G.,** Induction of specific food preference in lepidopterous larvae, *Entomol. Exp. Appl.,* 11, 211, 1968.

116. **Jermy, T., Horváth, J., and Szentesi, A.,** The role of habituation in food selection of lepidopterous larvae: the example of *Mamestra brassicae* L. (Lepidoptera, Noctuidae), in *Insects-Plants, Proc. 6th Int. Symp. on Insect-Plant Relationships,* Pau, Labeyrie, V., Fabres, G., and Lachaise, D., Eds., W. Junk, Dordrecht, 1986, 231.

117. **Jermy, T., Lábos, E., and Molnár, I.,** Stenophagy of phytophagous insects — a result of constraints on the evolution of the nervous system, in *Organizational Constraints on the Dynamics of Evolution,* Maynard-Smith, J. and Vida, G., Eds., Manchester University Press, Manchester, in press.

118. **Johansson, A. S.,** The food plant preference of the larvae of *Pieris brassicae* L., *Nor. Entomol. Tidsskr. B,* 8, 187, 1951.

119. **Johnston, T. and Gottlieb, G.,** manuscript.

120. **Kamil, A. C.,** Optimal foraging theory and the psychology of learning, *Am. Zool.,* 23, 291, 1983.

121. **Kamil, A. C. and Roitblat, H. L.,** The ecology of foraging behavior: implications for animal learning and memory, *Annu. Rev. Psychol.,* 36, 141, 1985.

122. **Kamil, A. C. and Sargent, T. D., Eds.,** *Foraging Behavior. Ecological, Ethological and Psychological Approaches,* Garland STPM Press, New York, 1981, 534.

123. **Kandel, E. R. and Spencer, W. A.,** Cellular neurophysiological approaches in the study of learning, *Psychol. Rev.,* 48, 65, 1968.

124. **Kozhantshikov, I. V.,** Biologitsheskie osobennosti evropeyskikh vidov roda *Galerucella* i usloviya obrazovaniya biologitsheskikh form u *Galerucella lineola* F., *Trudy Zool. Inst. Akad. Nauk. SSSR,* 24, 271, 1958.

125. **Krasne, F. B.,** Physiological analysis of learning in invertebrates, in *Cortical Integration,* Reinoso-Suarez, F. and Ajmone-Marsan, C., Eds., Raven Press, New York, 1984, 53.

126. **Krebs, J. R.,** Behavioural aspects of predation, in *Perspectives in Ethology,* Bateson, P. P. G. and Klopfer, P. H., Eds., Plenum Press, New York 1973, 73.

127. **Krebs, J. R. and Davis, N. B.,** *An Introduction to Behavioural Ecology,* Blackwell Scientific, Oxford, 1981, 292.

128. **Kuznetzov, V. I.,** Voprosy prisposoblenii tsheshuekrylykh k novym pishtshevym usloviyam, *Trudy Zool. Inst. Akad. Nauk SSSR,* 11, 166, 1952.

129. **Lawrence, E. S. and Allen, J. A.,** On the term 'search image', *Oikos,* 40, 313, 1983.

130. **Lee, J. C., Bernays, E. A., and Wrubel, R. P.,** Does learning play a role in host location and selection by grasshoppers?, in *Insects-Plants, Proc. 6th Int. Symp. on Insect-Plant Relationships,* Pau, Labeyrie, V., Fabres, G., and Lachaise, D., Eds., W. Junk, Dordrecht, 1986, 125.

131. **Levins, R. and MacArthur, R.,** An hypothesis to explain the incidence of monophagy, *Ecology,* 50, 910, 1969.

132. **Lewis, A. C.,** Memory constraints and flower choice in *Pieris rapae, Science,* 232, 863, 1986.

133. **Lewontin, R. C.,** Fitness, survival and optimality, in *Analysis of Ecological Systems,* Horn, D. J., Stairs, G. R., and Mitchell, R. D., Eds., Ohio State University Press, Columbus, 1979, chap. 1.

134. **Lorenz, K.,** Der Kumpan in der Umwelt des Vogels, *J. Ornithol.,* 83, 137, 289, 1935.

135. **Ma, W. C.,** Dynamics of feeding responses in *Pieris brassicae* Linn., as a function of chemosensory input: a behavioural, ultrastructural and electrophysiological study, *Meded. Landbouwhogesch. Wageningen,* 72-11, 1, 1972.

136. **Ma, W. C. and Visser, J. H.,** Single unit analysis of odour quality coding by the olfactory antennal receptor system of the Colorado beetle, *Entomol. Exp. Appl.,* 24, 520, 1978.

137. **Mark, G. A.,** Induced oviposition preference, periodic environments, and demographic cycles in the bruchid beetle *Callosobruchus maculatus, Entomol. Exp. Appl.,* 32, 155, 1982.

138. **Marlin, N. A. and Miller, R. R.,** Associations to contextual stimuli as a determinant to long-term habituation, *J. Exp. Psychol. Anim. Behav. Process.,* 7, 313, 1981.

139. **Mayr, E.,** Evolution and Verhalten, *Verh. Dtsch. Zool. Ges. Koeln.* 64, 322, 1970.

140. **McFarland, D., Ed.,** *The Oxford Companion to Animal Behaviour,* Oxford University Press, Oxford, 1981, 657.

141. **McGuire, T.,** Learning in three species of Diptera: the blowfly *Phormia regina,* the fruit fly *Drosophila melanogaster,* and the house fly *Musca domestica, Behav. Genet.,* 14, 479, 1984.

142. **McGuire, T.,** Further evidence for learning in Diptera: a reply to Holliday and Hirsch, *Behav. Genet.,* 16, 457, 1986.

143. **Meijsser, F. M.,** Internal report, Department of Animal Physiology, Agricultural University, Wageningen, 1983.

144. **Merzheevskaya, O. I.,** O pishtshevoi spetzializatzii sovok - ogorodnoy (*Polia oleracea* L., *P. dissimilis* Knoch), *Zool. Zh.,* 40, 707, 1961.

145. **Michely, M.,** Extrait d'une lettre parvenue par Sacc., *Bull. Soc. Imper. Zool. d'Acclimat.,* 8, 563, 1861.

146. **Mitchell, R.,** The evolution of oviposition tactics in the bean weevil, *Callosobruchus maculatus* (F.), *Ecology,* 56, 696, 1975.

147. **Müller, F. P.,** Genetic and evolutionary aspects of host choice in phytophagous insects, especially aphids, *Biol. Zbl.,* 104, 225, 1985.

148. **Murbach, R.,** Effect on plein champ de fongicides à base de fentin-acétate, de manèbe et d'oxychlorure de cuivre sur la densité de population du doryphore de la pomme de terre (*Leptinotarsa decemlineata* Say.), *Rech. Agron. Suisse,* 6, 345, 1967.

149. **Murbach, R. and Corbaz, R.,** Influence de trois types de fungicides utilisés en Suisse contre le mildiou de la pomme de terre [*Phytophthora infestans* (Mont.) de Bary] sur la densité de population du doryphore (*Leptinotarsa decemlineata* Say), *Phytopathol. Z.,* 47, 182, 1963.

150. **Nelson, M. C.,** Classical conditioning in the blowfly, (*Phormia regina*): associative and excitatory factors, *J. Comp. Physiol. Psychol.,* 77, 353, 1971.

151. **Nelson, M. C.,** The blowfly's dance: role in the regulation of food intake, *J. Insect Physiol.,* 23, 603, 1977.

152. **Oshima, K., Honda, H., and Yamamoto, I.,** Isolation of an oviposition marker from the azuki bean weevil, *Callosobruchus chinensis* (L.), *Agric. Biol. Chem.,* 37, 2679, 1973.

153. **Palmerino, C. C., Rusiniak, K. W., and Garcia, J.,** Flavor-illness aversions: the peculiar roles of odor and taste in memory for poison, *Science,* 208, 753, 1980.

154. **Papaj, D. R.,** Conditioning of leaf-shape discrimination by chemical cues in the butterfly, *Battus philenor, Anim. Behav.,* 34, 1281, 1986.

155. **Papaj, D. R.,** Leaf buds, a factor in host selection by *Battus philenor* butterflies, *Ecol. Entomol.,* 11, 301, 1986.

156. **Papaj, D. R.,** Shifts in foraging behavior by a *Battus philenor* population: field evidence for switching by individual butterflies, *Behav. Ecol. Sociobiol.,* 19, 31, 1986.

157. **Papaj, D. R.,** Interpopulation differences in host preference and the evolution of learning in the butterfly, *Battus philenor, Evolution,* 40, 518, 1986.

158. **Papaj, D. R. and Prokopy, R. J.,** Phytochemical basis of learning in *Rhagoletis pomonella* and other herbivorous insects, *J. Chem. Ecol.,* 12, 1125, 1986.

159. **Papaj, D. R. and Prokopy, R. J.,** Learning of host acceptance in the apple maggot fly, *Rhagoletis pomonella:* the role of fruit size, in *Insects-Plants, Proc. 6th Int. Symp. on Insect-Plant Relationships,* Pau, Labeyrie, V., Fabres, G., and Lachaise, D., Eds., W. Junk, Dordrecht, 1987, 408.

160. **Papaj, D. R. and Rausher, M. D.,** Individual variation in host location by phytophagous insects, in *Herbivorous Insects: Host-Seeking Behavior and Mechanisms,* Ahmad, S., Ed., Academic Press, New York, 1983, 77.

161. **Phillips, P. A. and Barnes, M. M.,** Host race formation among sympatric apple, walnut, and plum populations of the codling moth, *Laspeyresia pomonella, Ann. Entomol. Soc. Am.,* 68, 1053, 1975.

162. **Phillips, W. M.,** Modification of feeding "preference" in the flea-beetle, *Haltica lythri* (Coleoptera, Chrysomelidae), *Entomol. Exp. Appl.,* 21, 71, 1977.

163. **Pierce, G. J. and Ollason, J. G.,** Eight reasons why optimal foraging theory is a complete waste of time, *Oikos,* 49, 111, 1987.

164. **Pimbert, M. P.,** Reproduction and oviposition preferences of *Zabrotes subfasciatus* stocks reared from two host plant species, *Entomol. Exp. Appl.,* 38, 273, 1985.

165. **Plotkin, H. C.,** Learning in a carabid beetle (*Pterostichus melanarius*), *Anim. Behav.,* 27, 567, 1979.

166. **Polyakova, D. I.,** Obutshenie pri vybore kormovogo rasteniya u palotshnika *Bacillus rossius* (Phasmatoidea, Phyllinae), *Zool. Zh.,* 65, 1088, 1986.

167. **Prokopy, R. J., Averill, A. L., Cooley, S. S., and Roitberg, C. A.,** Associative learning in egglaying site selection by apple maggot flies, *Science,* 218, 76, 1982.

168. **Prokopy, R. J., Averill, A. L., Cooley, S. S., Roitberg, C. A., and Kallet, C.,** Variation in host acceptance pattern in apple maggot flies, in *Proc. 5th Int. Symp. Insect-Plant Relationships,* Wageningen, Visser,

J. H. and Minks, A. K., Eds., PUDOC, Wageningen, 1982, 123.

169. **Prokopy, R. J., Papaj, D. R., Cooley, S. S., and Kallet, C.,** On the nature of learning in oviposition site acceptance by apple maggot flies, *Anim. Behav.,* 34, 98, 1986.

170. **Pyke, H.,** Optimal foraging theory: a critical review, *Annu. Rev. Ecol. Syst.,* 15, 523, 1984.

171. **Quinn, W. G., Harris, W. A., and Benzer, S.,** Conditioned behavior in *Drosophila melanogaster, Proc. Nat. Acad. Sci., U.S.A.,* 71, 708, 1974.

172. **Rausher, M. D.,** Search image for leaf shape in a butterfly, *Science,* 200, 1071, 1978.

173. **Rausher, M. D.,** Variability for host preference in insect populations: mechanistic and evolutionary models, *J. Insect Physiol.,* 31, 873, 1985.

174. **Rausher, M. D. and Papaj, D. R.,** Foraging by *Battus philenor* butterflies: evidence for individual differences in searching behaviour, *Anim. Behav.,* 31, 341, 1983.

175. **Roitberg, B. D., van Lenteren, J. C., van Alphen, J. J. M., Galis, F., and Prokopy, R. J.,** Foraging behaviour of *Rhagoletis pomonella,* a parasite of hawthorn *(Crataegus viridis),* in nature, *J. Anim. Ecol.,* 51, 307, 1982.

176. **Rozin, P.,** The selection of foods by rats, humans, and other animals, *Advan. Study Behav.,* 6, 21, 1976.

177. **Rowell, C. H. F.,** Incremental and decremental processes in the insect central nervous system, in *Short-term Changes in Neural Activity and Behaviour,* Horn, G. and Hinde, R. A., Eds., Cambridge University Press, Cambridge, 1970, 237.

178. **Rowell-Rahier, M.,** The food plant preferences of *Phratora vitellinae* (Coleoptera, Chrysomelidae). B. A laboratory comparison of geographically isolated populations and experiments on conditioning, *Oecologia (Berlin),* 64, 375, 1984.

179. **Royama, T.,** Factors governing the hunting behaviour and selection of food by the great tit *(Parus major* L.), *J. Anim. Ecol.,* 39, 619, 1970.

180. **Saxena, K. N.,** Some factors governing olfactory and gustatory responses of insects, in *Olfaction and Taste II,* Hayashi, T., Ed., Pergamon Press, Oxford, 1967, 799.

181. **Saxena, K. N. and Schoonhoven, L. M.,** Induction of orientational and feeding preferences in *Manduca sexta* larvae for an artificial diet containing citral, *Entomol. Exp. Appl.,* 23, 72, 1978.

182. **Saxena, K. N. and Schoonhoven, L. M.,** Induction of orientational and feeding preferences in *Manduca sexta* larvae for different food sources, *Entomol. Exp. Appl.,* 32, 173, 1982.

183. **Schoonhoven, L. M.,** Loss of host plant specificity by *Manduca sexta* after rearing on an artificial diet, *Entomol. Exp. Appl.,* 10, 270, 1967.

184. **Schoonhoven, L. M.,** Sensitivity changes in some insect chemoreceptors and their effect on food selection behaviour, *Proc. K. Ned. Akad. Wet., Ser. C.,* 72, 491, 1969.

185. **Schoonhoven, L. M.,** Long-term sensitivity changes in some insect taste receptors, *Drug. Res.,* 28, 1978, (reprint).

186. **Schoonhoven, L. M. and Jermy, T.,** A behavioural and electrophysiological analysis of insect feeding deterrents, in *Crop Protection Agents — Their Biological Evaluation,* McFarlane, N.R., Ed., Academic Press, New York, 1977, 133.

187. **Schoonhoven, L. M. and Meerman, J.,** Metabolic cost of changes in diet and neutralization of allelochemics, *Entomol. Exp. Appl.,* 24, 689, 1978.

188. **Schweissing, F. C. and Wilde, G.,** Predisposition and nonpreference of the greenbug for certain host plants, *Environ. Entomol.,* 8, 1070, 1979.

189. **Scriber, J. M.,** The effects of sequentially switching foodplants upon biomass and nitrogen utiization by polyphagous and stenophagous *Papilio* larvae, *Entomol. Exp. Appl.,* 25, 203, 1979.

190. **Scriber, J. M.,** Sequential diets, metabolic cost, and growth of *Spodoptera eridania* feeding upon dill, lima bean and cabbage, *Oecologia (Berlin),* 51, 175, 1981.

191. **Scriber, J. M.,** The behaviour and nutritional physiology of southern armyworm larvae as a function of plant species consumed in earlier instars, *Entomol. Exp. Appl.,* 31, 359, 1982.

192. **Shettleworth, S. J.,** Learning and behavioral ecology, in *Behavioral Ecology, an Evolutionary Approach,* Krebs, J. B. and Davies, N. B., Eds., Sinauer Assoc., Sunderland, MA, 1984, 170.

193. **Siegel, R. W. and Hall, J. C.,** Conditioned responses in courtship behavior of normal and mutant *Drosophila, Proc. Natl. Acad. Sci., U.S.A.,* 76, 3430, 1979.

194. **Simmonds, M. S. J. and Blaney, W.M.,** Some neurophysiological effects of azadirachtin on lepidopterous larvae and their feeding responses, in *Natural Pesticides from the Neem Tree, (Azadirachta indica A. Juss) and Other Tropical Plants,* Proc. 2nd Int. Neem Conf., Rauischholzhausen, Schmutterer, H., and Ascher, K. R. S., Eds., Deutsche GTZ GmbH, Eschborn, 1983, 587.

195. **Smith, M. A. and Cornell, H. V.,** Hopkins host-selection in *Nasonia vitripennis* and its implications for sympatric speciation, *Anim. Behav.,* 27, 365, 1979.

196. **Solodovnikov, V. B.,** Ismeneniya povede-

niya gusenitz kitayskogo dubovogo shel-kopryada *(Antheraea pernyi)* na kormovom gradiente, *Dokl. Akad. Nauk SSSR,* 60, 321, 1948.

197. **Spatz, H.-Ch.,** Visuelles Lernen bei *Drosophila.* Wege zu einer Molekularbiologie des Lernens?, *Funkt. Biol. Med.,* 5, 276, 1985.

198. **Städler, E. and Hanson F. E.,** Influence of induction of host preference on chemoreception of *Manduca sexta:* behavioral and electrophysiological studies, *Symp. Biol. Hung.,* 16, 267, 1976.

199. **Städler, E. and Hanson, F. E.,** Food discrimination and induction of preference for artificial diets in the tobacco hornworm, *Manduca sexta, Physiol. Entomol.,* 3, 121, 1978.

200. **Stanton, M. L.,** Short-term learning and the searching accuracy of egg-laying butterflies, *Anim. Behav.,* 32, 33, 1984.

201. **Strebel, O.,** Biologische Studien an einheimischen Collembolen. II. Ernährung und Geschmackssinn bei *Hypogastrura purpurascens* (Lubb.), *Z. Wiss. Insektenbiol.,* 23, 135, 1928.

202. **Stride, G. O. and Straatman, R.,** The host plant relationship of an Australian swallowtail *Papilio aeqeus,* and its significance in the evolution of host plant selection, *Proc. Linn. Soc. NSW,* 87, 69, 1962.

203. **Szentesi, A. and Bernays, E. A.,** A study of behavioural habituation to a feeding deterrent in nymphs of *Schistocerca gregaria, Physiol. Entomol.,* 9, 329, 1984.

204. **Tabashnik, B. E., Wheelock, H., Rainbolt, J. D., and Watt, W. B.,** Individual variation in oviposition preference in the butterfly *Colias eurytheme, Oecologia (Berlin),* 50, 225, 1981.

205. **Takata, N.,** Studies on the host preference of common cabbage butterfly, *Pieris rapae crucivora* (Boisduval). XII. Successive rearing of the cabbage butterfly larva with certain host plants and its effect on the ovipositional preference of the adult, *Jpn. J. Ecol. (Nippon Seitai Gakkaishi),* 11, 147, 1961.

206. **Thibout, E., Auger, J., and Dakkouni, M.,** Conservation à court terme de l'information chimique issue de la plante-hôte lors de la ponte chez la Teigne du Poireau (Lépidoptère), *C. R. Séances Acad. Agric. Fr.,* 300, 489, 1985.

207. **Thompson, R. F. and Spencer, W. A.,** Habituation: a model phenomenon for the study of neuronal substrates of behaviour, *Psychol. Rev.,* 173, 16, 1968.

208. **Thon, B.,** Acquisition and retention of habituation as a function of intertrial interval duration during training in the blowfly, *Behav. Process.,* 15, 47, 1987.

209. **Thon, B. and Pauzie, A.,** Differential sensitization, retention, and generalization of habituation in two response systems in the blowfly *(Calliphora vomitoria), J. Comp. Psychol.,* 98, 119, 1984.

210. **Thorpe, W.H.,** Further studies on pre-imaginal olfactory conditioning in insects, *Proc. R. Soc. London, Ser. B.,* 127, 424, 1939.

211. **Thorpe, W. H.,** *Learning and Instinct in Animals,* Methuen and Co. Ltd, London, 1963.

212. **Thorpe, W. H. and Jones, F. G. W.,** Olfactory conditioning and its relation to the problem of host selection, *Proc. R. Soc. London, Ser. B,* 124, 56, 1937.

213. **Tinbergen, L.,** The natural control of insects in pinewoods. I. Factors influencing the intensity of predation by songbirds, *Arch. Neerl. Zool.,* 13, 265, 1960.

214. **Ting, A. Y.,** The Induction of Feeding Preference in the Butterfly *Chlosyne lacinia,* Unpublished thesis, University of Texas, Austin, 1970.

215. **Traynier, R. M. M.,** Long-term changes in the oviposition behaviour of the cabbage butterfly, *Pieris rapae,* induced by contact with plants, *Physiol. Entomol.,* 4, 87, 1979.

216. **Traynier, R. M. M.,** Associative learning in the ovipositional behaviour of the cabbage butterfly, *Pieris rapae, Physiol. Entomol.,* 9, 465, 1984.

217. **Traynier, R. M. M.,** Visual learning in assays of sinigrin solution as an oviposition releaser for the cabbage butterfly, *Pieris rapae, Entomol. Exp. Appl.,* 40, 25, 1986.

218. **Traynier, R. M. M.,** Learning without neurosis in host finding and oviposition by the cabbage butterfly, *Pieris rapae,* in *Insects-Plants, Proc. 6th Int. Symp. on Insect-Plant Relationships,* Pau, Labeyrie, V., Fabres, G., and Lachaise, D., Eds., W. Junk, Dordrecht, 1987, 243.

219. **Tully, T. P.,** Behavior-genetic Analysis of the Black Blowfly, *Phormia regina,* Using the Central Excitatory State (CES), Unpublished doctoral dissertation, University of Illinois at Urbana-Champaign, 1981.

220. **Tully, T.,** *Drosophila* learning: behavior and biochemistry, *Behav. Genet.,* 14, 527, 1984.

221. **Tully, T.,** Measuring learning in individual flies is not necessary to study the effects of single-gene mutations in *Drosophila:* a reply to Holliday and Hirsch, *Behav. Genet.,* 16, 449, 1986.

222. **Vargo, M. and Hirsch, J.,** Central excitation in the fruit fly *(Drosophila melanogaster), J. Comp. Physiol. Psychol.,* 96, 452, 1982.

223. **Vet, L. E. M. and van Opzeeland, K.,** The influence of conditioning on olfactory microhabitat and host location in *Asobara tabida* (Nees) and *A. rufescens* (Foerster) (Braconidae: Alysiinae) larval parasitoids of Drosophilidae, *Oecologia,* 63, 171, 1984.

224. **Visser, J.H. and Avé, D. A.,** General green leaf volatiles in the olfactory orientation of the Colorado beetle, *Leptinotarsa decemli-*

neata, Entomol. Exp. Appl., 24, 738, 1978.

225. **Volkonsky, M.,** Observations sur le comportement du Criquet Pèlerin (*Schistocerca gregaria* Forsk.) dans le Sahara algéro-nigérien, *Arch. Inst. Pasteur Alger.*, 20, 236, 1942.

226. **Waldbauer, G. P. and Bhattacharya, A. K.,** Self-selection of an optimum diet from a mixture of wheat fractions by the larvae of *Tribolium confusum, J. Insect Physiol.*, 19, 407, 1973.

227. **Waldbauer, G. P., Cohen, R. W., and Friedman, S.,** Self-selection of an optimal nutrient mix from defined diets by larvae of the corn earworm, *Heliothis zea* (Boddie), *Physiol. Zool.*, 57, 590, 1984.

228. **Waldbauer, G. P. and Friedman, S.,** Dietary self-selection by insects, *Proc. Int. Conf. Endocrin. Frontiers Physiol. Insect Ecol.*, Szklanska Poreba, Poland, Sept. 1987, in press.

229. **Wallace, R. A.,** *Animal Behavior. Its Development, Ecology and Evolution,* Goodyear Publishing Co., Santa Monica, 1979, 590.

230. **Walsh, B. D.,** On phytophagic varieties and phytophagous species, *Proc. Entomol. Soc. Phila.*, 3, 403, 1864.

231. **Walsh, B. D.,** On the phytophagic varieties of phytophagous species, with remarks on the unity of coloration of insects, *Proc. Entomol. Soc. Phila.*, 5, 194, 1865.

232. **Wasserman, S. S.,** Host-induced oviposition preferences and oviposition markers in the cowpea weevil, *Callosobruchus maculatus, Ann. Entomol. Soc. Am.*, 74, 242, 1981.

233. **Wasserman, S. S.,** Gypsy moth (*Lymantria dispar)* induced feeding preferences as a bioassay for phenetic similarity among host plants, in *Proc. 5th Int. Symp. Insect-Plant Relationships,* Wageningen, Visser, J. H. and Minks, A. K., Eds., PUDOC, Wageningen, 1982, 261.

234. **Wehner, R. and Raeber, F.,** Visual spatial memory in desert ants, *Cataglyphis bicolor,* (Hymenoptera: Formicidae), *Experientia,* 35, 1569, 1979.

235. **Whelan, R. J.,** Response of slugs to unacceptable food items, *J. Appl. Ecol.*, 19, 79, 1982.

236. **Wiklund, C.,** Host plant suitability and the mechanism of host selection in larvae of *Papilio machaon, Entomol. Exp. Appl.*, 16, 232, 1973.

237. **Wiklund, C.,** Oviposition preferences in *Papilio machaon* in relation to the host plants of the larvae, *Entomol. Exp. Appl.*, 17, 189, 1974.

238. **deWilde, J., Sloof, R., and Bongers, W.,** A comparative study of feeding and oviposition preference in the Colorado beetle (*Leptinotarsa decemlineata* Say), *Meded. Landbouwhogesch. Opzoekingsstn. Staat Gent,* 25, 1340, 1960.

239. **Wiseman, B. R. and McMillian, W. W.,** Feeding preferences of *Heliothis zea* larvae preconditioned to several host crops, *J. Ga. Entomol. Soc.*, 15, 449, 1980.

240. **Wright, R. H.,** Housefly memory, *Can. Entomol.*, 106, 223, 1974.

241. **Yamamoto, R. T.,** Induction of host plant specificity in the tobacco hornworm, *Manduca sexta, J. Insect Physiol.*, 20, 641, 1974.

242. **Yamamoto, R. T. and Fraenkel, G.,** The physiological basis for the selection of plants for egg-laying in the tobacco hornworm, *Protoparce sexta* (Johan.), *Verh. XI. Int. Kongr. Entomol.*, Wien, 3, 127, 1960.

243. **Zahorik, D. M. and Houpt, K. A.,** Species differences in feeding strategies, food hazards, and the ability to learn food aversions, in *Foraging Behaviour. Ecological, Ethological and Psychological Approaches,* Kamil, A. C. and Sargent, T. D., Eds., Garland STPM Press, New York, 1981, 289.

3

Phenolglucosides and Interactions at Three Trophic Levels: Salicaceae — Herbivores — Predators

M. Rowell-Rahier
Zoologisches Institut der Universitat Basel
Basel, Switzerland

J. M. Pasteels
Laboratoire de Biologie Animale
Université Libre de Bruxelles
Brussels, Belgium

TABLE OF CONTENTS

I.   Introduction..76
II.   Phenolglucosides 76
        A.   Chemical Structure and Methods of Analysis...........................76
        B.   Distribution 77
              1.   Interspecific Variation......................................77
              2.   Intraspecific Variations and Abiotic Factors 78
III.   Phenolglucosides in Host Plant Leaves and Distribution Patterns of
        Herbivores..80
        A.   Laboratory Studies..81
        B.   Field Studies 82
        C.   Effect of Altitude 84
        D.   Effect of Leaf Age 84
        E.   Effect of Environmental Stress...85
        F.   Summary..85
IV.   Specialization on Leaves Rich in Phenolglucosides...........................86
        A.   Antagonistic Effects of Phenolglucosides on Herbivores: Detoxification
              Mechanism in the Tiger Swallowtail Butterfly 86
        B.   Agonistic Effects of Phenolglucosides on Leaf-Beetles: Chemical
              Defense 88
V.   Phenolglucosides and Natural Enemies of Leaf-Beetles.....................90
VI.   Conclusions 91
Acknowledgments 92
References..92

I. INTRODUCTION

The significance of plant secondary metabolites for animal-plant interactions is a widely studied area.[47] Indeed, biochemical and behavioral adaptations of herbivores are often key factors in determining their distribution patterns and level of dietary specialization.

Among plant secondary metabolites, phenolics are the most common.[64] More particularly, the role of flavonoids and especially tannins has been discussed at length in the context of both the "apparency plant defense theory"[8,46] and the "resource availability hypothesis".[4] We will not discuss plant defensive strategies further here.[2,7]

In this review we will summarize our knowledge of the role of one group of simple phenols — the phenolglucosides, in a set of interactions between three trophic levels. For this purpose, we will use the example of the host-plant family Salicaceae, the herbivores feeding on them, and their natural enemies. This system is particularly interesting, since it involves antagonistic and agonistic interactions such as toxic, deterrent, or phagostimulant effects, and more complex effects such as the sequestration of plant toxins for defense of the beetles against their natural enemies.

Price[38,39] and Bernays and Graham[1] have stressed the potential importance of natural enemies for the speciation and dietary specialization of herbivores. They suggest that the use of specific plant toxins by herbivores to reduce predation could be one of the strategies leading to their dietary specialization.

II. PHENOLGLUCOSIDES

A. CHEMICAL STRUCTURE AND METHODS OF ANALYSIS

The Salicaceae include (within the genera *Salix* and *Populus*) many of the shrub and tree species of the boreal and montane regions of the northern hemisphere. Because of their relatively rapid growth rate, they are often planted for wood production.

Their chemistry is rather well known thanks to work of Thieme and Benecke,[62] Pearl and Darling,[37] and Steele et al.[57] and has been reviewed by Hegnauer,[11] Palo,[29] and Reichardt et al.[45] The Salicaceae as a family is characterized by the presence of phenolic glucosides in the bark of all species and the leaves of some species; no secondary metabolites other than phenolics have been identified so far in the leaves of Salicaceous plants. In northern temperate European species of *Salix* there is a separation between those with phenolglucosides in their leaves and those with condensed tannins (see below).

The structures of several of the more common phenolglucosides found in the Salicaceae are given in Figure 1. Salicin and salicortin are the simplest and most frequently identified ones. Salicin is well known in pharmacology for its analgesic and antipyretic effects.[26] The active metabolite is salicylic acid obtained after removal of the glucose moiety and oxidation of the remaining aglycone saligenin. Salicin is also toxic to some insects (such as ants,[30] and some Lepidoptera).[22,23] Phenolglucosides inhibit the feeding of some mammals also such as the Finnish mountain hare *(Lepus timidus)*[58] and the opossum *Trichosurus vulpecula* in New Zealand.[6]

The instability of many phenolglucosides has been reported by Thieme[60] and Steele et al.[56] Lindroth and Pajutee[21] summarize various forms of hydrolytic degradation and interconversion of phenolglucosides (Figure 2). These seriously affect the results of quantitative and qualitative chemical analysis and are dependent on the sample preparation technique (fresh, freeze-dried, or oven-dried leaves) and the extraction method utilized (water or methanol as a solvent, variation in pH, variation of relative volume of solvent, variation in extraction time).

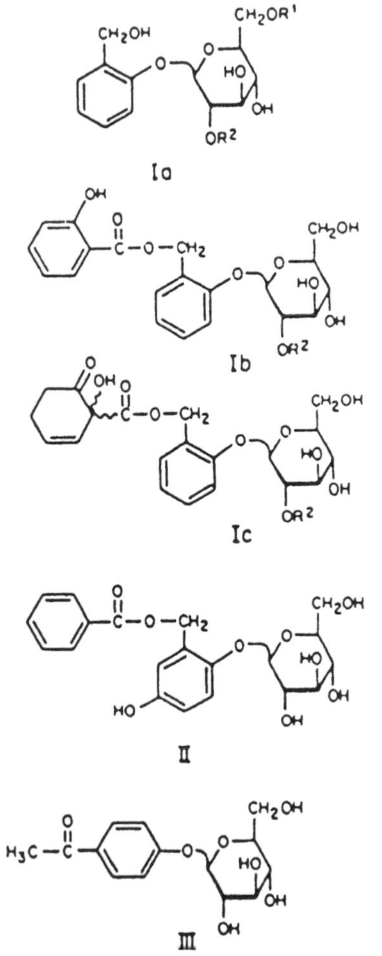

FIGURE 1. Structures of different phenolglu-cosides.[28] Ia: Salicin (R^1 = H, R^2 = H), Fragilin (R^1 = CH$_3$CO, R^2 = H); Ib: Salicoylsalicin (R^2 = H); Ic: Salicortin (R^2 = H), 2-O-Acetylsalicortin (R^2 = CH$_3$CO), Tremulacin (R^2 = C$_6$H$_5$CO); II. Salireprosid; III: Picein. (From Meier, B., Sticher, O., and Bettschart, A., *Dtsch. Apoth.*, 7, 341, 1985. With permission.)

In light of these chemical artefacts, great caution should be exercised when comparing the results from different studies where the chemical analyses are performed with different methods. Moreover, it is important to consider which one of the many possible sample preparation and extraction techniques is the more appropriate for the question being asked. For example, what a herbivorous insect gets from a plant might well differ from what is found in a methanol extraction.[52]

B. DISTRIBUTION
1. Interspecific Variation

The barks of all Salicaceae species are rich in phenolglucosides. This is not the case for the leaves; there is usually a larger quantity and a greater diversity of phenolglucosides in the bark than in the leaves.[29] In fact, the leaves of some *Salix* species

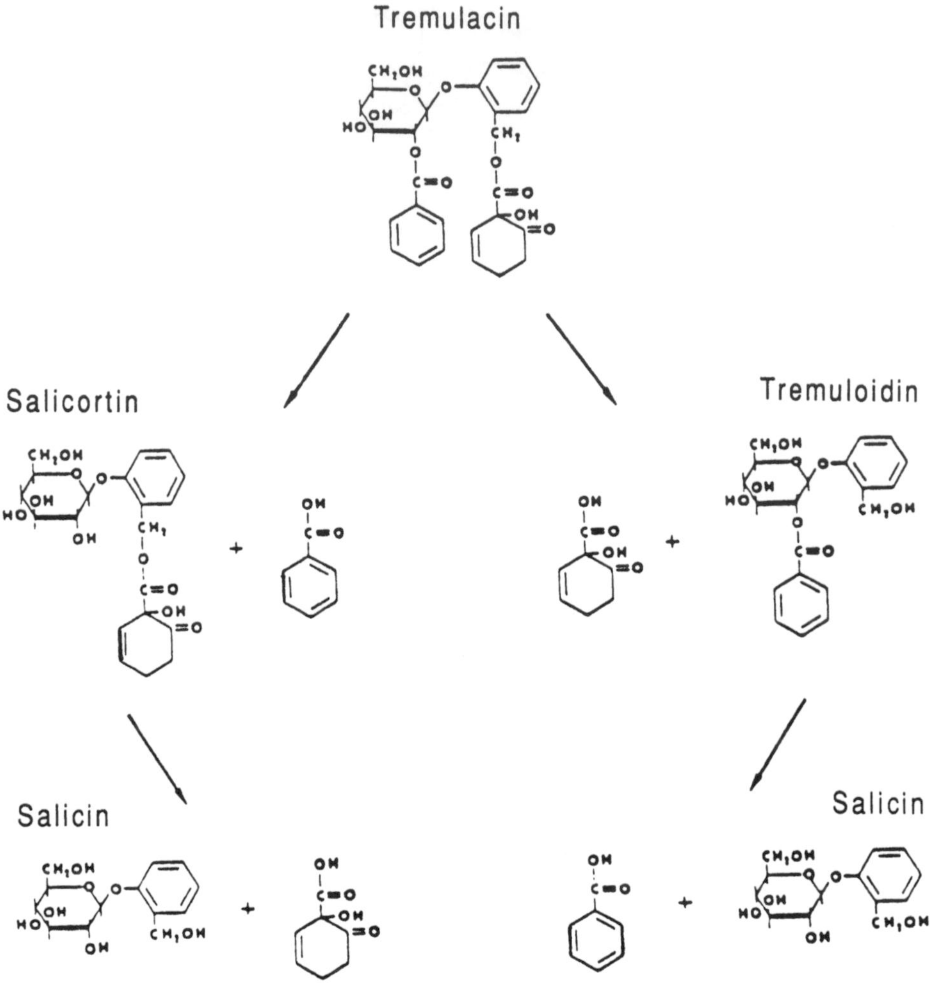

FIGURE 2.   Pathways of hydrolytic degradation and interconversion of phenolic glycosides. (From Lindroth, R. L. and Pajutee, R. S., *Oecologia*, 74, 144, 1987. With permission.)

do not contain any phenolglucoside at all, although the level of total phenolics in *Salix* and *Populus* leaves is generally high. The occurrence of phenolglucosides in the leaves of different *Salix* and *Populus* species is reviewed by Palo.[29] Additional information is found in Julkunen-Tiito,[16] who gives a chemotaxonomic survey of phenolics in the leaves of Northern species of Salicaceae. Meier et al.,[28] Rowell-Rahier,[49] Horn,[13] and Lindroth et al.[24] also analyzed the leaves of different *Salix* and *Populus* species. The profile of the different glucosides in *Salix* and *Populus* leaves appears to be relatively species specific despite intraspecific variations described in the next section.

2. Intraspecific Variations and Abiotic Factors

There are many other sources of variation in the phenolglucoside content of leaves, other than species differences. Content also varies with sex of the plant (the Salicaceae are dioecious); there is usually a higher level of glucosides in the leaves of female plants than in those of males.[60] Further, the phenolglucoside content of leaves tends to exhibit temporal variation. In general, it decreases later in the season (Figure 3)[13,37,49,61] but this is not always the case (e.g., *P. tremuloides* and *P. deltoides*).[20] Denno et al.[5] showed that mature leaves of *S. viminalis, S. dasyclados,* and *S. fragilis* are richer in

% DW

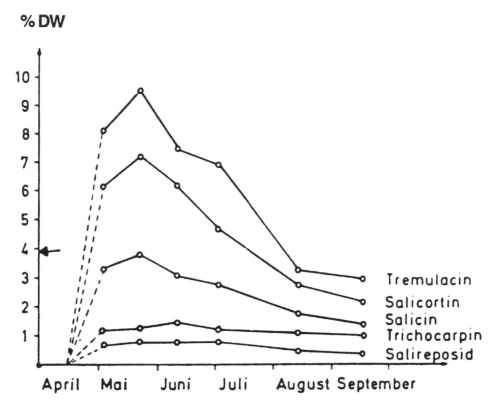

FIGURE 3. Seasonal variation in phenolglucoside concentration in the leaves of *P. trichocarpa* Hopk.[61] (From Thieme, H. and Benecke, R., *Pharmazie*, 25, 228, 1970a. With permission)

phenolglucosides than the young ones closer to the apex of the shoot. The reverse was observed in *P. tremuloides*, but this observation needs to be confirmed by additional measurements.[20] Since the Salicaceae produce new leaves during most of the growing season, two factors have to be differentiated when comparing leaves of different "ages": first, developmental variations (i.e., position on the shoot at a given moment in time),[5] and secondly, seasonal variations (i.e., same developmental stage at different time).[20] Diurnal variations have also been observed, showing decreasing quantity from a maximum in the early morning to a minimum in the evening.[60]

Environmental stress (water, nutrients, light, air pollution) are known to lead to mobilization of nitrogen and should reduce the production of carbon based secondary metabolites.[14] Fast growers should show much plasticity in their production of secondary metabolites, and give priority to growth. A reduction in phenolglucoside is observed in *P. deltoides* when the leaves are submitted to ozone stress (Jones, personal communication) while Larsson et al.[18] produced threefold variation in the phenolics of sapling *S. dasyclados* by varying the light intensity and nutrient supply. Relative carbon availability is linked to phenolic synthesis. Additionally, Bryant et al.[3] demonstrate that nitrogen fertilization results in decreased concentration of condensed tannins and phenolglucosides in the leaves of *P. tremuloides*. Thus if the C/N ratio is decreased, the content of the leaves in carbon-based secondary metabolite also decreases.

Elevation is another possible source of variation; salicin and salicortin are positively correlated with elevation in *S. lasiolepis, S. orestera*, and *S. planifolia*.[13]

Lastly there have been several reports of individual variation between leaves of one branch or of different plants and clones.[20,53-55] Summarizing, the causes of intraspecific variation in the phenolglucoside content of the leaves are numerous and each plant or

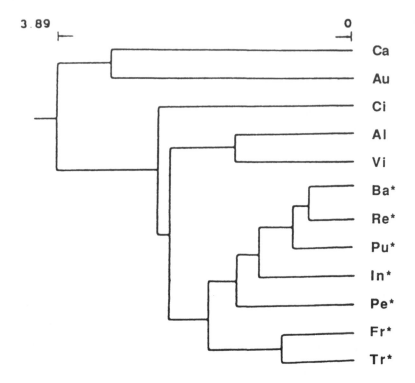

FIGURE 4.   Cluster analysis showing similarity of *Salix* species using presence/ absence data of oligophagous sawflies and weevils.[48] The units of the abscissa give the maximum diameter of a cluster in Euclidean metric. CA = *S. caprea;* Au = *S. aurita;* Ci = *S. cinerea;* Al = *S. alba;* Vi = *S. viminalis;* Ba = *S. babylonica;* Re = *S. repens;* Pu = *S. purpurea;* In = *S. incana;* Pe = *S. pentandra;* Fr = *S. fragillis;* Tr = *S. triandra.* ⋆ = species of Salix containing phenolglucosides in their leaves.

set of leaves probably presents at any given point in time a particular mixture of phenolglucosides to a potential herbivore. Intraspecific variation in the phenolglucosides content of the leaves suggests that the Salicaceae should be considered as an heterogenous habitat sensu Whitham.[66]

III. PHENOLGLUCOSIDES IN HOST PLANT LEAVES AND DISTRIBUTION PATTERNS OF HERBIVORES

Several recent studies (bibliographical and experimental) suggest that herbivore distribution is affected by the phenolglucoside content of the leaves of different *Salix* species, of which there are more than 30 in Europe alone.

In a bibliographical survey, Rowell-Rahier[48] shows that for British moths, French weevils, and German sawflies, the presence or absence of phenolglucosides (or alternatively the absence or presence of condensed tannins) in *Salix* leaves is related to the degree of dietary specialization of the insects feeding on these leaves. The results show that *Salix* species with phenolglucosides tend to be eaten by specialized herbivores and avoided by generalists. Conversely, *Salix* species without phenolglucosides tend to be the food of more polyphagous insects and tend to be avoided by more specialized ones. Moreover, the faunas of the different *Salix* species with phenolglucosides in their leaves are more similar to each other than are the faunas of the *Salix* species having no phenolglucosides in their leaves; this is particularly true for oligophages (Figure 4).

This affords a corroboration of the protective function of phenolglucosides in willow leaves. Indeed, if the dietary specialization of *Salix* herbivores is linked (proximately or ultimately) to the presence or absence of phenolglucosides, the faunas of the willow species with similar chemistry should be more similar to each other than the faunas of *Salix* species with different secondary chemistry.

In a study of the aphid *Pemphigus betae* on *Populus angustifolia* Zucker[68] shows that the leaves with lower level of total phenolic are more infested than those with higher level. It is also clear from that study that, as suggested by Whitham,[66] the leaves should be considered as a heterogenous habitat; the level of phenol between and within leaves is extremely variable, showing for example a gradient of concentration between the tip and the apex of a single leaf.

Among the herbivores on willows, the leaf beetles (Chrysomelidae) have been intensively investigated by several research groups. In northern and central Europe more than 70 such species are described.[9,17] The leaf beetles are particularly interesting because they offer a model to study not only the impact of plant secondary metabolites on herbivory but also the significance of these compounds for the interaction between herbivores and their enemies. We will now briefly summarize what is known about the relationship between phenolglucosides and (1) laboratory experiments on food plant choice by beetles, (2) abundance of beetles, and (3) herbivory level in the field.

A. LABORATORY STUDIES

Tahvanainen and co-workers[59] showed that the concentration and composition of the different phenolglucoside blends are species specific in willows. For four common leaf beetle species *(Phratora vitellinae, Plagiodera versicolora, Lochmea caprea, Galerucella lineola)*, the patterns of food plant selection observed in multiple choice preference experiments are closely related to the phenolglucoside spectra of the willows tested. Indeed, the second choice of the beetles was always the willow species which was chemically the most similar in its phenolglucoside content to the preferred host plant (Table 1). Rowell-Rahier[49-50] showed that *Ph. vitellinae* in laboratory food plant choice prefers *Salix* species rich in phenolglucosides *(S. nigricans, S. purpurea, S. fragilis)* to those whose leaves do not contain phenolglucosides *(S. caprea, S. cinerea, S. alba)*. However, geographically isolated populations can be local specialists and have markedly different preferences within those *Salicaceae* species which have phenolglucosides in their leaves. Especially in disturbed habitats (a small number of indigenous species and/or a large number of introduced species), the preferred host in the field might not be the same as the one established in laboratory tests. Denno et al.[5] studied *Ph. vitellinae* and *G. lineola* on *Salix* species (*S. viminalis* — low phenolglucoside content and native; *S. fragilis* — high phenolglucoside content and native; *S. dasyclados* — high phenolglucoside content but introduced). *Ph. vitellinae* oviposits preferentially on *S. fragilis* whereas *G. lineola* prefers to oviposit on *S. viminalis* followed closely by *S. fragilis*. For both beetle species *S. dasyclados* is the least preferred oviposition host. The performances (larval survival, pupal size, developmental time) of the two beetles on *S. fragilis* and *S. viminalis* were quite similar. We will discuss this apparent discrepancy between performance and oviposition in the next section.

Matsuda and Matsuo[27] indicate that for *Chrysomela vingitipunctata costella* and *Pl. versicolora distinctata*, salicin and populin in combination with sucrose have a phagostimulant effect, whereas for *Ch. v. costella* populin alone is active. These compounds have no effect on *L. caprea scribata*.

However, the feeding preference of *Ph. vitellinae* for *Salix* spp. rich in phenolglucosides is not directly caused by a phagostimulatory effect of salicin but is rather an indirect consequence of the pilosity of the leaves of many of the species which do not

Table 1
**FOR FOUR SPECIES OF LEAF BEETLES, THERE
IS A GOOD CORRESPONDENCE BETWEEN THE
RANKS OF PREFERENCE OBTAINED WITH
DIFFERENT *SALIX* SPECIES IN MULTIPLE CHOICE
FEEDING EXPERIMENTS (RF) AND THE RANKS
OF CHEMICAL SIMILARITY (RCS) OF THOSE
SALIX SPECIES COMPARED TO THE MOST
PREFERRED SPECIES (1)**

Salix sp.	*Ph. vitelli-nae* RCS	*Ph. vitelli-nae* RF	*Pl. versico-lora* RCS	*Pl. versico-lora* RF	*G. lineola* RCS	*G. lineola* RF	*L. caprea* RCS	*L. caprea* RF
Wild								
S. ni	1	1	4	4	4	4	4	4
S. ph	2	2	2	2	2	2	2	2
S. ca	3	4	1	1	1	1	1	1
S. pe	4	3	3	3	3	3	3	3
Cultivated								
S. aq	1	1	1	1	3	3	3	3
S. da	2	2	2	2	4	4	4	4
S. vi	3	3	3	3	2	2	2	2
S. tr	4	4	4	4	1	1	1	1

S. ni = *S. nigricans;* S. ph = *S. phylicifolia;* S. ca = *S. caprea;* S.
pe = *S. pentandra;* S. aq = *S. aquatica;* S. da = *S. dasyclados;* S.
vi = *S. viminalis;* S. tr = *S. triandra.*

Adapted from Tahvanainen, J., Julkenen-Tiito, R., and Kettunen, J.,
Oecologia, 67, 52, 1985.

contain phenolglucosides. For example, shaved *S. caprea* leaves are preferred to un-
shaved ones and, *Ph. vitellinae* does not show any preference between shaved *S.
caprea* leaves with or without salicin painted on the surface.[50] *Ph. vitellinae* is not the
only leaf beetles species for which the presence or absence of trichomes on the leaf
surface plays an important role in feeding choice of the adults. *Ph. laticollis* feeds on
different species of *Populus* and prefers *P. nigra italica. P. alba* has leaves covered
with trichomes and is rejected by *Ph. laticollis.* However, in laboratory test, silica discs
coated with methanol and dichloromethane extracts of *P. alba* are preferred to those
coated with the extract of *P. nigra.* This suggests that factors other than those present
in the extracts are important for the food plant choice of *Ph. laticollis,*[63] In preliminary
studies of another beetle belonging to the same genus, *Ph. vulgatissima,* Rowell-Rahier
(unpublished) was able to show that this species prefers *Salix* with leaves poor in
phenolglucosides (e.g., *S. caprea* and *S. cinerea*) and that salicin is deterrent for the
adults of this species.

B. FIELD STUDIES

In a field study, involving two of the same beetle species as in the previous examples
(*Ph. vitellinae* and *Pl. versicolora*), as well as a common sawfly *(Pontania proxima),* the
correlation between herbivore abundance and amount of salicin and sailcortin in the
leaves of individual *S. alba* (poor in phenolglucosides), *S. fragilis* (rich in phenolglucosides), and several of their hybrids containing intermediate levels of phenolglucosides

Table 2
RESULTS OF A MULTIPLE REGRESSION
BETWEEN THE ABUNDANCE OF EACH OF
THREE HERBIVORE SPECIES IN THE FIELD
(DEPENDENT VARIABLE) AND THE
PHENOLGLUCOSIDE CONTENT OF THE *SALIX*
LEAVES ON WHICH THEY WERE RECORDED

Herbivore species	% Variation explained	Overall significance
Ph. vitellinae		
Adults	4	ns
Larvae	8	p<.10
Pl. versicolora		
Adults	24	p<.005
Larvae	1	ns
P. proxima		
Galls	30	p<.005

Adapted from Rowell-Rahier, M., Soetens, Ph., and Pasteels, J. M., *Insects-Plants*, Labeyrie, V., Fabres, G., and Lachaise, D., Eds., W. Junk, Dordrecht, 1987, 91.

was tested.[52] Because the phenolglucosides influence larvae and adult beetles differently (see below), the abundances of both life stadia were recorded separately. The results show that phenolglucosides have no significant effect on the distribution of *Ph. vitellinae* adults and *Pl. versicolora* larvae. However, the phenolglucosides may partly explain the observed variation in the distribution of *Ph. vitellinae* larvae, *Pl. versicolora* adults, and *P. proxima* galls (Table 2).

The function and importance of the phenolglucosides for *Ph. vitellinae* larvae will be discussed below. The pattern of distribution of *Pl. versicolora* adults is rather well predicted by phenolglucoside alone (24% of variation explained), but the prediction is even better (multiple correlation coefficient 0.68, 47% of variation explained) when the combination of both phenolglucosides and abundance of other herbivores on the same plant is taken into account. The different herbivores studied prefer the same individual plants, and the observed distribution of *Pl. versicolora* adults can probably be explained by some plant characteristic positively correlated with phenolglucosides and influencing positively the different herbivores. For *Pl. versicolora*, the results obtained in the field and in the laboratory are different. In the laboratory study of Tahvanainen et al.,[59] *Pl. versicolora* preferred willow leaves with moderate to low total phenolglucoside content whereas Soetens[55] found the abundance of this beetle in the field is positively correlated with level of phenolglucosides, mostly salicortin. Obviously, factors other than food plant preferences influence the abundance of the beetles in the field (e.g., natural enemies), but we also have to keep in mind the possibility of local specialization (as demonstrated in *Ph. vitellinae*[49]).

The relationship between intraspecific variation in phenolglucosides in individual trees of *S. lasiolepis* and the damage due to herbivory by a leaf beetle, *C. aenicollis*, was established by Smiley et al.[54] in California. The level of herbivory is also positively correlated with level of phenolglucosides in the attacked leaves (Figure 5). Additionally, the relative success of *C. aenicollis* on *S. lasiolepis* and *S. orestera* is positively correlated with salicin content.

It is possible that the gall-forming sawfly, *P. proxima*, selects plants rich in phenolglucosides so that the larvae are better protected by the surrounding plant tissue.

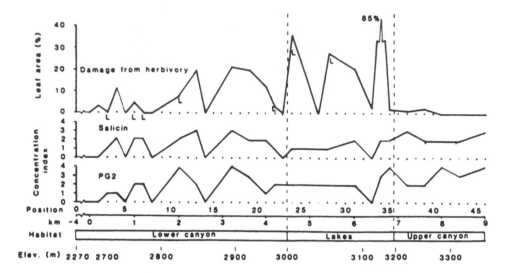

FIGURE 5.   Leaf damage and chemical content of willows growing along an elevation gradient in the east-central Sierra Nevada mountains. Sampling points were at 200 m intervals numbered 0 to 46; if no foliage was present none was collected. Data points indicated by L represent *S. lasiolepis*; otherwise they represent *S. orestera*. (From Smiley, J. T., Horn, J. M., and Rank, N. E., Ecological effects of salicin at three trophic levels: new problems from old adaptations, *Science*, 229, 649, 1985. © 1985 by the AAAS. With permission.)

Fritz et al.[10] showed that for four sawfly species (*Euura* spp.; *Pontania* sp.; *Phyllocolpa* sp.) found on *S. lasiolepis* there is a considerable variation among clone phenotypes in their susceptibility to sawfly oviposition; possible influences of phenolglucosides have not been examined.

C. EFFECT OF ALTITUDE

In Section II we have seen that several factors can influence the intraspecific distribution of phenolglucosides. Altitude was one of them. Indeed, the 100:1-fold variation in phenolglucosides observed by Smiley and Rank[53] within *S. orestera* and *S. lasiolepis* was observed along an elevation gradient (from 2270 m to 3300 m) in the Californian Sierra Nevada. It is difficult to assess the importance of altitude on performance of the beetle, because it affects not only host-plant secondary metabolites, but may also provide an important selection pressure which is experienced by the beetles, namely the direct effect of temperature. Larvae from *C. aenicollis* collected from high elevation site grow approximately 10% faster than individuals from a lower altitude population under controlled temperature and phenolglucoside conditions in the laboratory. Rapid growth is indeed to be expected from the high elevation population, whereas better defense of the beetles against more abundant predators is expected from the lower elevation populations. This seems to contradict the observation of lower concentration of salicin (which is used for insect defense; see following section) at lower elevation. Smiley and Rank[53] suggested that lower salicin content at low elevation could be an adaptation of the plants to escape heavy herbivory by specialists such as *C. aenicollis*, which are dependent on salicin for their own chemical defense.

D. EFFECT OF LEAF AGE

Leaf age is another factor known to affect the phenolglucoside content of leaves of Salicaceae (see Section II). Raupp and Denno[41] studied leaf age as a predictor of herbivore distribution and abundance using *Pl. versicolora* on *S. babylonica*. Raupp and

Denno did not mention phenolglucosides as a variable in their study, but they established that adult *Pl. versicolora* are mostly found on the young leaves and larvae and eggs on the older leaves of *S. babylonica*. Adults fed young leaves reproduce sooner and have bigger clutches of eggs than those fed old leaves. The larvae fed young leaves have a greater adult weight, have shorter developmental times and lower mortality than those on older leaves. In another study Raupp[40] shows that older leaves erode the cutting surface of the beetles mandibles more than tender leaves, and that this results in a slower consumption rate of older leaves, correlated with reduced fecundity. This mechanical factor is certainly not the only one explaining the better performance of *Pl. versicolora* on *S. babylonica* young leaves. These young leaves probably also have higher nutritive value and a qualitatively and quantitatively different phenolglucoside content than the older ones.

The larvae of two other beetle species (*Ph. vitellinae* and *G. lineola*) perform better (higher survival on *S. viminalis* and *S. dasyclados*, larger pupal size on *S. fragilis,* and faster development on *S. dasyclados*) on mature leaves than on younger ones.[5] In the case of these three willow species mature leaves contain larger amounts of phenolglucosides and other simple phenolics (see Section II). These mature leaves are also poorer in primary nutrients (such as proteinaceous amino acids) and are tougher than the terminal young ones.

E. EFFECT OF ENVIRONMENTAL STRESS

Adult *G. lineola*[18] have a reduced feeding rate on leaves with high phenolic content, obtained by increasing relative carbon availability (increased the light intensity under low nutrient condition).

Adult and larval *Pl. versicolora* prefer to feed on, and have a higher consumption rate of, ozone-stressed *P. deltoides* (possibly due to reduced levels of phenolglucosides, see Section II).[15] However, the females prefer to oviposit on nonstressed plants.

Adult *Pl. versicolora* avoid previously injured leaves of *S. babylonica*.[44] On damaged *S. babylonica* and *S. alba,* fecundity is reduced, the development of the larvae is slower, and they have a lower weight as adults.[42]

These results may seem contradictory; ozone stress seems to increase consumption by *Pl. versicolora* whereas stress originating from previous injury decreases it. However, the two types of stress probably have very different effects on the phenolglucoside content of the leaves; ozone stress probably decreases it, and injury of the leaves could have the opposite effect as observed by Reichardt et al.[45] in *P. tremuloides*.

F. SUMMARY

In summary, when looking at the pattern of distribution of herbivores and their relationships to phenolglucosides (Table 3), larvae and adults often have different responses. Studies of feeding and oviposition preference also give different answers. In the next section we will examine how this might be explained.

Most of the results summarized above were obtained by correlative rather than causal studies. Additionally, it is important to remember that in the relationships summarized above, the variation in phenolglucosides is linked to different origins (interspecific differences, age differences, environmental stress, etc.). These factors do not exclusively affect the level of phenolglucosides in the plants, but rather numerous other plant characteristics too, all of which might be important for the herbivore. In other words, the experiments are inadequately controlled in the purity of the prime variable.

Table 3
**FEEDING PREFERENCES AND GROWTH PERFORMANCE OF TWO LEAF
BEETLES: *PH. VITELLINAE* (LARVAE USE PHENOLGLUCOSIDES FOR
DEFENSE) AND *PL. VERSICOLORA* (LARVAE DO NOT USE
PHENOLGLUCOSIDES FOR DEFENSE)**

Studies	Ph. vitellinae	Pl. versicolora
In laboratory:		
49, 50	Feed on spp. rich in phenolglucosides	
59	Feed on spp. rich in phenolglucosides	Feed on spp. low to poor in phenolgluco-sides
5	Oviposit on spp. rich in phenolglucosides	
In the field:		
55	Abundance of adults not correlated with phenolglucoside content of the leaves; abundance of larvae positively corre-lated with phenolglucoside content of the leaves	Abundance of adults positively correlated phenolglucoside content of the leaves; abundance of larvae not correlated with phenolglucoside content of the leaves
Other studies:		
41		Prefer young leaves
5	Larvae perform better on mature leaves	
15		Prefer ozone-stressed leaves but oviposit on nonstressed plants
44		Avoid injured leaves, larvae perform less well on injured leaves

IV. SPECIALIZATION ON LEAVES RICH IN PHENOLGLUCOSIDES

A. ANTAGONISTIC EFFECTS OF PHENOLGLUCOSIDES ON HERBIVORES: DETOXIFICATION MECHANISMS IN THE TIGER SWALLOWTAIL BUTTERFLY

Although the evidence reported in the previous section strongly suggests that phen-olglucosides mediate the distribution of phytophagous insects on salicaceous plants, there are surprisingly very few data on the precise action of those phenols on behavior and physiology of herbivores. There are few direct demonstrations of the negative impact of the condensed tannins on the growth of herbivores, but Bryant et al.[3] have demon-strated that larval growth of large aspen tortrix *(Choristoneura conflictana)* is decreased by increasing concentration of condensed tannins and phenolglucosides in artificial diets. The toxicity of the phenolglucosides salicortin and tremulacin is due to their degradation product 6-hydroxycyclohexenane.[45]

There is no information on the detoxification of phenolglucosides in the guts of leaf beetles not dependent on these compounds for their own defense. However, recent studies by Lindroth et al.[19,25] provide insight into the way another insect, the tiger swallowtail (*Papilio glaucus canadensis* and *P. g. glaucus*) feeding on salicaceae (*P. tremuloides*) deals with the phenolic glycosides. *P. g. canadensis* is adapted to these toxins and grows well on *P. tremuloides*. *P. g. glaucus* does not grow so well on the same plants. Lindroth et al.[23,25] showed that the toxic phenolics are salicortin and tre-mulacin and not salicin. They cause gut lesions.[23] The active toxic compound is a cyclohexane saligenin ester.[25] The hypothetical degradation pathways of salicortin and tremulacin are summarized in Figure 6. The two species of tiger swallowtail differ at least in two ways in their enzymatic degradation potential. First, the midgut-β-glucosi-dase activity of the adapted *P. g. canadensis* is one third to one half that of the less adapted *P. g. glaucus*. The specificity of this β-glucosidase is still unknown. Ingestion

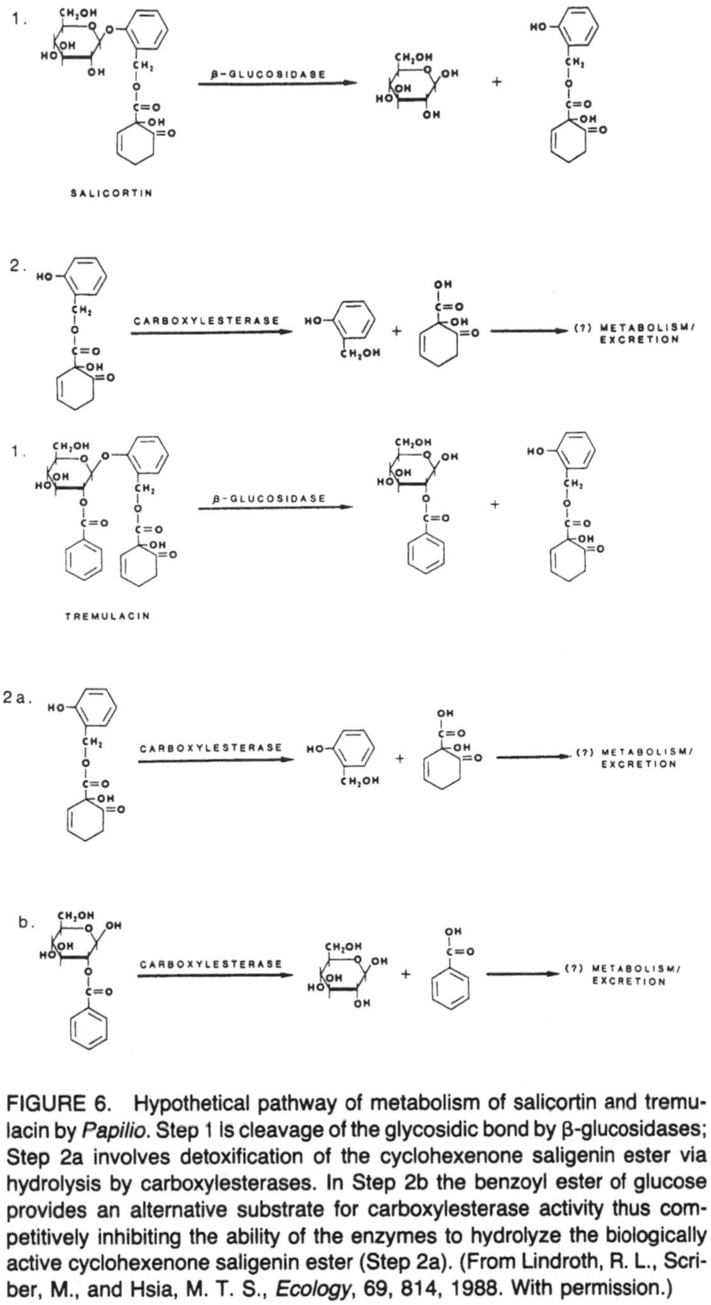

FIGURE 6. Hypothetical pathway of metabolism of salicortin and tremulacin by *Papilio*. Step 1 Is cleavage of the glycosidic bond by β-glucosidases; Step 2a involves detoxification of the cyclohexenone saligenin ester via hydrolysis by carboxylesterases. In Step 2b the benzoyl ester of glucose provides an alternative substrate for carboxylesterase activity thus competitively inhibiting the ability of the enzymes to hydrolyze the biologically active cyclohexenone saligenin ester (Step 2a). (From Lindroth, R. L., Scriber, M., and Hsia, M. T. S., *Ecology*, 69, 814, 1988. With permission.)

of phonolglucosides suppresses β-glucosidase activity in *P. g. canadensis* but induces it in *P. g. glaucus*.[19] Secondly, the actively toxic cyclohexane saligenin ester (liberated by the β-glucosidase) is further degraded (hydrolyzed) to harmless products by a carboxylesterase. The activity of the carboxylesterase is thus a determinant: this enzyme should act as fast as possible to avoid damage. Lindroth suggests that the evolution of an optimal carboxylesterase is a critical part of the adaptation to phenolglucosides. The biological activity of tremulacin is twice that of salicortin, since the benzoyl glucose moiety released by the β-glucosidase may provide an alternative substrate for the carboxylesterase.[25]

Table 4
DISTRIBUTION OF DEFENSE COMPOUNDS IN CHRYSOMELINAE LARVAE

Beetles	Host plants	Secretions
Subtribe Chrysomelina		
6 genera, 11 spp.	Salicacae	Iridoid monoterpenes
	Betulaceae	
	Ranuncala-	
	ceae	
	Apiaceae	
	Polygonacae	
	Scrophulari-	
	aceae	
	Cruciferae	
Chrysomela interrupta	Betulaceae	Phenylethyl esters
6 spp. of *Chrysomela*	Salicaceae	Salicylaldehyde
		Benzaldehyde
Gastrolina depressa	*Juglandaceae*	Juglone
Subtribe Phyllodectina		
3 spp. of *Phratora*	Salicaceae	Iridoid monoterpenes
Phratora vitellinae	Salicaceae	Salicylaldehyde

Adapted from Pasteels, J. M., Duffey, S., and Rowell-Rahier, M., *J. Chem. Ecol.*, 16, 1990.

B. AGONISTIC EFFECTS OF PHENOLGLUCOSIDES ON LEAF BEETLES: CHEMICAL DEFENSE

As shown in the data reviewed above, the distribution of leaf beetles on the Salicaceae seems to be linked, positively or negatively, to the phenolglucoside content of the leaves. Leaf beetles are known for their extremely diverse defensive mechanisms, which include chemical defense. Both adults and larvae can be so defended. Chemical defense of adult leaf beetles has been recently reviewed.[33] Chemical defense of adult leaf beetles feeding on the Salicaceae bears no direct relationship to the phenolglucosides of the leaves. The phenolglucosides are not directly sequestered by adults for defensive purpose. On the other hand, the phenolglucosides are known to influence the nature of the defensive secretion of the larvae of several species of the subtribe Chrysomelina and *Phratora* in the subtribe Phyllodectina.[35,51] Defensive compounds are identical in larval and adult stages of the subtribe Chrysolinina and the subfamily Galerucinae, which do not possess glands at all in the larval stage. This contrasts sharply with the Chrysomelina and Phyllodectina, which possess nine pairs of segmental exertile larval defensive glands. The compounds generally liberated by the larval defensive glands are listed in Table 4. The larval secretions from these glands are volatile and highly reactive compounds detectable at a distance.

In Table 4, we can see that the larvae of some species of *Chrysomela* and *Phratora* secrete not monoterpenes, as expected from their taxonomic position, but rather aromatic compounds such as salicylaldehyde and benzaldehyde. It has long been suggested that some chrysomelid larvae utilize salicin and possibly other phenolglucosides as precursors of salicylaldehyde produced in specialized defensive glands[12,36,65] and this utilization is now well documented. These species feed exclusively on trees belonging to the Salicaceae. The secretion of salicylaldehyde is an adaptation to the secondary chemistry of the Salicaceae. To confirm the role of salicin and salicortin as precursors of the salicylaldehyde secreted by the larvae of *Ph. vitellinae*, the consequence of a salicin-free diet on the defensive secretion of the larvae was examined. In nature, adult *Ph. vitellinae* are never seen on *S. caprea* and the larvae do not normally accept these hairy leaves which are salicin free. After being denuded of trichomes, however, the leaves of *S. caprea* were readily accepted by the larvae. The larvae developed normally on shaved leaves but did not produce any secretion. Addition of salicin or salicortin (from Dr. B. Meier) to the shaved leaves of *S. caprea* restored the secretion of salicylaldehyde. Salicin and salicortin contained in the food plant are thus the precursors of the larval secretion of salicylaldehyde.[50] Experimental feeding of *Ch. tremulae* larvae with labeled salicin confirmed the role of salicin as a precursor of salicylaldehyde. The transformation of salicin into salicylaldehyde occurs in the defensive glands; the β-glucosidase activity was found to be four times higher in the glands than in the guts.[34] Both β-glucosidase and oxidase needed to transform salicin in salicylaldehyde occur in the secretion itself, and the derivation of salicylaldehyde from salicin occurs in the gland reservoirs[31] (Duffey and Pasteels, in preparation). The concentration of glucose and aldehyde in the secretion are far from equimolar and this indicates that the glucose formed by the salicin hydrolysis is in great part recovered by the larvae and transferred back into the blood.[34] The recovered glucose should benefit the beetles. Thus the cost of defense is expected to be maximal in those species synthesizing defensive secretion *de novo* and minimal or negative in those which profit from the use of the plant phenolglucosides. Rowell-Rahier and Pasteels[51] showed experimentally that the autochthonously produced secretion of some beetle species entails an appreciable cost, expressed as a loss of weight. These costs can be avoided by the use of an appropriate plant precursor such as salicin, in which case the recovery of the glucose moiety of the salicin contributes significantly to the larval energy budget.

At hatching the larvae are clustered and immobile on the leaves. An early defense would therefore seem to be critical. However, only those species secreting salicylaldehyde have functional glands on hatching. Species producing *de novo* secretion are not able to secrete at birth. Paradoxically, then, only those larvae which depend on salicin normally found in their food seem to be able to produce a secretion before feeding. The obvious explanation is that the salicin is sequestered in the eggs and used as a precursor by the neonate larvae. This hypothesis has been confirmed experimentally; the eggs of some of the Salicaceae feeders contain salicin, and those species are also those whose larvae secrete salicylaldehyde. The adults do not sequester salicin for their own defense (they produce *de novo* isoxazolinone glycosides[33]), but females are able to sequester salicin in their eggs for the benefit of both eggs and neonate larvae.[30]

Based on evidence from *Plagiodera versicolora*, Pasteels et al.[31] suggested that the biochemical determinant of the chemical nature of the larval defensive secretion is the oxidase, rather than the β-glycosidase necessary for the transformation of salicin into salicylaldehyde. In other words, a change in the specificity of the oxidase, say the acquisition of the ability to oxidize saligenin, would permit an insect to develop a switch from *de novo* biosynthesis of iridoid monoterpenes to sequestered defense via the utilization of the plant glucoside salicin. The larval gland reservoirs of *Pl. versicolora* contain besides the iridoid aldehyde and lactone, glucose and a nonspecific β-glucosidase able to hydrolyze salicin. It was thus suggested that species producing monoterpenes also contain in the glandular reservoirs a β-glucosidase and possibly an oxidase, yet to be demonstrated, needed in the final steps of the biosynthesis of the oxidized iridoid monoterpenes from a reduced iridoid glucoside.[31]

Thus, the significance and function of phenolglucosides such as salicin and salicortin varies between species of beetles. *Ph. vitellinae* and several *Chrysomela* spp. are dependent on the glucosides for the defense of their larvae and eggs against natural enemies. Other species (*Pl. versicolora, G. lineola*) do not directly use salicin and salicortin. Accordingly, we expect the first set of species to be extremely well adapted to handling phenolglucosides compared with the second group of species. Beetle larvae using salicin and salicortin for defensive purpose should be particularly good at transporting these toxins into the glands, whereas the latter set should be adapted to detoxify the phenolglucosides. This remains to be documented. Additionally, we should expect that the beetles producing salicylaldehyde in the larval stage will select Salicaceae species rich in salicin and other phenolglucosides to lay their eggs. This was confirmed (see previous section).

V. PHENOLGLUCOSIDES AND NATURAL ENEMIES OF LEAF BEETLES

In view of the data viewed in the second section of this paper it is clear that the phenolglucosides from host plants influence, directly or otherwise, the distribution and performance of leaf beetles feeding on the Salicaceae. It is also clear (see Section IV) that the phenolglucosides are utilized by some beetle species for their own defense, whereas in other species detoxification mechanisms for these same compounds are present.

Although several studies compare the performance of different beetle species on *Salix* with different levels of phenolglucosides in their leaves, few, to our knowledge, try to explain the observed patterns by the interaction of the herbivores with their predators.[5] Larvae of *Ph. vitellinae* fed on leaves rich in salicylaldehyde precursors are immediately regurgitated alive by predaceous coccinellid larvae (a predator present in the field). On the other hand, the larvae of *Ph. vitellinae* fed on leaves lacking the precursor of their defensive secretion do not survive attack by the same predaceous coccinellid. This kind of interaction might explain why *Ph. vitellinae* avoid species of Salicaceae poor in salicin as oviposition hosts although the larvae develop well on these plants in the absence of predation! *G. lineola*, on the other hand, is not chemically defended by exocrine secretion and does not rely on phenolglucosides for its defense. In this species the oviposition preference of the adult matches the larval performance.

Smiley et al.[54] measured the effect of host derived defense on larval survival in the field, and showed a differential survival of *C. aenicollis* on willow clones containing different levels of salicin. Larvae feeding on salicin-rich clones were able to produce salicylaldehyde and had significantly greater survivorship than those larvae reared on salicin-poor clones, not producing salicylaldehyde.

There is, to our knowledge, no evidence demonstrating whether the beetle larvae defended by secretion synthesized *de novo* or the larvae of species which use plant phenolglucosides as a precursor of their defensive secretion are better protected from natural enemies. But it is clear that the efficiency of one type of compound is not a fixed characteristic. For example, Pasteels and Grégoire[32] showed that, when given a choice between larvae of *Pl. versicolora* and *Ph. vitellinae*, predaceous sawflies select the type of prey they have previously eaten. Clearly, chemical defense of larvae is not foolproof: predation on *Pl. versicolora* larvae by ladybird larvae has been observed[67] and predation on *Ph. vitellinae* larvae by spiders (*Xysticus* sp.) and neuropteran larvae *(Anisochrysa prasina)* has also been observed.[35] Some reports, still anecdoal, suggest that predatory wasps are "temporally" specialized on larvae producing salicylaldehyde.[35]

Raupp and Denno[41] showed that the generalist ladybird *Hippodamia convergens* has an enhanced search behavior on leaves of willows *(S. babylonica)* covered with feces of adult *Plagiodera versicolora*. The plant phenolglucosides could influence this behavior indirectly, since the distribution of the *Plagiodera* adults seems to be correlated with the phenolglucoside content of the plants.[55] Additionally, it is unknown what the cue for the change in the predator behavior is, but one cannot exclude a chemical factor, such as, for example, the presence of active compounds resulting from the degradation of phenolglucosides in the feces.

Defense is not the only reported function of the larval exocrine secretion of the leaf-beetles. At least in some cases, it has been demonstrated that the secretions have an impact on the distribution of other herbivores. Raupp et al.[43] showed that the iridoid monoterpenes secreted by the larvae of *Pl. versicolora* negatively influence the distribution of conspecific adults as well as the leaf consumption by another herbivore, *Nymphalis antiopa* larvae. *Prenelopis* ants tending aphids are repelled by salicylaldehyde[35] and stop tending the aphids.

VI. CONCLUSIONS

Phenolglucosides clearly influence the distribution and food seleciton of herbivores in a complex way. Indeed the interactions between plant and herbivores are part of a dynamic process of which we are only able to observe a frozen sequence in evolutionary time. On the basis of this single observation, we try to deduce the logical course of evolution designating it as coevolution, diffuse or not. It is outside the scope of this review to discuss the terminology which should be utilized to describe the dynamic changes in the interaction between two trophic levels (the host plants and their herbivores). Rather we have stressed that this process involves not only plants and herbivores but also the third trophic level which consists of the herbivores' natural enemies.

Our present understanding of the impact of phenolglucosides on herbivores is mostly speculative since many of the data are descriptive or correlative and very few studies include causal analysis. With these restrictions in mind we would like to divide the interaction between phenolglucosides and herbivores in four rough evolutionary "steps".

- First: The phenolglucosides in the leaves of the Salicaceae have a defensive function against nonadapted herbivores.
- Secondly: Some herbivores have developed adaptation to escape the negative impact of the phenolglucosides, for example, the detoxification mechanism present in *Papilio* spp.
- Thirdly: Some herbivores are able to use the phenolglucosides for their own benefit, for example, the sequestration and utilization which has evolved several times in larvae of leaf beetles.
- Fourthly: At the third trophic level, some natural enemies of the herbivores have adapted to handle the defense which the insects derive from the host-plant phenolglucosides. It is not excluded that some predators can use these plant-derived defensive volatiles as cues to find their prey.

Why are leaves of some northern temperate *Salix* defended by phenolglucosides against herbivory, whereas those of others produce tannins and trichomes? Although it has been suggested[54] that some *Salix* species could escape specialized herbivores which search for plants rich in phenolglucosides by reducing the phenolglucoside content of their leaves, it would be highly hazardous to speculate that it is only the selective pressure of particular herbivores that has prompted such a shift in plant defense. Phenolglucosides could be classified as qualitative defense (sensu Feeny[8]). However, the two types of leaf defense present in *Salix* are not obviously correlated to the apparency of the *Salix* species. Whether or not these two defensive mechanisms fit the resource allocation theory of Coley et al.[4] remains an open question. Additionally, it is important to remember that caution is necessary when dividing the *Salix* species in two categories.

There are contradictory data in the literature on the presence of phenolglucosides in some species which might be explained by geographic variation or by the ease with which hybridization occurs in this group.

ACKNOWLEDGMENTS

We thank H. Rowell and E. Bernays for their helpful comments on previous versions of this paper. M. Rowell-Rahier thanks the Swiss National Fund for financial support.

REFERENCES

1. **Bernays, E. and Graham, M.,** On the evolution of host specificity in phytophagous arthropods, *Ecology,* 69, 886, 1988.
2. **Bryant, J. P., Chaplin, F. S., Reichardt, P. B., and Clausen, T. P.,** Adaptation to resource availability as a determinant of chemical defense strategies in woody plants, in *Chemically mediated interactions between plants and other organisms,* Cooper-driver, G. A., Swain, T., and Conn, E., Eds., Plenum Press, New York, 1985, 219.
3. **Bryant, J. P., Clausen, T. P., Reichardt, P. B., McCarthy, M. C., and Werner, R. A.,** Effect of nitrogen fertilization upon the secondary chemistry and nutritional value of quaking aspen (*Populus tremuloides* Michx) leaves for the large aspen tortrix (*Choristoneura conflictana,* Walker), *Oecologia,* 73, 513, 1987.
4. **Coley, P. D., Bryant, J. P., and Chapin, F. S., III,** Resource availability and antiherbivore defence, *Science,* 230, 895, 1985.
5. **Denno, R. F., Larsson, S., and Olmstead, K. L.,** Host plant selection by Scandinavian willow-feeding leaf beetles (Coleoptera: Chrysomelidae): role of plant quality and predation, *Ecology,* submitted.
6. **Edwards, W. R. N.,** Effect of salicin content on palatability of *Populus* foliage to opossum *(Trichosurus vulpecula). N. Z. J. Sci.,* 21, 103, 1978.
7. **Fagerstrm, T., Larsson, S., and Tenow, O.,** On optimal defence in plants, *Functional Ecol.,* 1, 73, 1987.
8. **Feeny, P.,** Plant apparency and chemical defense, in *Biochemical Interaction Between Plants and Insects, Vol. 10, Recent Advance in Phytochemistry,* Wallace, J. W. and Mansel, R. L. Eds., Plenum Press, New York, 1976, 1-g.,
9. **Freude, H., Harde, K. W., and Lohse, G. A.,** *Die Kafer Mitteleuropas,* Vol. 9, Goecke und Evers Verlag Krefeld, 1966.
10. **Fritz, R. S., Sacchi, C. F., and Price, P. W.,** Competition versus host plant phen-otype in species composition: willow sawflies, *Ecology,* 67, 1608, 1986.
11. **Hegnauer, R.,** *Chemotaxonomie der Pflanzen,* VI, Birkhauser, Basel, 1973.
12. **Hollande, A. C.,** Sur la fonction d'excrétion chez les insectes salicicoles et en particulier sur l'éxistence de derivés salicylés, *Ann. Univ. Grenoble,* 21, 457, 1909.
13. **Horn, J.,** Distribution of phenolglycosides in willow (*Salix* spp.) leaves, in *White Mountain Research Symp. Proc.,* 1985.
14. **Jones, C. G. and Coleman, J. E.,** Plant "stress" and insect herbivory: toward an integrated perspective, *Symposium: An Integrated Approach to Study of Environmental Stress on Plant Growth,* submitted.
15. **Jones, C. G. and Coleman, J. E.,** Plant stress and insect behavior: cottonwood, ozone and the feeding and oviposition preference of a beetle, *Oecologia,* 1988.
16. **Julkunen-Tiito, R.,** A chemotaxonomic survey of phenolics in leaves of Northern salicaceae species, *Phytochemistry,* 25, 663, 1986.
17. **Krüssmann, C.,** *Handbuch der Laubgehize,* Vol. 3, Paul Parey, Berlin, 1962.
18. **Larsson, S., Wiren, A., Lundgren, L., and Ericsson, T.,** Effects of light and nutrient stress on leaf phenolic chemistry in *Salix dasyclados* and susceptibility to *Galerucella lineola* (Coleoptera), *Oikos,.* 47, 205, 1986.
19. **Lindroth, R. L.,** Hydrolysis of phenolic glycosides by midgut — glucosidases in *Papilio glaucus* subspecies, *Insect Biochem.,* in press, 1988.
20. **Lindroth, R. L., Hsia, M. T. S., and Scriber, J. M.,** Seasonal patterns in the phytochemistry of three *Populus* species, *Biochem. Syst. Ecol.,* 15, 681, 1987.
21. **Lindroth, R. L. and Pajutee, M. S.,** Chemical analysis of phenolic glycosides: art, facts, and artifacts, *Oecologia,* 74, 144. 1987.
22. **Lindroth, R. L. and Peterson, S. S.,** Effect of plant phenols on performance of southern armyworm larvae, *Oecologia,* 75, 185, 1988.

23. **Lindroth, R. L., Scriber, J. M., and Hsia, M. T. S.,** Differential response of tiger swallowtail subspecies to secondary metabolites from tulip tree and quaking aspen, *Oecologia,* 70, 13, 1986.

24. **Lindroth, R. L., Scriber, J. M., and Hsia, M. T. S.,** Characterisation of phenolic glycosides from quaking aspen *(Populus tremuloides), Biochem. Syst. Ecol.,* 15, 677, 1987.

25. **Lindroth, R. L., Scriber, M., and Hsia, M. T. S.,** Chemical ecology of tiger swallowtail: mediation of host use by phenolic glycosides, *Ecology,* 69, 814, 1988.

26. **Marks, V., Smith, M. J. H., and Cunliffe, A. C.,** The mechanism of the anti-inflammatory activity of salicylate, *J. Pharm. Pharmacol.,* 13, 218, 1960.

27. **Matsuda, K. and Matsuo, H.,** A flavonoid, luteolin-7 glucoside, as well as salicin and populin, stimulating the feeding of leaf-beetles attacking salicaceous plants, *Appl. Entomol. Zool.,* 20, 305, 1985.

28. **Meier, B., Sticher, O., and Bettschart, A.,** Weidenrinden-Qualitt, *Dtsch. Apoth.,* 7, 341, 1985.

29. **Palo, R. T.,** Distribution of birch *(Betula* spp.), willow *(Salix* spp.), and poplar *(Populus* spp.) secondary metabolites and their potential role as chemical defense against herbivores, *J. Chem. Ecol.,* 10, 499, 1984.

30. **Pasteels, J. M., Daloze, D., and Rowell-Rahler, M.,** Chemical defense in chrysomelid eggs and neonate larvae, *Physiol. Entomol.,* 11, 29, 1986.

31. **Pasteels, J. M., Duffey, S., and Rowell-Rahier, M.,** Toxins in Chrysomelid beetles: possible evolutionary sequence from de novo synthesis to derivation from food plant chemicals, *J. Chem. Ecol.,* 16, 1990.

32. **Pasteels, J. M. and Grégoire, J. C.,** Selective predation on chemically defended chrysomelid larvae. A conditioning process, *J. Chem. Ecol.,* 10, 1693, 1984.

33. **Pasteels, J. M., Rowell-Rahier, M., Braekman, J. C., Daloze, D., and Duffey, S.,** Evolution of exocrine defense in leaf-beetles (Coleoptera: Chrysomelidae), *Experientia,* 45, 295, 1989.

34. **Pasteels, J. M., Rowell-Rahier, M., Braekman, J. C., and Dupont, A.,** Salicin from host plant as precursor of salicylaldehyde in defensive secretion of Chrysomelinae larvae, *Physiol. Entomol.,* 8, 307, 1983.

35. **Pasteels, J. M., Rowell-Rahier, M., and Raupp, M. J.,** Plant derived defense in Chrysomelid beetles, in *Novel Aspects of Insect-Plant Interactions,* Barbosa, P. and Letourneau, D., Eds., John Wiley & Sons, New York, 1988, 235.

36. **Pavan, M.,** Studio sugli antibiotici e insetticidi di origine animale. I. Sul principio attivo della larva di *Melasoma populi* L., *Arch. Zool. Ital.,* 38, 159, 1953.

37. **Pearl, I. A. and Darling, S. F.,** The structures of salicortin and tremulacin, *Tetrahedron,* 44, 3827, 1970.

38. **Price, P. W.,** Relevance of ecological concepts to practical biological control, in *Biological Control in Crop Production,* Papavisas, G. C., Ed., Allenheld, Osmun, 1980, 3.

39. **Price, P. W., Bouton, P., Groos, B. A., and McPherson, J.N.,** Interactions among three trophic levels: influence of plants on interactions between insect herbivores and natural enemies, *Annu. Rev. Ecol. Syst.,* 11, 41, 1980.

40. **Raupp, M. J.,** Effects of leaf toughness on mandibular wear of the leaf beetle, *Plagiodera versicolora, Ecol. Entomol.,* 10, 73, 1985.

41. **Raupp, M. J. and Denno, R. F.,** Leaf age as a predictor of herbivore distribution and abundance, in *Variable Plants and Herbivores in Natural and Managed Systems,* Denno, R. F. and McClure, M. S., Eds., Academic Press, New York, 1983, 91.

42. **Raupp, M. J. and Denno, R. F.,** The suitability of damaged willow leaves as food for the leaf beetle, *Plagiodera versicolora, Ecol. Entomol.,* 9, 443, 1984.

43. **Raupp, M. J., Milan, F. R., Barbosa, P., and Leonhart, B. A.,** Methylcyclopentanoid monoterpenes mediate interactions among insect herbivores, *Science,* 232, 1408, 1986.

44. **Raupp, M. J. and Sadof, C. S.,** Behavioral responses of a leaf beetle to injury related changes in its salicaceous host, *Oecologia,* submitted.

45. **Reichardt, P. B., Clausen, T. P., and Bryant, J. P.,** The role of phenol glycosides in plant/herbivore interactions, in *Natural Toxins,* Vol. 6, in press.

46. **Rhoades, D. F. and Cates, R. G.,** Toward a general theory of plant antiherbivore chemistry, in *Biochemical Interaction Between Plants and Insects, Vol. 10, Recent Advance in Phytochemistry,* Wallace, J. W. and Mansel, R. L., Eds., Plenum Press, New York, 1976, 168.

47. **Rosenthal, G. and Janzen, D.,** *Herbivores: Their Interaction with Secondary Metabolites,* Academic Press, New York, 1979, 718.

48. **Rowell-Rahier, M.,** The presence or absence of phenolglycosides in *Salix* (Salicaceae) leaves and the level of dietary specialisation of some of their herbivorous insects, *Oecologia,* 62, 26, 1984.

49. **Rowell-Rahier, M.,** The food plant preferences of phratora vitellinae (Coleoptera: Chrysomelinae). B. A laboratory comparison of geographically isolated populations and experiments on conditioning, *Oecologia,* 64, 375, 1984.

50. **Rowell-Rahier, M. and Pasteels, J. M.,** The significance of salicin for a Salix-feeder, Phratora (Phyllodecta) vitellinae, in *Proc 5th Int. Symp. Insect-Plant Relationships,* Pudoc, Wageningen, 1982.

51. **Rowell-Rahier, M. and Pasteels, J. M.,** Economics of chemical defense in Chrysomelinae, *J. Chem. Ecol.,* 12, 1189, 1986.

52. **Rowell-Rahier, M., Soetens, Ph., and Pasteels, J. M.,** Influence of phenolglucosides on the distribuiton of herbivores on willows, in *Insects-Plants,* Labeyrie, V., Fabres, G., and Lachaise, D., Eds., W. Junk, Dordrecht, 1987, 91.

53. **Smiley, J. T. and Rank, N. E.,** Predator protection versus rapid growth in an alpine leaf beetle, in *White Mountain Res. Symp. Proc.,* 1985.

54. **Smiley, J. T., Horn, J. M., and Rank, N. E.,** Ecological effects of salicin at three trophic levels: new problems from old adaptations, *Science,* 229, 649, 1985.

55. **Soetens, P.,** Sensibilité differentielle de *Salix alba, S. fragilis* et leurs hybrides aux insectes phyllophages *Phratora vitellinae, Plagiodera versicolora* et *Pontania proxima,* Travail de fin d'étude, Université Libre de Bruxelles, 1986.

56. **Steele, J. W., Bolan, M., and Sudette, R. C. S.,** Phytochemistry of the family Salicaceae. II. The effect of extraction procedures on the apparent free phenolic glycosides content of Salix species, *J. Chromatog.,* 40, 370, 1969.

57. **Steele, J. W., Weitzel, P. F., and Audette, R. C. S.,** Phytochemistry of the Salicaceae, *J. Chromatogr.,* 71, 435, 1972.

58. **Tahvanainen, J., Helle, E., Julkunen-Tiito, R., and Lavola, A.,** Phenolic compounds of willow bark as deterrent against feeding by mountain hare, *Oecologia,* 65, 319, 1985.

59. **Tahvanainen, J., Julkunen-Tiito, R., and Kettunen, J.,** Phenolic glycosides govern the food selection pattern of willow feeding leaf beetles, *Oecologia,* 67, 52, 1985.

60. **Thieme, H.,** Die phenolglykoside der Salicaceae, *Planta Med.,* 13, 431, 1965.

61. **Thieme, H. and Benecke, R.,** Die Phenolglycosides der Salicaceae: Untersuchungen ber die Glykosidakkumulation in einigen mitteleuropischen *Populus*-Arten, *Pharmazie,* 25, 228, 1970a.

62. **Thieme, H. and Benecke, R.,** Die Phenolglykoside der Salicaceen. Mitteilung ber die Glykosidfuhrung einheimischer bzw. mitteleuropischen Kultivierter *Populus* Arten, *Pharmazie,* 25, 780, 1970b.

63. **Verboomen, M.,** Specificité alimentaire de *Phratora laticollis,* chrysomle infode aux Salicaceae, Memoire de fin d'étude, Université Libre de Bruxelles, 1982.

64. **Vickery, M. L. and Vickery, B.,** *Secondary Plant Metabolism,* Macmillan, London, 1981, 334.

65. **Wain, R. L.,** The secretion of salicylaldehyde by the larvae of the brassy willow beetle (*Phyllodecta vitellinae* L.), *Annu. Rep. Agric. Natl. Res. Stn.,* 108, 1943.

66. **Whitham, T. G.,** The theory of habitat selection: examined and extenting using *Pemphigus* aphids, *Am. Nat.,* 115, 449, 1980.

67. **Whitehead, D. R. and Duffield, R. M.,** An unusual specialised predator-pry association (Coleoptera, Coccinelidae, Chrysomelidae): failure of a chemical defense and possible practical application, *Coleopt. Bull.,* 36, 96, 1982.

68. **Zucker, W. V.,** How aphids choose leaves: the roles of phenolics in host selection by a galling aphid, *Ecology,* 63, 972, 1982.

4

Learning and Flower Use in Butterflies: Hypotheses from Honey Bees

Alcinda C. Lewis and Gloria A. Lipani
Department of Environmental, Population and Organismic Biology
University of Colorado
Boulder, Colorado

TABLE OF CONTENTS

I.   Introduction..96
II.   Hypotheses 97
       A.   Hypothesis 1: Butterflies Learn to Recognize Rewarding Flowers If
            Flower Constancy Increases Net Reward.............................97
            1.   Flower Constancy..97
            2.   Color/Light Intensity Learning 100
            3.   Short- and Long-Term Memory 102
            4.   Odor/Taste Learning..102
            5.   Shape and Pattern Learning 103
            6.   Unrewarding or Deleterious Flowers 104
       B.   Hypothesis 2: Flower Choice Will Be Influenced by Time to Learn
            Nectar Extraction 104
       C.   Hypothesis 3: Butterflies Learn the Location of Rewarding Plants
            When Resources are Stable, Reusable, and Hyperdispersed........107
III.   Summary 107
Acknowledgments 108
References 108

I. INTRODUCTION

A casual observation of insects collecting nectar in a field of flowers reveals striking differences among the species in behavior. In particular, honey bees *(Apis mellifera)* seem to earn their reputation for industrious and purposeful behavior while butterflies give the impression of aimless flight.[4,14] This view has persisted throughout years of research on the two groups. Richards,[56] a century after the first studies were reported, writes:

Anyone who, like me, has chased foraging butterflies for the purpose of capture or photography will indeed testify to the maddeningly *indecisive*, long distance and essentially unidirectional flight patterns they show. Perhaps for this reason, butterflies have been very little studied as foragers and pollen dispersants yet they are undoubtedly very important visitors (p. 149, italics mine).

Butterflies may appear to be indiscriminant foragers but we should find it surprising if two insect groups using a similar resource employ very different methods of finding and extracting the reward. We suggest that these differences between bees and butterflies are more apparent than real. Butterflies do show selectivity when foraging.[39] Selectivity in generalists implies the ability to learn at least some aspects of the resource. We explore here the role of learning in flower use; we suggest that butterflies have many of the same capabilities as bees, and that the observed behavioral differences are due to other constraints on choice.

The honey bee has become a model system for studying learning[44] while virtually no work has been done on learning related to flower use in butterflies. The honey bee therefore presents an unusual opportunity; future work on butterflies can be structured using the honey bee as a source of hypotheses. Comparing the same behavior in the two organisms will shed light on factors influencing the evolution of learning capabilities. Menzel et al.[43] suggest that various "life strategy adaptations" in the honey bee resulting form eusociality are more important in the evolution of learning than aspects of the environment. For example, they argue that the need to have a nest in which to rear brood leads to central place foraging. This in turn selects for location learning. But Galun and Fitzgerald[22] suggest that spatial learning evolves simply in proportion to navigational needs; therefore, sociality may be sufficient but not necessary. This is discussed in more detail below.

We present here specific hypotheses concerning butterfly flower use. We are less concerned with mechanisms of learning than with its role in foraging; we therefore discuss learning in functional rather than mechanistic categories. For each hypothesis, we review the existing data on butterflies, briefly summarize results on the honey bee, and suggest approaches to research. We do not present an extensive review of the literature on learning in other phytophagous insects or in the honey bee as this has been thoroughly done in several recent articles and books.[26,29,31,44,51,64,73,80] We also have not reviewed studies on learning related to the oviposition plant of the butterfly except where results suggest similar learning potential for flower cues.[49,52,67,70,71] Other aspects of adult butterfly feeding have been thoroughly reviewed by Boggs.[8]

Because of the relatively small amount of data available on butterflies, we are comparing butterflies, a diverse group, with a single insect, the honey bee. For purposes of this comparison, we divide butterflies grossly into two groups. Group 1 consists of short-lived, often temperate-zone, butterflies that are generalist flower feeders with unstable populations. They are often quite vagile, not necessarily inhabiting a particular site throughout their lifetimes. They are primarily nectar feeders and usually relatively small. The example cited most often below is *Pieris rapae*, the cabbage white.[13,16,35] Group 2 consists of longer-lived, often tropical, butterflies that have relatively stable populations, home ranges, and somewhat more specialized adult resources. These butterflies are usually larger than those in Group 1. These are typified by *Heliconius* butterflies, a neotropical pollen-feeding genus.[5,9,11,19,23,25,40,41,47,48,72] We acknowledge that these represent ends of a continuum of butterfly life history and many butterflies exhibit characteristics of both groups. Ultimately, an examination of the intermediate taxa should shed further light on the evolutionary ecology of learning.

II. HYPOTHESES

A. HYPOTHESIS 1: BUTTERFLIES LEARN TO RECOGNIZE REWARDING FLOWERS IF FLOWER CONSTANCY INCREASES NET REWARD

1. Flower Constancy

Honey bees are remarkable for the constancy of their flower visits; individuals will visit flowers of a single species, bypassing other rewarding species in their flight path. At the same time, other individuals are constant to the bypassed species, confirming its potential acceptability. Constancy to one species can last for a few hours, days, or weeks depending on nectar supply.[64,80] This seemingly maladaptive behavior can have multiple causes, but one of the oldest and simplest hypotheses is that of Darwin.[18] He suggested that the memory of the insect for the appearance and handling of flowers is limited; therefore, searching for and attempting to use several species simultaneously is inefficient. Learning a new flower will interfere with recall of the previously learned species.

Lewis[38] tested this hypothesis using the cabbage butterfly, *Pieris rapae*. Caged, naive butterflies were given a flower for nectaring. When a butterfly first lands on a flower, it searches the sepals and corolla, eventually finding the nectar. The time elapsed from first landing on the flower until contact with the nectar is termed discovery time. Discovery times for several consecutive attempts revealed rapid improvement in performance and followed a classic learning curve on the two species tested (Figure 1).

To test for memory capacity, members of one goup of butterflies were given one species to learn, followed by a second species. They were then re-tested on the first species. A control group was not given the second species but was also retested on the first species. Butterflies given the second species had to re-learn the first whereas members of the control group did not need to re-learn (Figure 2). Control group butterflies were held without food for a time equal to that required on average for the test group to find and learn their flowers, 20 min. Therefore, passage of time alone did not cause memory loss. The results may be explained by increased hunger of control butterflies, but it is unlikely that hunger alone would explain more rapid nectar discovery. In addition, butterflies in the experimental group that chose not to feed on the second flower, presumably because they were not hungry, had test times similar to those in the control group. This indicates that hunger was not necessary to maintain low discovery times.

Butterflies that switch among species in the field will suffer a cost in time to learn. Some evidence for similar constraints on memory of recognition cues have been found in certain bees.[79] That discovery time is an important variable, subject to selection, is supported by Laverty and Plowright's results.[37] Naive specialist bumblebees learned nectar extraction more quickly on their *Aconitum* host than did generalist bees. Learning may increase fitness by reducing energetic costs of nectar location.[32] By reducing the time during which the insect must remain motionless and conspicuous on a flower, learning may also reduce susceptibility to predation.

Despite the need for constancy suggested by these results, butterflies as a group

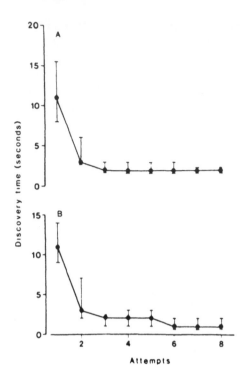

FIGURE 1. Discovery times (medians with first and third quartiles) for native butterflies given either (A) *Campanula rotundifolia* (n = 24) or (B) *Lotus corniculatus* (n = 18). (From Lewis, A. C., in *Science,* 232, 863, 1986. Copyright 1986 by the AAAS. With permission.)

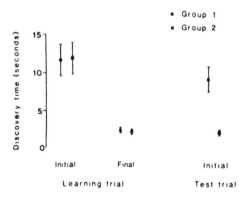

FIGURE 2. Interference of learning experiment. Group 1 (n = 17) given *Lotus corniculatus* between learning and test trials with *Campanula rotundifolia;* Group 2 (n = 20) given no flcwers during a similar period. Difference between discovery times for the learning and test trials: Group 1, $p < 0.01$, paired t test; Group 2, NS. (From Lewis, A. C., in *Science,* 232, 863, 1986. Copyright 1986 by the AAAS. With permission.)

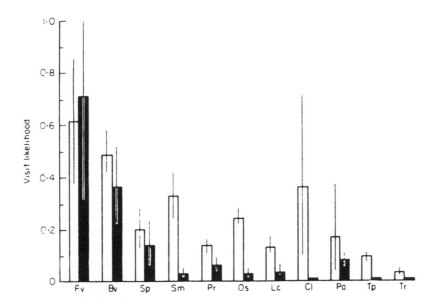

FIGURE 3.   Visit rates (visits/encounters) with 95% confidence limits when the encountered species was the same as last visited (open bar) and different from the last visited (closed bar). Species and number of encountered flowers for left and right bar: Pv, *Prunella vulgaris* (23, 7); Bv, *Barbarea vulgaris* (220, 59); Sp, *Specularia perfoliata* (136, 66); Sm, *Stellaria media* (123, 473); Pr, *Potentilla recta* (488, 174); Os, *Oxalis stricta* (691, 663); Lc, *Lotus corniculatus* (728, 812); Cl, *Chrysanthemum leucanthemum* (11, 529); Pa, *Potentilla argentata* (18, 263); Tp, *Trifolium pratense* (992, 3330); Tr, *Trifolium repens* (312, 4600). (From Lewis, A. C., in *J. Anim. Ecol.*, 58, 1, 1989. With permission.)

are considered to be less constant in their flower visiting than honey bees.[74,78] While casual observations do give this impression, extensive greenhouse and field observations of foraging butterflies have revealed that both males and non-ovipositing *P. rapae* females are capable of consistently visiting flowers of one species in the presence of other rewarding species.[39]

Field observations were made in an unmanipulated, species-rich habitat. Methods were borrowed from studies of oviposition: flight paths were flagged and every species visited and bypassed was recorded (reviewed in Lewis[39]). Switching rates (switches per visit) were 0.205 for males and non-ovipositing females and 0.497 for ovipositing females. A logit analysis was performed of visit likelihood when the encountered flower was the same as last visited and when it was different. Figure 3 gives visit rates (visits per encounters) and 95% confidence limits for 11 species when the species encountered in the flight path is the same (left bar) and different (right bar) from the last visited. The logit analysis indicates a highly significant main effect of recent visit history ($p < 0.0001$). The odds of visiting any flower are 0.095 but if the flower is the same species as last visited, these odds increase to 2.84.

Subsequent analyses suggest that the history effect is not due to differences in abundances among flower species, flower distribution patterns, or pooling of observations across individuals or time periods. Butterflies do, however, show preferences among species as can be seen in Figure 3. This was supported by the logit analysis: odds multipliers differ significantly among species. Preference differences among flowers probably have multiple causes as discussed below.

This study of *P. rapae* is the first quantitative description of flower constancy in a lepidopteran but many studies suggest that some degree of constancy is widespread (reviewed by Lewis[39]). A direct comparison with honey bees is not possible using the existing data but it is clear that butterflies as a group, particularly temperate-zone species with multiple nectar sources available (Group 1 butterflies), are less constant than honey bees when observations are made in the field. Honey bees may remain constant for days, while *P. rapae* may switch species within a foraging flight.

What is responsible for this difference in constancy? Despite the advantage of learning to recognize rewarding flowers, butterflies may not be capable of learning flowers with a precision comparable to that of honey bees. But this may itself arise from a need or ability to be less constant than honey bees, so additional hypotheses should be considered. For instance, butterflies do not have a division of labor; some species therefore mix behaviors at certain times of day, nectaring while searching for mates or host plants.[63,67] Only butterflies on morning nectaring flights were used in the study of *P. rapae* described above, but many species switch more frequently between activities. For instance, May[42] never observed mixing of behaviors by *Phoebis sennae* (Pieridae) but often observed *Agraulis vanillae* (Nymphalidae) alternate nectaring and oviposition. Mixing of behaviors can lead to interference of memory for the item being searched as suggested by Stanton;[67] ovipositing *Colias* butterflies appear to forget aspects of the larval hosts on which they oviposit following bouts of nectaring. This would reduce their ability to be constant to one flower species.

Some butterflies, particularly those lacking chemical protection or mimicry, may be more vulnerable to predators than honey bees. These butterflies may switch more often in response to a direct or suspected threat.

Another potential difference between bees and butterflies that might be related to learning ability is brain size. It is difficult to compare butterflies and bees directly because of the relative importance of odor for the two groups. Odor processing is probably much more important in the honey bee, and may require a proportionately larger brain volume. Aspects of sociality may also lead to larger brains in honey bees; larger size may in turn facilitate learning not directly related to sociality. The fact that certain butterflies do show learning of color, leaf shape, and chemical cues for oviposition (discussed below) argues against a brain size limitation on learning of many flower cues but upper limits must exist. Sivinski[66] compared brain sizes in seven species of nymphalid butterflies. Within the butterflies, those nearer the Group 2 end of the spectrum (e.g., *Heliconius charitonius* and *Dryas julia*) had much larger brains relative to body size and larger mushroom bodies relative to brain size than ones toward the Group 1 end (e.g., *Junonia coenia*). Sivinski suggests that this is related to the need to learn adult resource locations (see Hypothesis 3, p. 107).

2. Color/Light Intensity Learning

Butterflies have color vision adequate for discriminating among many flower species.[6,33,60,61,69] As with bees, there appear to be underlying preferences for certain colors although these can be overridden by conditioning.[2,33,36,68,69] These underlying preferences may vary among species. Scherer and Kolb[61] found that the feeding response occurred in the blue and orange regions in *Pararge aegeria* (Satyridae) and in the blue and yellow regions in *Aglais urticae* (Nymphalidae).

While there were a few early reports of flower color learning in *Heliconius*[69] and one of learning by *P. rapae* to associate green with a chemical stimulant,[70,71] Scherer and Kolb[60] failed in their attempts to train *Pieris brassicae* to associate colored lights of specific wave lengths with a reward. We recently attempted to train *P. rapae* to associate particular colors with a sugar solution reward. Naive butterflies were first given a chance to feed on a drop (0.25 μl) of 20% sucrose solution on a 1 cm square of either blue or yellow. These were matched for initial attractiveness and consisted of Scotch brand

Table 1
PROPORTION OF NAIVE BUTTERFLIES
CHOOSING BLUE OR YELLOW STIMULI
FOLLOWING ONE EXPOSURE TO EITHER
COLOR (N IN PARENTHESES)

	A. Rewarded training			**B. Unrewarded training**	
Choice color	**Training color**		**Choice color**	**Training color**	
	B	**Y**		**B**	**Y**
B	81 (22)	17 (4)	B	50 (12)	48 (12)
Y	19 (5)	83 (19)	Y	50 (12)	52 (13)
Total	100 (27)	100 (23)	Total	100 (24)	100 (25)

blue tape and Strathmore "Lido" yellow artist's cover stock. It was important to match the stimuli for their initial attractiveness to the butterflies so that innate preferences would not mask a learning potential; therefore, stimuli were not necessarily matched for brightness. For brevity, we refer to color in the following, acknowledging that butterflies may be using differences in light intensity as a cue. Pilot trials showed that butterflies were not using scent to distinguish between the squares.

For training, eight stimulus squares of the same color were placed equidistantly around the perimeter of a 15 cm diameter petri dish. Butterflies were given access to the plates until they had fed once. After a feed, the plate was removed and replaced by one containing squares of alternating blue and yellow. The first unambiguous feeding choice of the butterfly was recorded; that is, brief alightings not followed by proboscis extension were excluded. In the majority of cases (Table 1A), butterflies chose the color of their training (p <0.001, X^2 = 20.44). Error rates were similar on the two colors. These results suggest that butterflies learned to associate color with reward. Although not strictly comparable to work on honey bees, these results further suggest that *P. rapae* learns color faster than honey bees. Error rates for bees are higher, an average of 35% after the first feed.

A preliminary test revealed that error rates were considerably higher when butterflies had previous experience, over 1 or more days, of both colors. Butterflies that fed at both colors previously were given the standard training and test procedures. To determine if the relative number of feeds at the two colors influenced the outcome of a test, the difference between feeds at colors the same and different from the training color was taken. This difference is referred to as exposure heterogeneity (Figure 4). Sixty-seven percent of the trials in which an error was made occurred when the number of feeds prior to training to the alternate color exceeded that to the same color; this occurred in 22% of correct trials (p <0.05, Mann Whitney). Results are shown in Figure 4 as the proportion of responses which were correct for classes of exposure heterogeneity. A large positive number indicates that the butterfly previously fed primarily at the color used in training; a large negative number indicates multiple feeds at the color not used in training. Butterflies are increasingly likely to make a correct response in a test with increased prior feeding at the training color over the alternate.

Results from the multiple exposure test reveal that color memory is subject to interference as has been shown in bees.[45] Menzel[44] feels that short-term memory is an adaptive mechanism for tracking resources. But whether this result should be viewed as positive or negative, as interference or as increased ability to recognize rewarding stimuli, is as yet unclear.

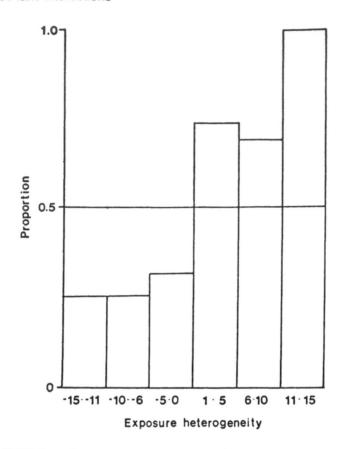

FIGURE 4. Proportion fo correct responses for classes of exposure heterogeneity. See text for details.

3. Short- and Long-Term Memory

Menzel and others (reviewed in Menzel[45]) have documented a short-term memory (STM) in honey bees of about 3 min during which time memory is subject to interference. Once consolidation into long-term memory (LTM) has occurred, interference is less likely. Menzel[45] speculates that the temporal dynamics of memory reflect the rate of encounters with food under natural conditions. Rate of learning is under genetic control in honey bees,[10] providing raw material for selection, but the relationship between, for instance, the length of STM and the rate of encounters with food needs to be tested across taxa differing in resource distribution. This will be a difficult task; for many species, the variance in resource distribution patterns within a species across its range probably equals or exceeds that found across many insect taxa.

Nothing is known about the existence of STM and LTM in butterflies. If butterflies have standard STM and LTM, and if Menzel's hypothesis is correct, then the duration of memory phases should vary in a predictable way among taxa with varying resource distribution. The more rapid switching of Group 1 butterflies suggests that they may have a longer STM, but a longer STM is not essential for rapid switching.

4. Odor/Taste Learning

Minnich[46] found that the red admiral butterfly is capable of detecting apple juice by olfaction and taste. Olfactory sensilla are hypothesized to occur in the antennae of *P. brassicae* and other butterflies[3] and to be used to detect an egg deterrent pheromone.[59] The existence of tarsal chemoreceptors has also been verified.[21] This work has largely been done in the context of oviposition.

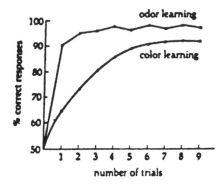

FIGURE 5. Typical learning curves for honey bees at feeders marked with odor or color. (For details see Gould, J., in *Foraging Behavior,* Kamil, A. C., Krebs, J. R., and Pulliam, H. R., Eds., Plenum Press, New York, 1987, 479. Reprinted with permission.)

Although some species can use color alone to initially find flowers, odors are essential for flower recognition by others. Pellmyr[53] found that fragrance was necessary for fritillary visitation of *Cimicifuga simlex* (Ranunculaceae); the unscented morph is visited only by bumblebees. The fragrance has two synergistic components which when both applied to the nonfragrant morph elicited landing by butterflies. Scherer and Kolb[61] found that *P. brassicae* will visit a color stimulus lacking an odor but will not visit odor alone, while the satyrid *Pararge aegeria* shows the opposite behavior.

We have not found any studies on butterfly learning of flower odor. We predict that such learning exists since odor is used as a cue by at least some species. Some butterflies can learn chemical cues. *Battus philenor* can learn to associate a chemical cue with leaf shape when searching for oviposition plants.[50] Traynier found that *P. rapae* can learn to associate the chemical sinigrin with the visual cue, the color green.[71]

Honey bees learn odor even more quickly than color (Figure 5) and odor is more important than color in the hierarchy of stimuli used for choice (reviewed by Gould[29]). This is possibly due to the greater specificity of odor as a sign stimulus. For instance, bees can distinguish sunflower cultivars, all of the species *Helianthus annuus*, based on odor; it is unlikely these cultivars could be distinguished on the basis of visual characters alone.[54] Flower odor is also brought back to the hive on the bodies of scouts, increasing its value as an unambiguous recognition cue.

The high switching rate of some butterflies, such as *P. rapae*, stemming from aspects of their life history, suggests that odor would be a less useful cue for butterflies. Learning a color shared by several species rather than a more specific odor cue will allow rapid recognition and switching to other rewarding species. This is particularly true if different flower species growing together facilitate pollination among one another by a form of Mullerian mimicry.[55]

This discussion has treated color and odor as separate cues, but honey bees have been trained to discriminate color-odor combinations.[17] While odor overshadows color as a cue for bees, the opposite may be true for butterflies if the suggestions about the relative importance of the two cues given above is correct.

5. Shape and Pattern Learning

As with odor learning, there has been no test made of the ability of a butterfly to learn to recognize flower shapes or patterns. Various authors have inferred this ability, in part from the UV patterns of flowers visited by butterflies (reviewed by Boggs[8]). Further support comes from results showing that *Battus philenor* can learn leaf shape.[49,50,52] The same reasons for learning other cues apply to shape and patterns; recognition increases the possibility of visit consistency. Honey bees can learn fine distinctions between complex patterns although this learning takes longer than color and odor learning.[27] The resolution of this learning is somewhat low, a finding Gould and Gould[30] suggest indicates a tradeoff between the number of patterns that can be stored and their precision. They further suggest that the bee can always fly closer to verify a pattern, tipping the balance in favor of storing more patterns with lower precision.

If this tradeoff does exist, then we expect the more generalist Group 1 butterflies to learn patterns with less precision than either honey bees or Group 2 butterflies. Gould's[66] test could be used to study shape and pattern learning in butterflies and to make a comparative test of the relative speed and precision with which the two taxa learn patterns.

6. Unrewarding or Deleterious Flowers

Using the same procedure as described above for rewarded color learning, we tested the ability of the butterfly to learn to avoid an unrewarded color. Table 1b shows that they cannot learn in one trial. This is not surprising; one color should not be excluded from visits based on one unrewarding encounter, particularly since a given color may be shared by many potentially rewarding species.

In honey bees, Gould[27] found that pattern learning is enhanced by the simultaneous presence of the unrewarded alternative. Data are not presented on the numbers of visits to the unrewarded stimulus, however. It is possible that butterflies would learn to avoid species-typical cues, and flowers, following one or more unrewarded encounters.

Butterflies do spend less time on unrewarding flowers in at least one case. Naive *Pieris rapae* butterflies were given inflorescences of a favored plant, the mint *Nepeta cataria*. The inflorescences contained both living and dead (nectarless) flowers. Visits to living and dead flowers were interspersed. The time spent investigating living flowers decreased with attempts and was followed by drinking, as on other species. The time spent investigating dead flowers also decreased (Figure 6). Further work is needed to determine if the reduced time spent on dead flowers is due to learning to recognize nectarless flowers or to other processes. For instance, butterflies may learn the morphology of the flower irrespective of reward and then use the information in different ways of depending on the presence or absence of reward. Butterflies do not learn to avoid these dead flowers; landing rates do not decrease even though time spent investigating the flowers does decrease. A similar behavior has been recently reported in hermit crabs; crabs spent less time investigating a previously rejected shell than a novel one (presumably due to learning) but were unable to distinguish novel and rejected shells by contact or sight alone.[34]

It may be advantageous to actively avoid some flowers of a species. Murawski[47] observed avoidance (skipping flowers in an inflorescence) of particular *Gurania* flowers by *Heliconius* in the field; these flowers had been infested with tephritid fly larvae. The nectar becomes viscous and possibly distasteful. Based on the butterflies' flight pattern, Murawski speculates that they learned to avoid the flowers and are not simply using a long distance cue, although this awaits testing. Honey bees are capable of aversion learning[1] but these observations were laboratory ones using unnatural stimuli; the ecological significance of the result is unclear.

B. HYPOTHESIS 2: FLOWER CHOICE WILL BE INFLUENCED BY TIME TO LEARN NECTAR EXTRACTION

The speed of information acquisition should influence decision making (Elwood and

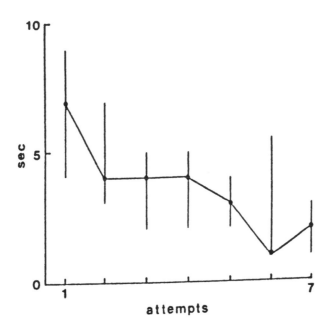

FIGURE 6. Time (median with first and third quartiles for n = 15) spent by naive butterflies investigating dead flowers of *Nepeta cataria* during first 7 visits to inflorescences.

Neil[20]). Honey bees base their choices among flower types on net reward where net reward is a function of time to extract nectar and its quantity, concentration, and variance.[57,62,75,76] Waddington[76] suggests that honey bees maximize energetic efficiency per foraging trip and minimize flight costs. May[42] speculates that butterfly choice will also be influenced by reward profitability. There have been no tests of this hypothesis. We predict that Group 1 butterflies will be influenced by net reward but that the decision to continue to feed on flowers of a given species will be made immediately, following a simple threshold rule, rather than by complex comparisons among species. Butterflies therefore will have, in the words of Shettleworth,[65] a short memory window where recent experience will be more important than past experiences. This prediction stems from our view, given above, of these butterflies as rapid decision makers, constrained by predator pressures and the need to switch frequently between nectaring, oviposition flight, and mate search.

Group 2 butterflies should behave more like bees. They require high quality adult food in habitats where resources are relatively rare. Selection will favor butterflies that choose and revisit exceptionally rewarding flowers. Choosing the most rewarding plants will be even more important at times of intense competition, a situation more likely experienced by these butterflies than those of Group 1. For instance, when plant populations are reduced by drought, *Heliconius* butterflies abandon traplining and defend individual plants and/or flowers.[47]

An important component of the assessment of reward will be the time required to learn within-flower nectar location and extraction techniques. Tests on 11 flower species, using the methods described above, reveal that initial discovery time is generally correlated with flower morphology. Discovery times are higher and more variable on more complex flowers (Figure 7). There is great variation among flower species which, according to the hypothesis, should be related to flower choice in the field. A test of this aspect of the hypothesis awaits the appropriate field study but preliminary results reveal that the species learned more quickly are indeed more likely to be visited.

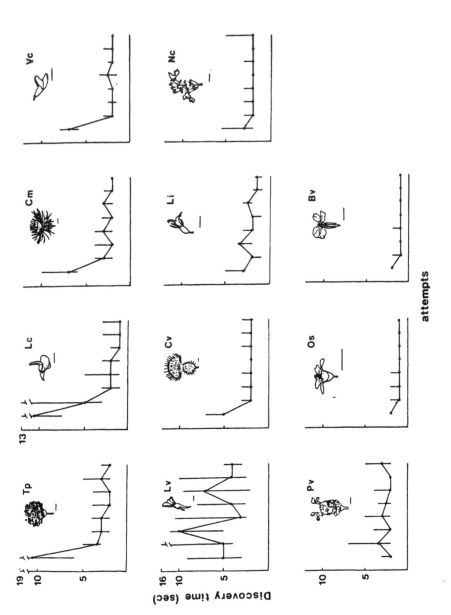

FIGURE 7.   Median discovery times with first and third quartiles for the first 8 attempts to locate nectar by naive butterflies. Horizontal lines next to flowers show 1 cm at the scale of each drawing. Species abbreviations and number of butterflies tested are: Tp, *Trifolium pratense* (20); Lc, *Lotus corniculatus* (20) Cm, *Centaurea maculosa* (22); Vc, *Vicia cracca* (8); Lv, *Linaria vulgaris* (15); Cv, *Cirsium vulgare* (18); Li, *Lobelia inflata* (8); Nc, *Nepeta cataria* (20); Pv, *Prunella vulgaris* (18); Os, *Oxalis stricta* (14); Bv, *Barbarea vulgaris* (15).

The time to learn nectar extraction should be a more important component of choice for Group 1 butterflies than for Group 2 butterflies or honey bees. For the reasons given above, the latter two groups probably require species with higher net rewards. Both Group 2 butterflies and bees can spend the time required to find particularly rewarding flowers: Group 2 butterflies by surveying and possibly remembering locations of specific plants, honey bees by marshalling scouts. This idea is consistent with the results of Laverty and Plowright[37] who found that generalist bumble bees are more likely to give up before finding nectar on complex flowers than are specialists when both groups are given the flower used by the specialist. Generalists will spend less time learning a flower even though the net reward, once learned, is potentially higher than in other available flowers. Instead, they search for another species that offers rewards more readily.

C. HYPOTHESIS 3: BUTTERFLIES LEARN THE LOCATION OF REWARDING PLANTS WHEN RESOURCES ARE STABLE, REUSABLE, AND HYPERDISPERSED

Menzel[43] speculates that eusociality in honey bees, through the central place foraging that is a consequence of the nest, facilitates and is facilitated by location learning. We argue that the stability and distribution of resources can place the same demand on other insects. Biebach et al.[7] suggest that time-and-place learning should be found in animals that return to a renewing resource, although they maintain that it had not been proven in any taxon until their recent study of garden warblers. Bees do learn locations using landmarks,[12,15,28] and there is strong circumstantial evidence for an ability to link time and place.

Group 2 butterflies live in relatively stable habitats with perennial adult host plants. Search time would be significantly reduced by learning the locations of plants they revisit. Field observations suggest that this butterfly learns nectar and pollen plant locations; butterflies visit individual plants repeatedly in a set sequence referred to as a trapline. Some *Heliconius* butterflies apparently learn the sites of their communal nocturnal roosts. Others learn to avoid sites where they have been captured, not returning for 2 days following capture.[41] Testing for location learning and then landmark learning could be adapted for the butterfly from the methods developed for honey bees,[12,15] and for time-and-place learning from the above cited study.[7]

It is unlikely that Group 1 butterflies will be capable of location learning. They do not have home ranges similar to those found in Group 2 butterflies or bees. Individual plants of both the larval and adult hosts are unlikely to be reused repeatedly; they are often small plants, and in the case of the adult hosts, offer small rewards per plant. When these butterflies find a rewarding patch, they remain in the patch by simply increasing their turning rate.[58] Other aspects of the behavior and ecology of Group 1 butterflies are unlikely to select for location learning. For example, these butterflies do not have stable, communal nocturnal roosts or restricted home ranges as are found in some Group 2 butterflies.

III. SUMMARY

We present hypotheses on butterfly learning of flower cues and location inspired by research on honey bees. We argue that butterflies as a group should have all the learning abilities found in honey bees except ones directly related to sociality. Both within butterflies and between butterflies and bees, taxa will differ in specific learning abilities, in the precision of learning, and in the weight given to various cues. Differences will be due in a predictable way to variation in life history characteristics and in ecology.

New data bearing on the hypotheses are given for the cabbage white butterfly, *Pieris rapae*, and existing data for other groups is reviewed.

ACKNOWLEDGMENTS

We thank the following for assistance of various kinds: E. A. Bernays, W. Getz, L. E. Gilbert, C. Boggs, F. S. Chew, S. Courtney, R. E. Jones, The Hon. M. Rothschild, C. Jones, J. Sivinsky, and C. H. Lewis. We thank the staffs of the IES and the EPO Biology Department, University of Colorado, for providing space for this research. Contribution to the program of the Institute of Ecosystem Studies of the New York Botanical Garden. Supported by NSF grant BSR 85-06072 and the American Philosophical Society.

REFERENCES

1. **Abramson, C. I.,** Aversive conditioning in honeybees, *J. Comp. Physiol.,* 100, 108, 1986.
2. **Backhaus, W., Werner, A., and Menzel, R.,** Color vision in honey bees: metric, dimensions, constancy and ecological aspects, in *Neurobiology and Behavior of Honeybees,* Menzel, R. and Mercer, A., Eds., Springer-Verlag, New York, 1987, 172.
3. **Behan, M. and Schoonhoven, L. M.,** Chemoreception of an oviposition deterrent associated with eggs in *Pieris brassicae, Entomol. Exp. Appl.,* 24, 163, 1978.
4. **Bennett, A. W.,** On the constancy of insects in their visits to flowers, *Zool. J. Linn. Soc.,* 17, 175, 1883.
5. **Benson, W. W.,** Natural selection for mullerian mimicry in *Heliconius erato, Science,* 176, 936, 1972.
6. **Bernard, G. D.,** Red-absorbing visual pigment of butterflies, *Science,* 203, 1125, 1979.
7. **Biebach, H., Gordjin, M., and Krebs, J. R.,** Time-and-place learning by garden warblers, *Sylvia borin, Anim. Behav.,* 37, 353, 1989.
8. **Boggs, C. L.,** Ecology of nectar and pollen feeding in Lepidoptera, in *Nutritional Ecology of Insects, Mites and Spiders,* Slansky, F. and Rodriguez, J. G., Ed., John Wiley & Sons, New York, 1987, 369.
9. **Boggs, C., Smiley, J. T., and Gilbert, L. E.,** Patterns of pollen exploitation by *Heliconius* butterflies, *Oecologia,* 48, 284, 1981.
10. **Brandes, Ch., Frisch, B., and Menzel, R.,** Time-course of memory formation differs in honey bee lines selected for good and poor learning, *Anim. Behav.,* 36, 981, 1988.
11. **Brown, K. S.,** The biology of *Heliconius* and related genera, *Annu. Rev. Entomol.,* 26, 427, 1981.

12. **Cartwright, B. and Collett, T. S.,** Landmark learning in bees, *J. Comp. Physiol.,* 151, 521, 1983.
13. **Chew, F. S.,** Coexistence and local extinction in two pierid butterflies, *Am. Nat.,* 118, 655, 1981.
14. **Christy, R. M.,** On the methodic habits of insects when visiting flowers, *Zool. J. Linn. Soc.,* 17, 186, 1883.
15. **Collett, T. S. and Kelber, A.,** The retrieval of visuo-spatial memories by honeybees, *J. Comp. Physiol. A.,* 163, 145, 1988.
16. **Courtney, S. P.,** The ecology of pierid butterflies: dynamics and interaction, *Adv. Ecol. Res.,* 15, 51, 1986.
17. **Couvillon, P. A. and Bitterman, M. E.,** Compound-component and conditional discrimination of colors and odors by honeybees: further tests of a continuity model, *Anim. Learn. Behav.,* 16, 67, 1988.
18. **Darwin, C.,** *On the Effects of Cross and Self Fertilisation in the Vegetable Kingdom,* John Murray, London, 1876.
19. **Ehrlich, P. and Gilbert, L. E.,** Population structure and dynamics of the tropical butterfly *Heliconius ethilla, Biotropica,* 5, 69, 1973.
20. **Elwood, R. and Neil, S.,** Information and motivational change, in *Quantitative Models in Ethology,* Colgan, P. and Zayan, R., Eds., Privat, Toulouse, 1986, 97.
21. **Feeny, P., Rosenberry, L., and Carter, M.,** Oviposition behavior in butterflies, in *Herbivorous Insects: Host-seeking Behavior and Mechanisms,* Ahmad, S., Ed., Academic Press, New York, 1983, 76.
22. **Galun, S. J. and Fitzgerald, R. W.,** Sexual selection for spatial learning ability, *Anim. Behav.,* 37, 322, 1989.
23. **Gilbert L. E.,** Pollen feeding and reproduc-

tive biology of *Heliconius* butterflies, *Proc. Natl. Acad. Sci. U.S.A.*, 69, 1403, 1972.

24. **Gilbert, L. E.**, Ecological consequences of a coevolved mutualism between butterflies and plants, in *Coevolution of Animals and Plants*, Gilbert, L. and Raven, P., Eds., University of Texas Press, Austin, 1975, 100.

25. **Gilbert, L. E.**, The biology of butterfly communities, in *Biology of Butterflies*, Vane-Wright, R. I. and Ackery, P. R., Eds., R. *Entomol. Soc. London Symp.*, Academic Press, London, 1984, 41.

26. **Gould, J.**, Natural history of honeybee learning, in *The Biology of Learning*, Marler, P. and Terrace, H. S., Eds., Springer-Verlag, New York, 1984, 149.

27. **Gould, J.**, Pattern learning by honey bees, *Anim. Behav.*, 34, 990, 1986.

28. **Gould, J.**, Landmark learning in honeybees, *Anim. Behav.*, 35, 26, 1987.

29. **Gould, J.**, The role of learning in honeybee foraging, in *Foraging Behavior*, Kamil, A. C., Krebs, J. R., and Pulliam, H. R., Eds., Plenum Press, New York, 1987, 479.

30. **Gould, J. and Gould, C.**,*The Honey Bee*, Scientific American, New York, 1988.

31. **Gould, J. L. and Towne, W. F.**, Honey bee learning, *Adv. Insect Physiol.*, 20, 55, 1988.

32. **Heinrich, B.**, Learning in invertebrates, in *The Biology of Learning*, Marler, P. and Terrace, H. S., Eds., Springer Verlag, New York, 1984, 135.

33. **Isle, D.**, Uber den farbersinn der tagfalter, *Z. Vergl. Physiol.*, 8, 658, 1928.

34. **Jackson, N. and Elwood, R.**, Memory of information gained during shell investigation by the hermit crab, *Pagurus bernhardus Anim. Behav.*, 37, 529, 1989.

35. **Jones, R. E.**, Behavioral evolution in the cabbage butterfly *Pieris rapae*, *Oecologia*, 72, 69, 1987.

36. **Kay, Q.**, Preferential pollination of yellow-flowered morphs of *Raphanus raphanistrum* by *Pieris* and *Eristalis* supp., *Nature, (London)*, 261, 230, 1976.

37. **Laverty, T. and Plowright, R. C.**, Flower handling by bumblebees: a comparison of specialists and generalists, *Anim. Behav.*, 36, 733, 1988.

38. **Lewis, A. C.**, Memory constraints and flower choice in *Pieris rapae*, *Science*, 232, 863, 1986.

39. **Lewis, A. C.**, Flower visit consistency in *Pieris rapae* the cabbage butterfly, *J. Anim. Ecol.*, 58, 1, 1989.

40. **Mallet, J.**, Dispersal and gene flow in a butterfly with home range behavior: *Heliconius erato*, *Oecologia*, 68, 210, 1986.

41. **Mallet, J., Longino, J. T., Murawski, D., Murawski, A., and Simpson de Gamboa, A.**, Handling effects in *Heliconius:* where do all the butterflies go?, *J. Anim. Ecol.*, 56, 377, 1987.

42. **May, P. G.**, Determinants of foraging profitability in two nectarivorous butterflies, *Ecol. Entomol.*, 13, 171, 1988.

43. **Menzel, R.**, Biology of invertebrate learning, group report, in *The Biology of Learning*, Marler, P. and Terrace, H. S., Eds., Springer-Verlag, New York, 1984, 249.

44. **Menzel, R.**, Learning in honeybees in an ecological and behavioral context, in *Experimental Behavioral Ecology and Sociobiology*, Hölldobler, B. and Lindauer, M., Eds., Sinauer, Sunderland, 1985, 55.

45. **Menzel, R.**, Memory traces in honeybees, in *Neurobiology and Behavior of Honeybees*, Menzel, R. and Mercer, A., Eds., Springer-Verlag, New York, 1987, 310.

46. **Minnich, D. E.**, An experimental study of the tarsal chemoreceptors of two nymphalid butterflies, *J. Exp. Zool.*, 33, 1921, 173.

47. **Murawski, D.**, Floral resource variation, pollinator response, and potential pollen flow in *Psiguria warscewiczii*, *Ecology*, 68, 1273, 1987.

48. **Murawski, D. and Gilbert, L. E.**, Pollen flow in *Psiguria warscewiczii:* a comparison of *Heliconius* butterflies and hummingbirds, *Oecologia*, 68, 161, 1986.

49. **Papaj, D. R.**, Shifts in foraging behavior by a *Battus philenor* population: field evidence for switching by individual butterflies, *Behav. Ecol. Sociobiol.*, 19, 31, 1986.

50. **Papaj, D. R.**, Conditioning of leaf-shape discrimination by chemical cues in the butterfly, *Battus philenor*, *Anim. Behav.*, 34, 1281, 1986.

51. **Papaj, D. and Prokopy, R.**, Ecological and evolutionary aspects of learning in phytophagous insects, *Annu. Rev. Entomol.*, 1989, in press.

52. **Papaj, D. R. and Rausher, M.**, Individual variation in host location by phytophagous insects, in *Herbivorous Insects: Host-Seeking Behavior and Mechanisms*, Ahmad, S., Ed., Academic Press, New York, 1983, 77.

53. **Pellmyr, O.**, Three pollination morphs in *Cimicifuga simplex;* incipient speciation due to inferiority in competition, *Oecologia*, 68, 304, 1986.

54. **Pham-degue, M., Masson, C., Etievant, P., and Azar, M.**, Selective olfactory choices of the honeybee among sunflower aromas, *J. Chem. Ecol.*, 12, 781, 1986.

55. **Ratchke, B.**, Competition and facilitation among plants for pollination, in *Pollination Biology*, Real, L., Ed., Academic Press, New York, 1983, 305.

56. **Richards, A. J.**, *Plant Breeding Systems*, George Allen and Unwin, London, 1986, 149.

57. **Real, L., Otte, J., and Silverfine, E.**, On the tradeoff between the mean and variance in foraging: an experimental analysis of bumblebees, *Ecology*, 63, 1617, 1983.

58. **Root, R. B. and Karieva, P. M.**, The search

for resources by cabbage butterflies *(Pieris rapae)*: ecological consequences and adaptive significance of Markovian movements in a patchy environment, *Ecology,* 65, 147, 1984.

59. **Rothschild, M. and Schoonhoven, L. M.,** Assessment of egg load by *Pieris brassicae,* *Nature (London),* 266, 352, 1977.

60. **Scherer, C. and Kolb, G.,** Behavioral experiments on the visual processing of color stimuli in *Pieris brassicae, J. Comp. Physiol. A.,* 160, 645, 1987.

61. **Scherer, C. and Kolb, G.,** The influence of color stimuli on visually controlled behavior in *Aglais urticae* and *Pararge aegeria* (Lepidoptera), *J. Comp. Physiol,* 161, 891, 1987.

62. **Schmid-Hempel, P., Kacelnik, A., and Houston, A. J.,** Honeybees maximize efficiency by not filling their crop, *Behav. Ecol. Sociobiol.,* 17, 61, 1985.

63. **Schmitt, J.,** Pollinator foraging behavior and gene dispersal in *Senecio, Evol.,* 34, 489, 1980.

64. **Seeley, T.,** *Honeybee Ecology,* Princeton University Press, Princeton, 1987.

65. **Shettleworth, S.,** Animal foraging in the laboratory: problems and promises, *J. Exp. Psychol. Anim. Behav. Processes,* 15, 81, 1989.

66. **Sivinski, J.,** Mushroom body development in nymphalid butterflies: a correlate of learning? *J. Insect Behav.,* 2, 277, 1989.

67. **Stanton, M.,** Short-term learning and the searching accuracy of egg-laying butterflies, *Anim. Behav.,* 32, 1984, 33.

68. **Stanton, M., Snow, A., and Handel, S.,** Floral evolution: attractiveness to pollinators increases male fitness, *Science,* 232, 1625, 1986.

69. **Swihart, C. and Swihart, S.,** Color selection and learned feeding preferences in the butterfly *Heliconius charitonius, Anim. Behav.,* 18, 60, 1970.

70. **Traynier, R.,** Associative learning in the ovipositional behavior of the cabbage butterfly, *Pieris rapae, Physiol. Entomol.,* 9, 465, 1984.

71. **Traynier, R.,** Visual learning in assays of sinigrin solution as an oviposition releaser for the cabbage butterfly *Pieris rapae, Entomol. Exp. and Appl.,* 40, 25, 1986.

72. **Turner, J. R. G.,** Experiments on the demography of tropical butterflies. II. Longevity and home-range behavior in *Heliconius erato, Biotropica,* 3, 21, 1971.

73. **von Frisch, K.,** *Bees: Their Vision, Chemical Senses and Language,* Cornell University Press, Ithaca, NY, 1971.

74. **Waddington, K.,** Floral-visitation sequences by bees: models and experiments, in *Handbook of Experimental Pollination Biology,* Jones, C. E. and Little, R. J., Eds., Scientific and Academic Editions, New York, 1983, 461.

75. **Waddington, K.,** Foraging behavior of pollinators, in *Pollination Biology,* Real, L., Ed., Academic Press, New York, 1983, 213.

76. **Waddington, K.,** Perception of foraging costs and intakes, and foraging decisions, In *Neurobiology and Behavior of Honeybees,* Menzel, R., and Mercer, A., Eds., Springer-Verlag, New York, 1987, 66.

77. **Waser, N.,** A comparison of distances flown by different visitors to flowers of the same species, *Oecologia,* 55, 251, 1982.

78. **Waser, N.,** The adaptive nature of floral traits: ideas and evidence, in *Pollination Biology,* Real, L., Ed., Academic Press, New York, 1983, 254.

79. **Waser, N.,** Flower constancy: definition, cause and measurement, *Am. Nat.,* 127, 593, 1986.

80. **Winston, M.,** *The Biology of the Honey Bee,* Harvard University Press, Cambridge, 1987.

5

The Mechanisms of Nutritional Compensation by Phytophagous Insects

S. J. Simpson and C. L. Simpson*
Department of Zoology and University Museum
Oxford University
Oxford, U.K.

TABLE OF CONTENTS

I.   Introduction... 112
II.   Compensatory Responses to Variations in Dietary Nutrients.............. 113
        A.   Compensating by Altering Consumption 113
                1.   The Response to Dilution of An Artificial Diet................... 114
                        a.   The Possible Effects of Dietary Dilution on Intermeal
                                Interval 114
                        b.   The Possible Effects of Dietary Dilution on Meal Size 115
                2.   The Response to Dilution of Specific Nutrients in the Diet...... 116
                        a.   Protein... 116
                        b.   Compensation by Nymphs of *Locusta migratoria* to
                                Changes in Dietary Protein: A Case Study 117
                        c.   Carbohydrate.. 121
                3.   The Response to Nutrient Imbalances 123
                4.   The Response to Previous Food Deprivation................... 124
                5.   The Response to Previous Deprivation of Specific Nutrients ... 128
        B.   Compensating by Dietary Selection 129
                1.   Studies Using Natural Foods 130
                2.   Studies Using Artificial Diets 131
                3.   The Underlying Mechanisms of Compensatory Selection....... 133
                        a.   The Initial Choice of Food................................. 133
                        b.   The Ability to Recognize Nutritional Inadequacy 133
                        c.   The Role of Learning 135
                        d.   Direct Feedbacks 137
        C.   Post-Ingestive Compensation................................... 138
III.   Compensating for Changes in Nutritional Requirements................. 141
        A.   Compensating for Changes Within a Larval Stadium 141
        B.   Compensating for Changes Between Larval Stadia.................. 144
        C.   Compensating for Changes Associated with Reproduction.......... 145
IV.   Constraints on Compensation 146
V.   Compensation and Allelochemicals.................................... 148
VI.   Concluding Remarks 150
Acknowledgments 151
References... 152

* The authors are not related.

I. INTRODUCTION

In terms of their nutritional composition, phytophagous insects are clearly not what they eat. Despite having evolved the morphological, physiological, and behavioral means to ensure that they obtain adequate supplies of nutrients from their hosts, phytophages are still faced with more immediate problems, in that the nutritional quality of plants (a function of the quantity, balance, and availability of nutritients) is variable both in time and in space.[78,198,227] Additionally, the nutritional needs of an insect are not constant, but vary with the requirements of growth and development, reproduction, and so on. Unless a phytophage is able to respond to the constantly changing mismatch between what it needs and what the plant can provide, it suffers consequences such as extended development, reduced fecundity, or premature death.[227] The study of compensatory mechanisms whereby an insect reduces this discrepancy is therefore central to an understanding of insect-plant relations. It also has potential consequences for the control of phytophagous insect pests[78,154] and is relevant to theories about the optimality of feeding behavior.[168,170,235]

There are three basic types of compensatory response (Figure 1). First, the insect can alter its rate of consumption in order to ingest sufficient of whatever nutrients are most limiting. Secondly, it can leave its present food and select something else, either another part of the same plant or another host of the same or a different species. Thirdly, it can alter its digestive and metabolic efficiencies to make best use of ingested nutrients. During recent years, with the enormous interest among ecologists in what has become known as the paradigm of nutritional ecology,[198,226,227] many apparent examples of all three compensatory responses have emerged. Relatively few studies, though, have investigated compensation as a subject in its own right, and even fewer have been concerned with the underlying controlling mechanisms. This has made our task rather difficult, since a discussion of the mechanisms of compensation inevitably ends up being a patchy mix of hearsay and the superficial, on the one hand, and detailed analyses of particular examples on the other. The importance of the subject and the fact that at least some detailed examples exist make the exercise worthwhile, however. In view of the paucity of relevant data, we have taken certain liberties and included work on those celebrated phytophages the cockroach and the blowfly.

In this review we have aimed to assess critically the available detailed studies and to take a representative, but by no means exhaustive, selection of the rest. We hope that this account complements the excellent reviews of Scriber and Slansky[198] and Slansky and Scriber[227] on nutritional ecology, and those of Barton Browne,[13] Bernays and Simpson,[29] and Bernays[22] on the regulation of feeding. The former provide the ecological context, while the latter outline the physiological and proximate principles involved.

There are several important areas which we have omitted, most notably the role of water as a nutrient, for which the reader is directed to the accounts of Scriber[197] and Bernays.[22] We have also been selective in our treatment of compensation for changing needs; in particular, we do not consider nutritional compensation by a host phytophage for the effects of parasitism. The recent review by Slansky[223] is a useful entry into this exciting area, in which the key problem is whether compensation is a passive response to the presence of a parasitoid or whether parasitoids actively regulate host physiology and feeding behavior. Finally, allelochemicals have been discussed only in the way in which they interact with compensatory feeding for nutrients.

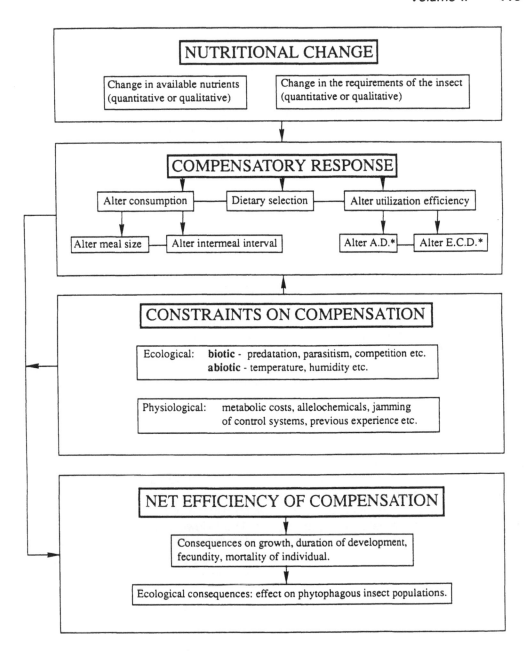

FIGURE 1.   A diagram summarizing the various relationships between the components of nutritional compensation. (*A.D. refers to the efficiency at which nutrients are digested and absorbed from the gut, while E. C. D. is the efficiency at which nutrients absorbed from the gut are converted into body tissue).

II. COMPENSATORY RESPONSES TO VARIATIONS IN DIETARY NUTRIENTS

A. COMPENSATING BY ALTERING CONSUMPTION

A commonly used approach to the study of dietary compensation has been to provide animals with one of several foods varying in nutrient quantity or quality and to monitor subsequent feeding behavior, either in detail or by simply measuring total consumption.

Another experimental tactic has been to deprive an animal of food or specific nutrients for a period and to compare its feeding behavior once given food with that of undeprived animals. Both types of approach have provided valuable insights into the nature and control of compensatory feeding.

1. The Response to Dilution of An Artificial Diet

A number of studies have investigated the effect of diluting the entire nutrient complement of an artificial diet on the rate of food consumption by various species. Some of these are discussed in detail in the review by Barton Browne.[13] Certainly the most impressive compensatory response was that demonstrated in fifth-instar nymphs of *Melanoplus sanguinipes* by McGinnis and Kasting.[140] The grasshoppers were fed a diet of wheat sprout meal diluted to various degrees with cellulose. When the sprout meal was diluted by two-, four-, and eight-fold, consumption over the first 5 days of the stadium increased by 2, 4, and 7 times respectively. A further dilution to $^1/_{16}$ of the control diet exceeded the compensatory abilities of the nymphs under the experimental conditions used. Nevertheless, such insects are six times as much as those provided with the undiluted food. Compensatory increases in consumption of artificial diets diluted with cellulose have been found in a number of other acridids, including *Locusta migratoria*,[70,127] *Schistocerca gregaria*,[70] and *Melanoplus bivittatus*.[139] Adult males of the cockroach *Blattella germanica* were shown by van Herrewege[107] to eat more of an artificial diet diluted with aluminum oxide, while House[114] demonstrated increased fresh weight consumption by fifth-instar larvae of the sphingid *Celerio euphorbia* when their artificial diet was diluted up to two-fold with water. Peterson et al.[167] found that adding up to 20% dry weight of cellulose or silica to the artificial diet of *Spodoptera eridania* resulted in the larvae exhibiting increased consumption. Slansky and Wheeler[228,229] have also reported increased consumption in velvetbean caterpillars, *Anticarsia gemmatalis*, after dilution of their food with either water or cellulose.

Most, if not all, insects feed in bouts or meals, separated by more or less extended interfeeds.[211] It follows that increased consumption of food can be achieved by two means: either by decreasing the time between meals or by eating more during meals. It is worth considering the various ways in which dilution of dietary nutrients might be expected to alter these behavioral parameters. The following account is based on data reviewed in several places during recent years.[13,22,25,29,79,213,214] It sets the scene for the more recent experimental evidence which is discussed elsewhere in this review, some of which runs counter to predictions based on the earlier work.

a. The Possible Effects of Dietary Dilution on Intermeal Interval

In those few species which have been studied in detail, intermeal intervals have been shown to consist of two phases. Soon after completing a meal the insect becomes relatively quiescent, resting either on or some distance from the food. This period of post-prandial quiescence is followed by an increase in activity, which leads to renewed stimulation of appropriate chemosensilla by food, whereupon feeding is initiated if excitation exceeds the behavioral threshold for the insect at that time. Hence, an insect provided with abundant food may feed more frequently either because it remains quiescent for a shorter time after a meal or because it has a lower threshold for the initiation of feeding once post-prandial quiescence is broken.

One of the most important regulators of the duration of intermeal intervals is the inhibitory influence of previous meals. Three sources of post-feeding inhibition have been described in locusts and blowflies, some or all of which are likely to occur in other species: tonic input from stretch receptors, associated with the gut in the locust and blowfly but found no the body wall in other species; hormones released as a result of taking a meal (locusts and possibly blowflies); and changed blood composition, particularly variation in blood osmolality and nutrients. These factors are known to interact. For example, stretching of the crop during a meal in the locust leads to the release of hormones (as yet unidentified) from the storage lobes of the corpora cardiaca which inhibit locomotor activity, reduce the sensitivity of gustatory sensilla on the mouthparts, and stimulate crop emptying. Crop emptying in the locust is also affected by the osmolality of the blood, which is in turn related to the concentration of nutrients in the diet. As in the blowfly, the lower the osmolality of the blood, the faster the rate of crop emptying. It could be that increased consumption of a diluted diet occurs simply because blood osmolality is lower than in insects feeding on undiluted diet, hence the gut empties more quickly and volumetric inhibition declines sooner after a meal.

b. The Possible Effects of Dietary Dilution on Meal Size

The amount of food eaten during a meal results from the interaction between excitation, generated primarily by input from gustatory receptors responding to phagostimulants in the food, and inhibition, which in most insects appears to come mainly from stretch receptors on the gut or body wall as well as from gustatory receptors responding to deterrents in the food. At first sight it seems difficult to explain in these terms how an insect feeding on a diluted diet might increase consumption by taking larger meals, since the amount of excitation generated while eating a diluted food would be expected to be less than on undiluted diet due to the lower concentration of phagostimulants. From the above discussion it seems reasonable to predict that compensation on a diluted diet would occur as a result of insects eating large numbers of small meals. Feeding patterns have been recorded in only a few species of insect, and the only study in which feeding patterns are available for insects fed diets in which the entire compliment of nutrients has been diluted is that of Timmins et al.[240] on fifth-instar larvae of the tobacco hornworm, *Manduca sexta*.

Timmins et al. gave larvae one of four artificial diets, containing either 0, 50, 75 or 90% cellulose. The two-fold dilution elicited an increase in consumption rate throughout the stadium of 1.5 times. An increase in consumption of 2.25 times accompanied the four-fold dilution, while the dilution to $^1/_{10}$ of the control diet only resulted in the same increase infeeding as did the two-fold dilution. The percentage of time spent feeding increased with dilution in the sameway as did total consumption. As might have been predicted, the mean intermeal interval decreased as the diet was diluted up to four-fold, after which it increased again on the most dilute, and thus least stimulating, diet. Although not measured directly, retention time of food in the gut apparently declined with dilution, as would be expected in terms of reduced blood osmolality. However, there was one major surprise; rather than getting smaller, the duration of the average meal increased with dilution, being longest on the diet diluted tenfold. The authors suggested a hypothesis to explain these results which involved both volumetric and nutritional feedbacks. A meal continues until some combination of gut stretch and nutrient feedback from the meal terminates it. If nutrients are in low concentration in the food then more gut stretch is required to end a meal. Thus a greater volume is eaten during an average meal of the more dilute diets.

Timmins et al. also altered the amount of water in the diet. When water content was reduced to provide twice the concentration of nutrients, larvae took longer meals about half as often, as would be expected from conventional explanations as outlined above. (To reiterate, the diet, being more concentrated, is more stimulating and larger meals are eaten, producing a larger increase in blood osmolality which slows the rate of gut emptying and causes extended intermeal intervals). As water content was increased, diluting the diet up to fourfold, intermeal interval declined somewhat and, as for cellulose dilution, meal duration increased considerably. However, unlike the situation for cellulose dilution, this increase in volume eaten per meal is consistent with a simple volumetric control mechanism. Because water is rapidly removed from the gut during feeding,[240,181] more of a water-diluted diet must be eaten in order to accrue sufficient solid food to stretch the gut. Retention time did not appear to be affected by increasing the water content of the diet as much as it was by diluting with cellulose. If osmotic control of gut emptying is occurring in *Manduca*, this might suggest that the osmolality of the gut contents rather than the blood is critical, as has been suggested for the cockroach.[74] Further work is needed in which parameters such as rate of gut emptying, blood and gut osmolality are measured directly.

Work on adult males of the Australian sheep blowfly, *Lucilia cuprina,* provides an alternative and more parsimonious explanation to the volumetric-nutritional feedback hypothesis, and is discussed in Section II.A.2.c. A similar model could explain the results of Timmins et al. for *Manduca,* although the possibility of a direct nutrient feedback is not discounted. In fact, Timmins[239] has tried to separate nutritional from osmotic and volumetric feedbacks in the control of meal size, with some suggestive results.

2. The Response to Dilution of Specific Nutrients in the Diet

When attempting to explain the mechanisms underlying an instance of compensation to dilution of a diet, an obvious question is whether certain nutrients are more important in eliciting compensatory changes in feeding than are others. In particular, are specific nutrient feedbacks involved?

The nutrients required by insects include proteins, carbohydrates, lipids, vitamins, minerals, and salts.[71] The only nutrients for which there are published data on the mechanisms of compensatory feeding are proteins and carbohydrates. In fact, even reports of compensatory feeding by altered consumption for other nutrients are rare. Navon et al.[158] found that larval *Manduca sexta* increased food consumption soon after removal of ascorbic acid from the diet (an essential nutrient for most insects that feed on fresh plant foliage). The mechanism whereby removal of vitamin C led to increased feeding is not known, although there are numerous possible routes for a feedback, since there are major changes in the titers of various blood amino acids and sugars, as well as in the fat body, following its removal.[44, 175] Timmins[239] has data suggesting compensatory feeding for lipids in *Manduca* larvae. Fernando-Warnakulasuriya et al.[91] found no difference in body weight or in the concentration of the blood lipoprotein, lipophorin, in *Manduca* larvae reared on diets containing 1.2 or 5.9% lipid by wet weight. Food consumption was not measured, however. Dwivedy[88] altered the dietary concentration of cholesterol from 0.02 to 0.22% wet weight and found that the sterol content of larval *Musca domestica* only increased by 1.5 times, but whether this resulted from changes in food consumption or in the efficiency at which ingested cholesterol was utilized is not known.

a. Protein

There has been considerable interest during recent years in the role of plant nitrogen as a limiting factor for insect herbivore populations. The nitrogen content of plant tissue ranges from about 0.0002% weight per volume in xylem sap up to 7.0% by dry weight in young, growing tissues, and in storage tissues such as seeds. Herbivore tissue, on the other hand, contains 7 to 14% N by dry weight.[149] The search by ecologists for evidence of nitrogen limitation of phytophagous insect communities has provided many instances of compensatory feeding behavior (see References 142, 149, 197, 198, 224, 232, and 251). Variation in the nitrogen content of plants fed to experimental insects in such studies has been achieved in a number of ways, including using different plant species, different tissues of the same plant, leaves of varying ages within a plant, plants of different age or stage in the season, primary or regrowth foliage, plants fertilized to different degrees, experimentally stressed or damaged plants, and plants raised in different concentrations of carbon dioxide (see above reviews, and References 104, 133, 163, 171, 194, and 243).

As pointed out by a number of authors, there are problems in the interpretation of results from studies such as these, in which plants varying in nitrogen content are used. Crude nitrogen level is presumably not what is regulated by insects, rather the levels and balance of specific proteins and amino acids are likely to be critical.[31,243] Much of what is measured as total nitrogen may not in fact be available to the insect for digestion.[149] In addition, it is virtually impossible to separate the effects of nitrogen content from correlated changes in other plant nutrients, water, and secondary compounds.[198,277] Two alternatives are to coat plant tissues with protein or to use artificial diets. Schroeder[194] coated the leaves of *Tilia americana* with a protein mixture and demonstrated reduced consumption but unchanged growth rate in final instar larvae of the lepidopteran, *Datana ministra.* He suggested that the reduced ingestion rate may be due to "feeding inhibition triggered by higher hemolymph concentrations of digestion products (e.g. amino acids...)." This has been recently demonstrated experimentally in locusts (see below).

Defined artificial diets usually bear little resemblance to the composition of food plants. This is especially the case for the protein component, since the most widely used artificial diets contain non-plant proteins such as casein, albumen, and peptone. Work by Horie and Watanabe[111] showed that the amount eaten and the degree of compensation exhibited by fifth-instar *Bombyx mori* larvae to dilution of the protein component with cellulose depended upon the nature of the protein source in an artificial diet. For example, when soybean meal was used there was no effect of diluting it from 60 to 20%. When casein was the protein source, the larvae ate less than when there was an equivalent quantity fo soybean meal, but showed a marked compensatory increase in feeding when the casein was diluted.

Studies in which insects fed artificial diets and plants have been compared demonstrate marked differences in feeding behavior and food utilization.[63,181,243] Nevertheless, the use of defined artificial diets has been essential in the search for the underlying mechanisms of compensation for variation in dietary protein. Once model mechanisms are established, their efficacy can be tested in more natural situations.

b. Compensation by Nymphs of Locusta migratoria to Changes in Dietary Protein: A Case Study

The mechanisms of compensation for altered levels of dietary protein have been studied in greatest depth in fifth-instar nymphs of *Locusta migratoria* by using detailed studies of feeding behavior as the basis for physiological and electrophysiological investigations.

The pattern of feeding behavior was recorded for nymphs given one of four artificial diets containing either 28 or 14% protein and 28 or 14% digestible carbohydrate.[212] The bulk of the diets and the concentration of all other nutrients were maintained using cellulose, which is indigestible to locusts. The nymphs provided with the lower protein diets ate nearly 1.5 times as much over the 12-h experimental period as those fed the high protein diets, while there was no compensation for digestible carbohydrate, or any interaction between protein and carbohydrate at the levels tested. (This does not mean that the same would apply at different levels, of course; see Section II.A.5.) The increased consumption of the lower protein diets was achieved by eating the same sized meals more frequently.

The next step was to monitor the composition of the blood and the rate of passage of food through the gut in insects fed the different diets.[2] Blood osmolality was found to be similar at the beginning of meals taken during *ad libitum* feeding, but about 10% higher after a meal of the high rather than the lower protein diets, with osmolality rising during the course of a high protein meal and remaining elevated for about 70 min afterwards, before falling back to the pre-meal level. Results from the previous experiments had shown that the mean intermeal interval on high protein diets was 70 min. Levels of digestible carbohydrate did not affect blood osmolality; apparently digested sugars are removed from the blood to the fat body as soon as they are absorbed from the gut.[241]

Since *in vitro* studies by Baines[8] had shown an effect of blood osmolality on gut motility, it was expected that the rate of passage of food through the gut would be faster in insects fed the lower protein diet. Such a mechanism would have provided the simplest explanation of the change in intermeal interval. This was not the case, however, no differences were found in the rate of gut emptying. Locusts were able to take meals of the same average size more frequently on the lower protein diets, without an increase in the rate of passage of food through the gut. This was because an amount of food equivalent to the mass of an average meal had cleared the gut by 40 min after the last meal, this being less than the average intermeal interval on either low or high protein diet. Beyond 40 min after a meal the rate of gut emptying slowed considerably, so that by the time the next meal was taken the gut contained a similar volume of food, irrespective of dietary protein.

Another possible explanation of the altered intermeal interval could also be discounted. The experiments of Bernays and Chapman and colleague,[24,28,32] showing effects of hormones released from the corpora cardiaca, both on the sensitivity of mouthpart taste receptors and on locomotor activity, demonstrated that it was the volume of food in the gut rather than its chemical composition which was important in eliciting hormone release. Since there was no difference in the quantity of food eaten during average meals on high and low protein diets, it seemed unlikely that differences in hormone release could explain the results. The only likely mechanism remaining was a direct influence of blood composition on feeding behavior.

There are two parameters of blood composition which could be influencing intermeal interval: osmolality and the concentration of specific nutrients. An analysis of blood composition showed that free amino acids contributed 40% of the osmolality difference between locusts fed diets containing either 14 or 28% protein, with 11 of the 16 amino acids measured occurring in significantly higher concentrations in the blood of nymphs fed a high protein diet. In order to test whether a general osmotic effect or a specific amino acid feedback was involved, nymphs were taken as they completed a meal of the lower protein diet during *ad libitum* feeding and left without food for 40 min (the average intermeal interval on the lower protein diets being 45 min). They were then injected with one of four solutions and given access to food, after which they were observed until they next fed. The injections were designed to alter the levels of free amino acids and/or blood osmolality to those occurring at the same time since a meal of the high protein diet. Injections that increased amino acid levels in the blood, but did not raise osmolality, significantly delayed feeding relative to the control injection of isotonic saline. An increase in blood osmolality without altering amino acid concentrations (xylose was used) also delayed feeding, while the injection which raised both amino acid concentration and osmolality had the greatest effect. It was clear, then, that both specific nutrient feedbacks and a general osmotic effect contributed to compensatory feeding behavior.

Subsequent experiments[1,205] showed that 8 of the 11 amino acids found in higher concentrations in the blood following a meal of the high protein diet are especially important in evoking compensatory feeding when their concentrations are low. Unless L-glutamine, serine, methionine, leucine, phenylalanine, lysine, valine, and alanine are all present in elevated levels in an artificial diet, a marked compensatory increase in feeding is exhibited, with the extent of the increase depending on the particular amino acid omitted. The response was most marked for lysine and least so for leucine.

Five of these (methionine, leucine, lysine, phenylalanine, and valine) are among the ten so-called "essential amino acids", which most animals are unable to synthesize. The eight are a representative mix of all the major synthetic pathways, as well as being a selection of acidic, basic and neutral, and polar and nonpolar amino acids. It is possible that together they provide a sign stimulus so that, by regulating compensatory feeding with respect to this suite of eight, a locust is assured of receiving a proper balance of all amino acids. Whether or not the same amino acids are involved in protein compensation during different stages of development, when protein requirements are known to vary (see Section III), is not yet known. Other experiments by Bernays and Woodhead[31] have shown that phenylalanine can be limiting to the locust *Schistocerca gregaria*. Supplementing a diet of lettuce with phenylalanine improved food utilization and, at least in one replicate experiment, led to a decrease in food comsumption.

The question remains as to how the effect of specific amino acid feedbacks and blood osmolality are mediated. There are several possibilities. Intermeal intervals may be shortened as a result of reduced post-prandial quiescence (i.e., the insect commences locomotion sooner after the last meal) and/or an increased probability of initiating ingestion on contacting food. In face, quiescence was lengthened and the probability of acceptance reduced by injections which increased blood amino acids and osmolality, with a stronger effect of amino acids on the latter.[1] Unless muscle excitability is influenced directly, an effect of blood composition on post-prandial quiescence is likely to be via the central nervous system, perhaps with blood amino acids being monitored by internal chemoreceptors or by direct effects on the CNS. Both mechanisms are known in vertebrates,[130,150] but not as yet in insects. An effect on the threshold for the initiation of feeding once post-prandial quiescence is broken could also be a central phenomenon, but there is the possibility that the sensitivity of mouthpart chemoreceptors is altered, since the initiation and maintenance of feeding is dependent upon input from gustatory receptors.[35,68]

A series of electrophysiological studies was carried out in order to test for an effect of dietary protein on the sensitivity of gustatory receptors on the maxillary palps of locust nymphs. In one experiment, locusts were subjected to the same protocol as used in the previously described injection experiments. Insects were taken after completing a meal of the lower protein diet and left for 35 min without food. The responses of taste sensilla on the palps to stimulation by a mix of the eight amino acids, a sucrose solution, and a salt solution were then recorded, before and at various times after injection with one of the four solutions used in the behavioral studies.[3]

There was a reduction of 70% in the number of spikes elicited by stimulation of sensilla with the amino acid mix within 5 min of an injection which increased the free amino acid concentration of the blood up to levels found in high protein fed nymphs. This was independent of the osmotic effect of the injection, and injections which only altered blood osmolality did not differ in their effect from the control. The response fo the same sensilla to stimulation with sucrose, the other major phagostimulant in the diets, was not influenced by altering the blood osmolality or amino acid content by injection, however. There was a statistically significant reduction in firing rate to salt stimulation, after injections which increased either blood amino acids or osmolality, although the reduction was only 20%: not nearly so marked as the specific reduction to stimulation with amino acids after injections which raised blood amino acid concentration.

These experiments provided the first evidence in any animal of a specific nutrient feedback influencing compensatory feeding behavior, and showed that this feedback may be partly mediated by "need specific" modulation of peripheral gustatory sensitivity. The effect on taste responsiveness was less clear in a comparison of locusts fed the high and low protein diets, where a decline in responsiveness to sugar and salt stimulation was also found in high protein-fed insects. More recent work by C. L. Simpson[203] has verified the results of injection experiments, but is not in accordance with the data comparing low and high protein-fed insects as reported by Abisgold and Simpson.[3]

It is important to know whether gustatory receptors within the oral cavity are similarly modulated, since it is these which are most likely to taste amino acids during feeding. Recent studies have demonstrated that changes in sensitivity with blood amino acid levels are not restricted to the maxillary palps; the labial palp and tarsal chemosensilla are also modulated.[203] Work is also needed to discover whether specific peripheral modulation occurs for individual amino acids or other nutrients. A start has been made in this direction.[205] Omission of lysine from an injection of amino acids (see above) results in *increasing* sensitivity of maxillary chemosensilla to stimulation with lysine and to stimulation with a mixture of amino acids. This is in marked contrast to control insects receiving the complete amino acid injection, which show a decline in responsiveness to both lysine and the amino acid mix. A similar, though less pronounced effect has been found for alanine. In behavioral experiments, reduced concentrations of these two amino acids in the diet produced the greatest increase in consumption. With leucine, the effect was less marked and corresponding electrophysiological studies show that omission of leucine from the amino acid injection does not result in a difference relative to controls.

Recent investigations into the way in which blood amino acids alter the sensitivity of gustatory receptors on the palps have shown that the response is not due to centrifugal control from efferent nerves, since sectioning the palp nerve as it leaves the subesophageal ganglion does not remove the effect of injecting amino acids into the abdomen.[203]

Reports of dietary quality affecting the responsiveness of taste receptors in other insects are scarce[40] but there are some suggestive results in caterpillars. The sensitivities of various receptors of *Manduca sexta, Spodoptera littoralis,* and *S. exempta* larvae are influenced by the nature of the rearing diet.[188,190,192,230] In particular, specific changes have been found in receptors responding to allelochemicals, with a reduction in sensitivity being found when diets contain the compounds. Unpublished experiments by Simmonds and Blaney,[201] in which sugars have been injected into the blood of larval *S. littoralis,* have shown a specific reduction in the responsiveness of mouthpart receptors to stimulation with sugar but not to amino acids or allelochemicals.

McClain and Feir[137] published data showing an effect of injecting amino acids into the hemolymph on the feeding behavior of fifth-instar nymphs of the milkweed bug, *Oncopeltus fasciatus.* The bugs were deprived of food for 24 h prior to being tethered and injected with one of a number of amino acid solutions. After this, the volume ingested of a 0.1 M solution of the same amino acid which had been injected was recorded during a test meal. Alternatively, bugs which had been injected or prefed for 15 min on one of the amino acid solutions were freed into an arena containing milkweed seeds and their behavior was recorded over an 8-h period. The amount of an amino acid solution ingested in the test meal was influenced by the volume of the injection, as well as by the amino acid injected. Neither alanine nor glycine injections influenced subsequent meal size on solutions of the same amino acids. Injections of glutamine enhanced subsequent intake of glutamine, while injections of histidine and α-aminoisobutyric acid had the opposite effect. The effects of injection of prefeeding on the feeding behavior of bugs given *ad libitum* access to seeds were also complex. Total time spent feeding was decreased following injections of 0.1 *M* glycine, 0.1 *M* alanine, 0.5 *M* glutamine, 0.5 *M* histidine, and a 0.5 *M* mix of these four, as well as by prefeeding on 0.1 *M* alanine, but feeding time was increased by the injection of 0.1 *M* histidine, 0.5 *M* glycine, and by prefeeding on 0.1 *M* glycine or glutamine. Changes in both mean meal duration and number of meals contributed to differences in total time spent feeding. There are several methodological problems with these experiments which make their interpretation difficult. No details were given of the levels of amino acids normally occurring in the blood, nor of blood volume, so it is not possible to say whether the injections provided realistic changes in blood composition. No osmotic controls were run either, making it impossible to distinguish specific nutrient effects from general osmotic ones. Finally, the insects used in the liquid feeding experiments were deprived of water as well as food for the 24 h preceding the experiment, so drinking and feeding were confounded.

A role of blood osmolality in the control of feeding in *Locusta migratoria* nymphs was reported by Bernays and Chapman.[26] Whereas Abisgold and Simpson[2] found that blood osmolality influenced intermeal interval but not meal size in nymphs feeding *ad libitum* on artificial diets, Bernays and Chapman showed that raising blood osmolality by injection caused a reduction in the size of grass meals eaten by previously deprived nymphs. They also proposed that there was a cumulative effect of blood osmolality on meal size, whereby the extent of the reduction in meal size depended not only on the osmolality of the blood at the start of a test meal but also on the time prior to the meal over which blood osmolality was elevated. There was no evidence of such an effect in the *ad libitum* studies of Simpson and Abisgold, since meal size was not reduced in high protein fed nymphs, despite the fact that blood osmolality is elevated more and for longer during intermeal intervals than in low protein fed nymphs.

It is clear from both experiments that increasing blood osmolality inhibits feeding. The differences between the two experiments may be due first to the fact that Bernays and Chapman raised blood osmolality to a greater extent than occurred following a meal during *ad libitum* feeding, and secondly to their use of previously deprived nymphs. When nymphs are feeding *ad libitum* inhibition provided by elevated blood osmolality manifests itself as extended interfeeds, while in previously deprived insects which have been injected and placed with food the inhibition is translated into a reduction in net excitation as the test meal begins. Under normal conditions insects would not experience a rise in blood osmolality immediately before being given food after a period of deprivation.

c. Carbohydrate

Compensatory increases in feeding following dilution of digestible carbohydrate in the diet have been demonstrated in a number of adult insects, including the blowflies, *Phormia regina*[84,98] and *Lucilia cuprina*,[216] the cockroaches, *Blattella germanica*[99] and *Periplaneta americana*,[33] the butterfly, *Pieris brassicae*,[75] the Mediterranean fruit fly, *Ceratitis capitata*,[161] and the mosquito, *Aedes taeniorhynchus*.[159] In each of these studies, except that on adult female *Periplaneta*,[33] insects were fed a diet consisting solely of digestible carbohydrate, diluted with water in the case of the fluid feeding flies and butterfly, and with cellulose for the cockroaches. In contrast, the cockroaches in Bignell's[33] experiments were fed a diet containing a fixed proportion of proteins and salts but variable proportions of dextrin and cellulose. Interestingly, compensation to dilution of the sugar component was less marked in Bignell's experiments than in the others mentioned above, probably due to the problems to the insects of increasing consumption on the nutritionally more complete diets (see Sections II.A.2, II.A.3, and IV).

The nature of the carbohydrate source has an important effect on amounts eaten and the degree of compensation to dilution that is exhibited, as was found for dietary proteins (see above). For example, work on male *B. germanica*[99] showed that compensation is more precise on preferred than less preferred sugars (preference being measured in choice tests). A 50% dilution of solid sucrose with cellulose caused an increase in consumption over several days of 1.8 times, as compared with 1.6 times for glucose and trehalose, 1.4 times for fructose and maltose, and a decrease to 0.8 times for raffinose. All of these sugars are nutritious and were eaten in similar quantities when undiluted. A similar effect was found in male *Phormia regina,*[98] where it was demonstrated that compensation is independent of the nutritive value of a sugar but is dependent on its stimulating power, as coded by the firing of gustatory receptors, and its effect on blood osmotic pressure, and therefore rate of crop emptying, once digested. Stimulating power varies with the nature of the sugar and with its concentration in the diet.

Gordon[100] proposed a mathematical model to account for compensatory feeding in the cockroach and blowfly which predicted that increased consumption of a diluted carbohydrate diet would occur as a result of insects taking frequent, small meals (see also Section II.1.b). Gelperin[97] and Barton Browne[13] came to similar conclusions. The only study to test this hypothesis was that carried out by Simpson et al.[216] on males of the Australian sheep blowfly, *Lucilia cuprina.* Flies were fed either 1.0 or 0.1 *M* glucose solutions and their feeding patterns recorded over 2 days. Like *Phormia, Lucilia* compensated on the lower sugar concentration by increasing consumption threefold. However, they did so by doubling the volume eaten during an average meal and feeding 1.5 times as often. At first sight it appeared that nutritional feedbacks might be controlling meal size, with flies continuing to feed until some standard amount of sugar has been imbibed (Section II.A.1.b). However, there is another possible explanation.

Previous work, showing a strong positive correlation between the volume of solution ingested in a single meal and the concentration of sugar in the solution, used flies which had previously been deprived of food for periods as long as 24 or even 72 h.[46,82,84] This means that the crop would have been at least partly emptied at the time of testing an insect's response to a sugar solution, and so any inhibitory volumetric input from previous meals would be reduced or absent. When crop volumes were measured in *Lucilia feeding ad libitum.* it was found that flies began a meal of 0.1 *M* glucose with an emptier crop than those flies fed 1.0 *M* glucose. Hence, a fly initiating a meal on 0.1 *M* glucose has less residual inhibition from gut stretch receptors, as well as less phagostimulatory excitation from the food than does a fly feeding on the more concentrated solution. The *net* excitation present as a meal begins (phagostimulation-induced excitation minus residual inhibition from the gut) will tend to converge in the two flies. It is this net excitation which will determine how much inhibition from crop stretch receptors will be tolerated before the meal is terminated. Because net excitation is similar, both flies will continue feeding until a similar level of crop distension is reached (experiments showed that crop fullness was similar at the end of meals taken on the two sugar concentrations) but the fly feeding on the 0.1 *M* solution will have to ingest more during the meal to reach the required fullness since it began with an emptier crop.

The shortening of intermeal interval on the 0.1 *M* glucose solution was much as expected. It was found that a meal of 1.0 *M* glucose caused more prolonged post-prandial quiescence than did a similarly sized meal of 0.1 *M* solution, although quiescence only contributed about 20% of an average intermeal interval. A fly would become active but not commence ingestion on contacting the solution until some time later. It has been shown that the threshold for the extension of the proboscis upon stimulating the tarsi with sugar is related to both the concentration of sugar and the fullness of the crop. Flies must, on average, have an emptier crop before stimulation of the tarsi with a dilute sugar solution will initiate proboscis extension.[89,233] As predicted from work on *Phormia*,[96] the rate of crop emptying was faster in *Lucilia* fed 0.1 *M* glucose than in those fed the 1.0 *M* solution. As a result, the threshold level of crop fullness was reached sooner after a meal in flies fed the more dilute solution, even though the crop had to be emptier before feeding was initiated than for flies given 1.0 *M* glucose.

As yet there is no published evidence for a specific nutrient feedback regulating carbohydrate ingestion in any insect, although it is apparent that carbohydrate and protein intake are under separate control. For example, although *Locusta migratoria* nymphs did not compensate for dilution of digestible carbohydrate in the experiments of Simpson and Abisgold,[212] other experiments show that locusts do regulate their intake of carbohydrate and protein independently when offered a choice of foods[59,217] (see Section II.A.5). The fact that there is no compensatory response to dilution of carbohydrate in a complete diet when the insects are given no choice suggests that a dilution from 28 to 14% is not sufficient to overcome the consequences of ingesting more of the other nutrients in order to increase carbohydrate consumption. Gordon[100,101] also performed experiments on *B. germanica* in which the proportion of protein (yeast) and digestible carbohydrate (sucrose) were varied independently in the diet. Compensation occurred for dilution of both protein and digestible carbohydrate, with the degree of compensation exhibited to one of the nutrient groups being limited by the consequences of ingesting an excess of the other. His data suggest that the cost of catabolizing and excreting excess protein inhibits compensatory feeding on diets which are high in protein but low in carbohydrate, while compensation for low dietary protein is limited when carbohydrate levels are high.

3. The Response to Nutrient Imbalances

Theoretically, as an insect starts a meal there will be a requirement for a precise mix of amino acids, sugars, lipids, salts, vitamins, and other nutrients. The nature of that mix will depend on the species of insect, its age and physiological state, and the prevailing environmental conditions (see Section III). An optimal food would provide the required mix in a readily digested form but, of course, food is never optimal. All plants and currently available artificial diets are to some extent nutritionally unbalanced and require the insect to compensate. In order to ingest enough of the most limiting nutrients an insect which has no alternative food will have to eat more of other nutrients than it needs, and will also have to ingest greater quantities of any harmful compounds present in the diet. If the mismatch between the optimal ratio of nutrients and that present in the food is sufficiently great, the consequences of ingesting unwanted nutrients and other compounds may be deleterious to the performance of the insect.

The first detailed studies into the importance of nutrient balance were those of House, and these experiments, as well as other earlier work, were reviewed by House in 1969.[117] He performed a series of experiments with fifth-instar larvae of the sphingid, *Celerio euphorbia*.[114] When the entire nutrient compliment of a nutritionally adequate artificial diet was diluted up to twofold by the addition of water, the caterpillars compensated for the dilution by increasing consumption and improving the efficiency with which they utilized ingested food (see Reese and Beck[178] who reanalyzed House's data and showed the latter to be true). However, if the diet was altered so as to contain half as much vitamins, a third more casein, and increased amounts of glutamic acid, glycine, and tryptophan (but the same proportion of sucrose, salts, and other ingredients), larvae appeared to reduce their food intake, utilize food less efficiently, and grow less well. Although none of these effects was statistically significant, House tentatively interpreted the results as a response to the "metabolic difficulty" of eating the unbalanced diet.[114] The critical experiment which House did not perform was to increase casein by a third but leave vitamins unchanged and omit the amino acid supplements. The apparent reduction in feeding on the unbalanced diet could have been due to a regulatory response to the increased protein rather than to the deleterious effects of ingesting the diet. There was no reason to expect larvae to increase feeding on the unbalanced food, as House seems to imply, unless they were attempting to compensate for the 50% dilution in dietary vitamins.

A better designed study was performed by House[115] on larvae of the sarcophagid fly, *Agria affinis*. Experiments were based on a factorial design, in which diets contained one of four levels of the amino acid mixture used in the standard diet, combined with one of four levels of the remaining nutrients (glucose, lipids, vitamins, etc.). In one experiment the salt content of all the diets was the same, while in another it varied as part of the non-amino acid nutrients. Results showed an interaction between levels of amino acids and other nutrients. When salt levels were constant, growth and development rates were similar for larvae fed diets representing all combinations of 0.75, 1.1, and 1.5 times the normal level of the amino acid mix with 1.1, 1.5, and 1.9 times the normal content of other nutrients. This indicates that compensatory feeding and/or food utilization must have occurred, but, unfortunately, these were not measured directly. If the amino acid content of a diet was increased to 1.9 times normal, however, other nutrients had to be present at 1.5 or 1.9 times the level in the standard diet for growth and development to be unaffected. The implication is that the larvae were unable to compensate for lower levels of the other nutrients because in so doing they would overload themselves with amino acids. This effect was exacerbated by allowing salt to vary as part of the non-amino acid nutrients.

There are a number of reports describing detrimental consequences on growth and development as a result of adding large amounts of single amino acids to an artificial diet, although it is not known whether such effects are because of toxicity of the added amino acids, antagonism with other amino acids, or some other effect of imbalance per se.[145] There are also reports of supplementing a plant or artificial diet with limiting amino acids resulting in improved food utilization and growth[31,111] and reduced food consumption.[31]

4. The Response to Previous Food Deprivation

Since all insects eat intermittently, it is true to say that every time an insect with constant access to nutritious, palatable food begins a meal, it does so after a period of deprivation. Even when such periods are entirely "self-imposed" there is evidence of subsequent compensation. The amount of food eaten during a meal has been shown to be positively correlated with the duration of the preceding intermeal interval in larval *Manduca sexta*[45,181] and *Pieris brassicae*,[143] as well as in adult males of the blowfly, *Lucilia cuprina*.[216] This is not the case for nymphs of *Locusta migratoria*, however, although there is a tendency for very short intermeal intervals to be followed by small meals and *vice versa*.[208] The lack of clear pre-prandial correlations in locusts feeding *ad libitum* is due to the effects of factors such as defecation and an endogenous short-term rhythm, which increase the probability of a meal beginning independently of time since the previous meal.[206,211,214]

Longer periods of deprivation are imposed when food runs out, when there is a decrease in the acceptability of available food, or when environmental factors prevent feeding. As deprivation continues most insects increase locomotor activity,[13] apparently in relation to the level of metabolic reserves held by the insect.[51,185] Within a matter of 6 h without food, locust nymphs show almost continuous marching, whereas there is no change, perhaps even a decrease, in locomotor activity exhibited by adult *Periplaneta americana* during 13 days without food.[185]

Numerous studies have measured the amount of food eaten in the first meal following a period of deprivation.[13,29] Generally, the size of the first meal rises with increasing deprivation, up to a point beyond which further time without food has no effect. (Extreme deprivation leads to a reduction in meal size due to morbidity, but this will not be considered here). In those cases where it has been investigated, the rate at which food is ingested during the meal also increases with deprivation. Simpson et al.,[220] working on nymphs of *L. migratoria*, showed that deprivation influenced meal size and ingestion rate in a similar way to excitatory influences such as increased levels of phagostimulants in the food and crowding by other nymphs. The amount of excitation sets both the rate of ingestion and the amount of inhibition needed to end the meal. Because most of this inhibition comes from stretch receptors on the gut (at least when levels of deterrents are low in the food), this means that ingestion rate and meal size increase together proportionally, and hence meal duration remains constant. Above a certain level of excitation (e.g., that present after 5-h deprivation when fed seedling wheat, or after 2 h without food when given seedling wheat coated in sugar) ingestion rate reaches its limit and meal duration increases proportionally with meal size. After 8 h without food, meal size also reaches its limit. Similarly, Bowdan[45] found a positive correlation between deprivation and ingestion rate in fifth-instar larvae of *M. sexta*. Although not measured directly, it was apparent that meal size as well as ingestion rate increased with deprivation periods of up to 5 h. The rise in ingestion rate was not sufficient to cause meal duration to remain constant, however. In both *Manduca* and *Locusta* the faster ingestion rate with deprivation was a result of increasing the rate of biting and spending less time pausing during the meal.

What, though, are the "deprivation-related factors" responsible for ensuring that the first meal taken after a period of deprivation is larger and eaten faster than normal? One likely candidate is a decline in inhibition resulting from previous meals: in other words, a simple extension of the same mechanisms that determine intermeal intervals in undeprived insects.[13] For instance, meal size on seedling grass reaches a maximum in nymphs of both *Schistocerca gregaria* and *L. migratoria* after periods of deprivation long enough to guarantee that the crop and the midgut/ileum are empty. Input to the CNS from stretch receptors on both these gut regions reduces net excitation at the start of meals taken by nymphs feeding *ad libitum*, since the gut normally contains some food at the time when meals begin.[184,210] Similarly, levels of humoral factors and blood nutrients will fall with continued deprivation beyond the normal intermeal interval and provide reduced inhibition of meal size.[2,25] Whether changes in blood composition or any other deprivation-related factor have a direct excitatory effect on the CNS rather than an indirect one via reduced inhibition, is not clear. There are some most suggestive results which implicate changed levels of the biogenic amine, octopamine, in deprivation-induced arousal. Davenport and Evans[73] showed that octopamine titer increased in the blood of adult male *S. gregaria* with food deprivation. Levels dropped once the insects were given access to food after the 9-h deprivation period, but remained elevated relative to control insects for 4 h afterwards. Whether or not the change in octopamine titer contributes to the increased locomotion and increased sensitivity to food which accompanies deprivation, or is just a consequence of them is yet to be established. It will be intriguing to discover what role changes in levels of this and other neuromodulators play in the control of "hunger" in insects (see also Section II.B.3.d).

There are several reports of changes in taste receptor sensitivity with food deprivation.[40] Such changes could result in raised excitation during feeding. Schoonhoven et al.[192] demonstrated increased firing from the medial sensilla stylochonica of larval *Spodoptera littoralis* to stimulation with sinigrin and nicotine. They found that responsiveness increased up to 8 h without food, then declined with further deprivation. Similar effects are shown in receptors responding best to sugars.[201] Omand[164] and Omand and Zabara[165] reported increased sensitivity of the sugar receptor in blowflies following deprivation. Other studies have failed to find such changes, although this could be because of methodological differences.[40,46,105,172] The work of Bernays and Chapman[24] and Bernays et al.[32] showed that the hormonally induced reduction in the sensitivity of palp taste sensilla in *Locusta* was largely removed by 2 h after a meal. The effect appears to be due to changes in the pore at the tip of each sensillum, rather than to intrinsic variation in receptor responsiveness or composition of the receptor lymph.

Apparently, none of these peripheral changes are specific to levels of particular nutrient reserves, although the experiments of Abisgold and Simpson[3] (see Section II.A.2.b) indicate that such changes can occur.

Compensatory feeding for previous deprivation does not necessarily stop after the first meal taken. This is hardly surprising since, as stated above, the size of the first meal becomes maximal in most cases after relatively short deprivation periods. Are insects able to make up the nutritional deficit incurred during deprivation once they are provided with *ad libitum* access to food?

When adult males of the cockroach, *B. germanica*, were left for 3 days without food their intake of a solid carbohydrate increased over the first day following deprivation. The extent of the increase depended on the nature of the sugar provided. Glucose intake was 3 times higher than in undeprived insects and consumption of sucrose was 2.5 times higher, but there was virtually no change in amounts of the nutritious but unstimulating sugar sorbose eaten.[99] These results indicate that increased responsiveness to phagostimulants occurs, in the CNS or at the periphery, and is maintained for some time after feeding recommences. A similar conclusion may be drawn from the work of Sinoir[221] on nymphs of *L. migratoria*. Insects were deprived for 8 or 30 h and the amount eaten of paper soaked in a weak sucrose solution recorded. The nymphs deprived for longer ate considerably more. In both these experiments, however, insects were not offered an opportunity to compensate for the full effects of previous deprivation since only carbohydrates were provided.

When Sinoir gave the locust nymphs grass rather than paper soaked in sucrose, the intake of 8- and 30-h deprived insects over the following 24 h was similar. Whether or not the 30-h deprived insects ate more on subsequent days was not measured, but a later study by Louveaux[136] demonstrated that 24-h deprived nymphs compensated almost exactly for the amount which would have been eaten during the deprivation period by increased consumption over several following days.

Like Gordon,[99] van Herrewege[107] deprived male *B. germanica* for 3 days, but then he provided them with a nutritionally adequate artificial diet. Intake was about three times that of undeprived insects during the first day following deprivation and was elevated for the next 2 days. As a result, insects almost precisely compensated for the missing intake.

Rollo[185] deprived adult male and female *Periplaneta americana* for 13 days. Subsequent daily intake of a complete artificial diet was initially 5 times higher than before starvation and feeding was significantly elevated for about 20 days. This enhanced feeding did not result in any loss in the efficiency at which ingested food was assimilated; in fact, this appeared to be elevated for the first day following the end of deprivation. Reports of work on lepidopteran larvae indicate that deprivation can have complex and subtle effects on various digestive and utilization efficiencies as well as on feeding behavior[102,103,157,193] (see Section II.C).

Another study in which increased consumption for some time following deprivation has been found is that of McLean and Kinsey[141] on the aphid, *Acyrthosiphon pisum*. Depriving aphids for 24 h resulted in an increase in the time spent ingesting during the following 16 h, with the effect being most marked soon after deprivation ended.

In adults of *Oncopeltus fasciatus* food intake is normally high during the first week after ecdysis and substantial lipid reserves are accumulated in readiness for migratory flight, which occurs in the second week. Slansky[225] found that bugs compensated for being deprived of food for the first week after adult ecdysis, by increasing their consumption during the following week by almost threefold relative to control insects. Bugs which were fed during the first week but deprived during the second did not compensate during the third week, however.

The detailed behavioral mechanisms of increased consumption following a period without food are not well known. Rollo[185] states for *Periplaneta* that "greater consumption was largely accomplished by longer, larger meals (unpublished data) and not so much by increased meal number." Jones et al.[123] performed experiments on adults of the Mexican bean beetle, *Epilachna varivestis,* in which feeding behavior on bean leaf discs was recorded intermittently (from once to 4 times per hour) for 8 h following deprivation times of 16 to 72 h. Since feeding was not recorded at more frequent intervals, it was not possible to derive measures for meal duration and intermeal interval. If feeding had occurred in a single or in several consecutive periods between observations these were considered to constitute an "interval with feeding", while single or consecutive periods during which nothing was eaten constituted an "interval without feeding". Deprivation times of between 16 and 64 h did not differ from each other in their effects, all causing increased consumption relative to control insects during the 8 h following deprivation. The latency to feeding was reduced by deprivation and there was a reduction in the number of intervals in which no feeding occurred, but there was no increase in the number of intervals with feeding. The amount of variability in each of these measures was also reduced relative to undeprived insects.

The most detailed study of behavior following deprivation is that of Bowdan[45] on fifth-instar *M. sexta* deprived for periods of from 1 to 5 h then given *ad libitum* access to an artificial diet. The major effect of deprivation was seen on the first meal (see above) but there was also a significant negative correlation between deprivation time and the duration of the intermeal interval following the first meal, with insects deprived longest feeding again soonest. Feeding behavior was recorded for 3 h after the end of deprivation and there was evidence of compensatory feeding throughout this time. For meals other than the first there was a positive correlation between deprivation time and both the duration of meals and the number of eating bouts during meals. Intermeal intervals were reduced and bite frequency increased in deprived as compared with control larvae, although neither measure was affected by the degree of deprivation. Presumably both had reached their limit within the shortest of the deprivation times tested (1 h). Hence, previously deprived larvae compensated by eating larger, longer, and more frequent meals.

Bowdan attempted to explain the increased feeding in terms of volumetric feedback from the gut (there is indirect evidence that gut stretch provides inhibition in larval *Manduca*[45,181,240]) and suggested that increased feeding following deprivation could be due to reduced inhibition as a result of partial emptying of the gut during the time without food. However, she found that the volume of food lost from the gut during a 5-h deprivation period was more than made up in the first meal, so could not explain why larvae continued to compensate for some time afterwards. In fact, if the actual deprivation deficit is calculated by reanalyzing Bowdan's data (i.e., the amount of food which larvae would have eaten were they not deprived), it is clear that this is only partly made up during the first meal, with 1-, 2-, 3-, 4-, and 5-h deprived insects compensating for only about 28, 33, 20, 30, and 49%, respectively, of their deficit by increasing the number of bites in the first meal relative to undeprived larvae.

Bowdan argues that *Manduca* larvae are able to monitor their nutritional deficit in a more sophisticated manner than by just monitoring gut fullness. This must also be the conclusion from the experiments on cockroaches, locusts, aphids, and bean beetles, where the animals were left without food for much longer periods than required to empty the gut, yet continued to compensate for the amount of nutrients which they "missed out on" during deprivation for a considerable period afterwards. How, then, is this achieved?

It is known that deprivation results in changes in blood composition and in the fat body, largely as a result of depletion of nutrient reserves. Newly digested and absorbed nutrients are likely to be removed from the blood more quickly in a previously deprived than in an undeprived insect and may continue to be so for as long as it takes to cause end-point inhibition of enzyme systems. This will presumably be related to the time for which the insect has been deprived, the state of its reserves prior to deprivation, and its behavioral and physiological response during deprivation. As discussed above, it is known that blood composition can cause altered rates of gut emptying, as well as have both specific nutrient and general osmotic effects on central and peripheral nervous excitability. Any of these could result in prolonged compensatory changes in feeding behavior. Experiments to test such ideas are clearly needed, as is work on the possible involvement of hormones and neuromodulators.

5. The Response to Previous Deprivation of Specific Nutrients

In the previous section we considered the case where an insect is deprived of food altogether. There are also instances where particular key nutrients may be missing or inaccessible in a diet, or even present in small amounts but, because of the relative concentrations of other nutrients, the insect is not able to adequately compensate for the deficiency by altering food consumption (see Sections II.A.3 and IV). The most detailed study of the behavioral mechanisms of compensation following such nutrient-specific deprivation in phytophagous insects is that of Simpson et al.[210,217] on larval *S. littoralis* and *L. migratoria*.

Spodoptera and *Locusta* were reared from the start of the final larval stadium on an appropriate, nutritionally adequate artificial diet, containing 20% protein and 10% digestible carbohydrate by dry weight. At a defined time before mid-stadium they were fed for 4, 8, or 12 h either on the same diet (PC) or on one of three variants of it: P, which contained the same concentration of protein but had the digestible carbohydrate replaced with cellulose; C, which had the protein but not the digestible carbohydrate substituted; and O, which had both protein and digestible carbohydrate replaced. At the end of this conditioning period the insects were given a choice of the P and C diets, and thus the opportunity to feed selectively for the nutrients missing from the conditioning diet. Their behavior was recorded every 10 s (locusts) or 30 s (caterpillars) for the first 60 min of choice, and then at 60-s intervals for the following 8 h.

Even after a conditioning period of 4 h there was a striking effect on feeding behavior during the first hour of choice. Locusts which had been conditioned on the P diet fed almost exclusively on the C diet when given a choice, the P diet usually being rejected either before taking a sample bite, or after a short period of ingestion. During the hour the nymphs ingested more than five times the amount of C diet than did PC-conditioned control insects. Locusts conditioned on the C diet showed an equally strong selection in the opposite direction. Those nymphs prefed on the O diet did not exhibit a preference for either P or C but increased consumption of both by about 5 times relative to controls. These compensatory responses continued for the following 8 h but were less marked, tending to be obscured by a preference shown by all nymphs for the C diet. Recent work[218] has shown that just a single meal on one of the deficient diets during *ad libitum* feeding is sufficient to evoke compensatory selection. *Spodoptera* larvae also exhibited marked compensatory feeding after conditioning on diets lacking protein or carbohydrate. The response to previous protein deprivation was not apparent during the first hour of choice but was very clear during the following 8 h. The reason for the delayed response was due to the fact that, unlike locusts, the caterpillars ate only small amounts of the C diet during the conditioning period, and so were effectively both carbohydrate and protein deprived at the time when they were provided with a choice. After making up the carbohydrate deficit they then fed selectively for protein. The implications of differences in response to deficient diets between the locust and caterpillar are considered in a later section (II.B.2).

The physiological basis of compensatory feeding following deprivation of protein and carbohydrate is not fully established as yet. Blood sugar and amino acid levels need to be measured in deprived insects, and a similar experimental approach to that used to establish the mechanisms of compensation for dietary protein in locusts by Abisgold and Simpson[2,3] should prove valuable. Preliminary studies[219] indicate that the responsiveness of mouthpart taste receptors varies in both insects, with receptors of P-conditioned larvae being more responsive to stimulation with sucrose than are receptors of larvae prefed the C diet. The responsiveness to various amino acids follows the opposite pattern. Such results suggest that nutrient-specific changes in peripheral sensitivity could play a role.

Geissler and Rollo[95] have recently tailored the design of the *Locusta-Spodoptera* study to the cockroach, *Periplaneta americana,* with similar results. Adult cockroaches were deprived of food for 2 weeks (in order to deplete fat body stores) then fed one of four diets, PC, P, C, and O, for 2 days. They were then provided with a choice of all four diets for 24 h. Insects conditioned on PC or P diets ate mainly PC and C; those conditioned on C diet ate mainly PC and P; and those prefed O diet ate PC, P, and C in approximately equal amounts.

A study of protein deprivation in adult *Phormia regina* by Belzer[16] showed that both male and female flies increase consumption of a protein diet after having been reared on only sugar for 4 days following emergence. If given *ad libitum* access to both sugar and protein, flies of both sexes normally exhibit a small peak of protein feeding during the first 2 days after emergence. Those flies deprived of protein for 4 days made up this deficit in only 5 h once given access to protein. The mechanisms of protein regulation in relation to vitellogenesis in blowflies are discussed further in Section III.C.

Some elegant work showing compensatory feeding for previous deprivation of vitamin A has been carried out by Dicke and colleagues[85] on predatory mites, but such experiments fall outside the scope of this review.

Studies such as these introduce the next of the major compensatory responses: dietary selection behavior.

B. COMPENSATING BY DIETARY SELECTION

Most, if not all, phytophagous insects are faced with nutritional heterogeneity in their food. The quantity and quality of available nutrients vary both temporally and spatially within individual plants, as well as between plants of the same species and between species.[78,142,198,227] Such variety provides the possibility that an insect can achieve a nutritionally appropriate diet, without having to face the problems associated with altering consumption rate, by selecting from the "cafeteria" available to it. This could be achieved by selecting the food which is closest to being nutritionally optimal, or, if no such single food is available, by selecting a mixture.

In order to demonstrate that an insect is compensating by selective feeding requires evidence, first, that the balance and quantity of nutrients which are ingested meet the current demands of the insect better than that provided by eating the same sources of nutrients randomly or in any other combination[249,250] and, secondly, that the behavior is directly modulated by the insect's nutritional state. Selection is not a compensatory response sensu stricto unless both are true, since the first proviso can be met, without any influence of current nutritional state, by an insect responding in a predetermined way to characteristics of the food or of the environment. For instance, an insect (hypothetical) may have evolved a diurnal locomotory rhythm which results in its feeding on foliage at the top of the plant during the day and at the bottom of the plant during the night. The balance of nutrients eaten as a consequence may be the best possible for growth and development, but in no proximate sense has the animal shown compensatory feeding. Similarly, an insect would not be exhibiting compensatory selection behavior if it had evolved to follow a simple functional rule independently of its nutritional state, such as negative frequency-dependent selection.[54]

1. Studies Using Natural Foods

There have been numerous reports of individual phytophagous insects feeding on a variety of plants and plant parts, or even showing occasional carnivory.[4,5,142,149] Some studies have demonstrated that insects allowed to feed selectively survive and grow better than those which are restricted to a single food,[11,12,109,138,169,249] but in no case has both the above requirements for the demonstration of compensatory selection been met in a study using real plants rather than artificial diets.

Waldbauer and Bhattacharya[248] reared larvae of the confused flour beetle, *Tribolium confusum,* either on one of three components of wheat kernal (bran, starchy endosperm, germ) or on a mix of all three (each finely ground) from which they could select. Larvae survived, utilized food and grew far better on the mixed diet than on the single components. From the mix they selected a 1:13:60 ratio of bran, starchy endosperm, and germ. While highly suggestive, these results do not provide evidence for compensatory selection. Although it was clear that each component alone was nutritionally suboptimal, the ratio selected by larvae was not tested against a series of other possible ratios, something which Waldbauer and his colleagues have subsequently done in studies on caterpillars and cockroaches (see below). Nor was it demonstrated that insects were responding to post-ingestive nutritional cues, rather than responding in a predetermined way to physical or chemical properties of the three components.

Cohen et al.[66] found that final instar larvae of the corn earworm, *Heliothis zea,* fed preferentially on the kernels, rather than husks, silk, or cob from maize ears. The germ of the kernel was preferred over the endosperm in a choice test. Larvae forced to enter a kernel from the endosperm end tunneled through the endosperm until the germ was reached. If larvae were able to enter the kernel from the germ end, however, they ate little endosperm. Cohen et al. suggested that "It seems that if a larva eats the germ when it first encounters a kernel, it received nutritional feedback that affects its feeding behavior, causing it to move on to another kernel before it has eaten a large proportion of the endosperm. Conversely, a larva that first encounters endosperm does not receive this feedback and often tends to continue eating — usually until it does encounter the germ. Only then is it likely to move on to another kernel." Larvae confined to germ alone, although not differing from larvae restricted to endosperm alone in growth rate, final size, or the efficiency at which they digested food, utilized digested nutrients more efficiently and ate less. Endosperm and germ differ in nutrient content, the germ containing more protein and oil but less carbohydrate. In an attempt to separate nutritional from physical properties, lyophilized germ and endosperm were incorporated into artificial diets and given to larvae in a choice test. Preference was still exhibited for germ over endosperm, although the effect was not so marked as with the real structures.

2. Studies Using Artificial Diets

The experiments of Cohen et al.[66] complemented a series of detailed papers by Waldbauer, Cohen, Friedman and colleagues in which dietary selection was investigated in larval *H. zea* fed on artificial diets. These studies provide compelling evidence that compensatory selection behavior occurs in this insect. The first paper in the series[250] describes experiments in which newly emerged final instar larvae were placed in a container with two blocks of defined artificial diet. Controls were given two blocks of a nutritionally complete diet, while experimental larvae were given one which lacked casein (the only protein source) and another which lacked sucrose (the only source of digestible carbohydrate). The concentration of casein in the sucrose-deficient block was the same as that of sucrose in the casein-deficient block, and this was twice the concentration in which both nutrients were present in the control diet. Experimental larvae ate the casein:sucrose blocks in the ratio 80:20 when intake across the stadium was summed. Overall, they ate a similar dry weight of food to the control larvae (although at a lower relative consumption rate) but ingested 1.5 times as much protein and 0.4 times as much sugar as the controls. Larvae fed the deficient cubes did not perform quite as well as the controls, gaining less weight, having a longer stadium duration, and digesting food less efficiently.

Having found that larvae selected an 80:20 ratio of casein to sucrose, Waldbauer et al. mixed casein and sucrose together in various ratios in the diet, in order to determine whether the selected ratio was the most appropriate one for sustaining growth and development. Ratios of 100:0, 80:20, 50:50, and 20:80 were tested. It turned out that the diet containing the 80:20 ratio was indeed the best of the four. It promoted a similar growth rate and stadium duration to the 50:50 diet, but larvae ate significantly less food and utilized what they ate for growth more efficiently on the 80:20 diet. This suggested that the slightly lower performance by larvae in the selection experiment was a consequence of their having to select as compared to feed on a ready mixed diet. Even though the 50:50 diet was not optimal, the consequences on growth and development of increasing carbohydrate intake in order to gain sufficient protein were not as great as those of having to select an appropriate protein:sugar ratio by feeding from two deficient diets. The diet containing the more extreme 20:80 ratio resulted in the stadium being 35% longer, since larvae did not increase their consumption rate above that shown at the 50:50 ratio, perhaps because of the effects of the excess carbohydrate.

Cohen et al.[186] compared the behavior of larvae provided with two blocks of diet, one lacking casein and the other sucrose, with that of controls given two complete blocks. Time-lapse filming was carried out over the whole stadium and the frequency at which larvae switched between cubes of diet was determined. Control and selecting larvae showed a high frequency of switching during the initial part of the stadium, before settling down to more continuous periods without switching. This unsettled phase lasted about 1 h in control larvae and 4 h in the experimental insects. Once settled, control larvae tended to stay on the same block, while selecting insects swapped blocks frequently and, over the whole stadium, spent 17% of their time on the sucrose block and 83% on the casein block. This suggests that the amount of feeding during periods on the two blocks was similar, since the ratio of 83:17 is in close agreement to the relative amounts of the diets eaten.[250] The ratio was not constant throughout the stadium, however, with sucrose feeding occurring more toward the end as a result of longer feeding bouts. Protein intake remained relatively constant during the stadium.

Schiff et al.[186] extended the study of selection in larval *H. zea* to include variations in dietary lipids and vitamins. Insects were provided with two blocks of diet each containing an 80:20 ratio of casein and sucrose. The blocks differed in that one contained lipids (corn oil) but no vitamins and the other contained vitamins but no lipids. In another experiment both blocks had lipids, but one had all vitamins except choline chloride and the other had only choline chloride. Neither of these deficient diets alone was able to support normal growth and development. Whereas control larvae given two similar, nutritionally complete diet blocks ate almost exclusively from only one, selecting larvae fed from both diet blocks. As a result they gained enough lipids and vitamins to complete their development. As in the previous work, selecting larvae gained less weight, grew more slowly and utilized food less efficiently than did the controls.

While experiments with larval *H. zea* supported the notion that insects are able to select an appropriate balance of nutrients, work by Cohen et al.[65] on the cockroach, *Supella longipalpa,* gave conflicting results. Cockroaches were subjected for their entire larval life to essentially the same procedure as that used by Waldbauer et al.[250] Nymphs given a choice of casein and sucrose diets ate them in the ratio of 16:84, with this ratio remaining relatively constant between instars. Selecting insects ate less, digested and assimilated food more efficiently, and grew at the same rate as controls. They ingested 0.3 times as much protein as did control nymphs and 1.5 times as much carbohydrate. All in all, the cockroaches did exactly the opposite of the caterpillars. In view of the differences between the insects in developmental biology, this is not unexpected. However, when nymphs were provided with a diet containing 20:80 casein to sucrose, close to the ratio they selected, only 5% survived to adulthood. In fact, of the various ready-mixed diets tested, only that containing a 50:50 ratio supported growth and development. In other words, the ratio selected by the nymphs was nothing like nutritionally optimal. Cohen et al. suggested that this was because the 20:80 ratio, although representing the mean ratio chosen throughout nymphal development, did not reflect changes occurring within each stadium. They showed that the amount of the sucrose diet selected varied throughout the first and final (and presumably the other) larval stadia, being highest at the beginning and then falling. Protein intake remained much more constant. Hence, the 20:80 diet was ideal early in a stadium but contained too much carbohydrate for later needs. The 50:50 diet provided a more suitable compromise. This explanation is not supported by the data from *H. zea,* however, since the caterpillars exhibited an equally marked variation in the ratio of casein to sucrose ingested throughout a stadium, albeit in the opposite direction, yet did perform best on a diet mixed to the selected ratio.[67]

The implication of the explanation proposed by Cohen et al. to explain the cockroach results is that the insects were regulating their carbohydrate intake and that they actually needed all the sucrose they ingested when selecting. This was questioned by Simpson et al.[217] who pointed out that a mobile, opportunistic omnivore kept in a small container with ample food might be expected to eat large amounts of the highly phagostimulatory sucrose diet. This might be especially true if post-ingestive inhibition is less following meals of sucrose diet than after feeding on the casein diet, due to the absence of amino acid and osmotic feedbacks (as occurs in locusts; see Section II.A.2.b). Simpson et al. found that larvae of another lepidopteran, *S. littoralis,* selected protein and carbohydrate in the ratio 67:33, while another orthopteroid, *L. migratoria,* chose 54:46. There is no doubt that lepidopteran and orthopteroid larvae have different nutritional requirements,[23] but this does not mean that the patterns of selection exhibited are directly controlled by nutritional feedbacks. Because of their greater need for carbohydrates, exopterygotes may have evolved such that sugars are much more stimulating than they are to endopterygotes. This could result in entirely different selection patterns without any nutritional feedback being involved.

There are two ways to establish beyond doubt whether or not an insect is selecting different foods as a result of nutritional feedbacks: first, by experimentally altering the nutritional state of the insect and seeing whether subsequent selection behavior is altered, and secondly, by altering the concentrations of nutrients in the selected diets and seeing whether the consumption ratio changes to effect compensation. The experiments of Simpson et al.,[217] which showed that short periods of conditioning on nutritionally deficient diets profoundly influenced subsequent choice behavior in *Spodoptera* and *Locusta,* provide the only evidence of the first sort for the existence of nutritional feedbacks (see Section II.A.5).

The key experiment to demonstrate beyond doubt that the basic patterns of selection found for caterpillars, cockroaches, and locusts are determined by nutritional feedbacks would be to alter the concentration of protein and carbohydrate in the two diets provided in the choice test. This was done by Simpson et al.[217] When larval *Spodoptera* and *Locusta* were given one diet with 20% protein and no digestible carbohydrate, and another with 10% carbohydrate but no protein, they ate them in the ratio 50:50 and 30:70 by total dry weight, respectively. This meant that the insects had ingested protein and digestible carbohydrate in the ratios 67:33 and 46:54. When larvae were given diets with 30% protein or 30% carbohydrate, the ratio eaten was 62:38 in the caterpillar and 43:57 in the locust — very close to what would be expected if the insects were regulating the balance of nutrients eaten, rather than just responding in a predetermined manner to the phagostimulants in the diets.

3. The Underlying Mechanisms of Compensatory Selection
a. The Initial Choice of Food

When an insect is provided with a variety of foods of similar physical properties it will choose the one which provides greatest phagostimulation, and feed less on those containing relatively lower levels of phagostimulants or higher levels of deterrents.[29] The phagostimulatory power of a food depends on its chemical composition, the firing rate of the insect's gustatory receptors, and the response of central nervous pathways to the input from these taste receptors. The first of these is probably out of the insect's immediate control (although rapid feeding-induced changes in plant nutrient levels have been reported[78,149]), while the other two can vary with the insect's physiological state. Because many nutrients as well as key secondary plant substances are phagostimulatory, the food selected is often nutritionally suitable. This is not always the case, however, and the initially chosen food, although highly phagostimulatory, may be nutritionally unsuitable. Compensatory selection behavior requires that the insect somehow "recognizes" a deficiency, ceases to respond to phagostimulants in the initially chosen food, leaves and selects a different food source.

House[116,118-120] found that larvae of the fly, *Agria affinis,* initially selected food on the basis of the level of amino acids present, even if the diet was seriously imbalanced, but that continued exposure led to a reversal of preference so that diets were chosen which provided best growth and development. A diet containing 6% amino acids and 0.5% glucose was initially chosen over one with the nutritionally more suitable mix of 2.25% amino acids and 0.5% glucose. After 4 days on the diets the preference was reversed, even though there was no evidence of a change in nutritional requirements of the larvae with development over this time. House[119] suggested that " . . . the larvae simply get misled by *always* seeking instinctively the richest level of amino acids (or protein), regardless of proportional relations with other nutrients — misled perhaps by the olfactory or gustatory stimuli of certain nutrients — only to be 'turned around' eventually by the dictates of normal nutrition . . . ." It is what constitutes this "turning around" that is at the heart of compensatory dietary selection.

b. The Ability to Recognize Nutritional Inadequacy

The ability of an insect to recognize a dietary deficiency in an initially acceptable food, and the time it takes to do so (the extent of "eventually" in House's statement) varies with the nature of the deficiency, and the species of insect. The experimental assay used also has some effect. For instance, if a *L. migratoria* nymph is left to feed on a diet lacking protein but containing all other necessary nutrients, it will continue to eat large amounts of food (more, in fact, than of a complete diet) up until about 8 h of exposure, after which intake falls rapidly.[217] However, it is apparent that the omission of protein was "recognized" much earlier, since a nymph will select a diet containing protein after having taken only a single meal of a protein deficient diet if given a choice.[218] Similarly, intake of a carbohydrate deficient diet continues at a relatively high level for more than 12 h, whereas selection can be exhibited within 4 h. This is not the case for larval *S. littoralis,* however, which reduce consumption of diets lacking either protein or carbohydrate after less than 4 h of exposure.[217]

Since most studies have not followed the detailed time-course of changes in feeding behavior with exposure to an inadequate diet, it is impossible to determine whether reduction in feeding occurs as a consequence of lack of phagostimulants in the food or to the post-ingestive consequences of eating the food. It might be expected that reduced consumption due to a dearth of phagostimulants should be apparent during the very first meal on a deficient diet, whereas post-ingestive effects would take longer to become apparent. The two are not quite so easily distinguished, however, as digestion products enter the blood within the course of a meal.[2,65] Nevertheless, detailed studies of behavior on diets lacking specific nutrients would undoubtedly prove enlightening.

Zucoloto[254] found that young larvae of the fruitfly, *Ceratitis capitata,* could recognize the deficiency of yeast, honey, or corn meal in an artificial diet when tested in a choice experiment against the complete diet. Selection occurred at some unknown stage during a 24-h exposure period, and could simply have been due to a reduction in phago-stimulatory power of the deficient diets rather than any post-ingestive consequences of eating them. Mittler[151] showed that larvae of the aphid, *Myzus persicae*, reduced their uptake over 2 days of exposure was shown if the diet lacked any other of the 17 amino acids in the diet. The effect of omitting methionine was fully apparent during the first 24 h, whereas removing histidine caused reduced consumption only on the second day. Other experiments showed that methionine is a powerful phagostimulant, so the decline in feeding might have been due to lack of phagostimulation. This would not explain the response to histidine, however. Another possibility was that the continued lack of histidine resulted in an inhibition of growth, which in turn caused reduced feeding relative to larger control insects. However, no reduction in feeding was found when isoleucine was removed, despite the fact that its omission causes a similar effect on larval growth to lack of methionine or histidine. Perhaps, then, histidine levels in the insect were providing a source of feedback. Leckstein and Llewellyn[128] performed similar experiments on young presumptive alate larvae of *Aphis fabae* and found that intake was reduced over the first 4 days of larval life when the diet lacked alanine, histidine, methionine, proline, or serine, but no reduction occurred if cysteine, phenylalanine, or tyrosine was missing. However, when aphid size was taken into account, it appeared that removal of methionine and histidine resulted in increased consumption. These amino acids were the only ones of those tested which were essential for protein synthesis, as evidenced by reduced growth in their absence. It seems that the undernourished aphids were actually attempting compensation for the deficiency by increasing consumption. That relative consumption rate but not growth was reduced by removal of alanine, proline, and serine was interpreted as indicating that these amino acids were acting as phagostimulants, not as sources of post-ingestive feedback. When aphids were provided with a choice of a diet deficient in a single amino acid and the complete diet, only for diets lacking methionine and proline was there any evidence of selective feeding on the complete diet. Mittler and Dadd[152] demonstrated that aphids select a diet containing a mix of various amino acids with sucrose to sucrose alone, but this is explicable in terms of differences in levels of phagostimulants. Dietary selection was also studied in the aphids *Acyrthosiphon pisum* and *Macrosiphon euphorbiae* by Cartier.[53] He allowed aphids access to six sachets of artificial diet, representing a graded series of sucrose, amino acids or pH, with all other constituents held constant. After $2^{1}/_{2}$ h, aphids had settled preferentially on diets containing 30 or 35% sucrose, 3.9 or 4.3% amino acids, and a pH of 7.0 or 7.3. These ranges were those which supported growth most successfully in longer trials. Once again, this does not prove that post-ingestive nutritional feedbacks are involved. Obviously more work is needed to clarify aphid compensatory selection, if it occurs.

The question arises as to how ingesting a nutritionally inadequate diet results in decreased consumption and a tendency to leave and feed on other foods. Of course, decreased consumption may occur because the animal becomes malnourished and ill, but this would not account for the subsequent increase in consumption that occurs when a more appropriate food is located. There seem to be two likely explanations. The first involves learning, while the second entails direct feedbacks.

c. The Role of Learning

There is a small amount of evidence for phytophagous insects that both nonassociative and associative learning can be involved in "recognizing" the nutritional inadequacy of a food that is initially acceptable. The best known nonassociative response in vertebrates is an alteration in the strength of the innate aversion which exists to novel foods (neophobia). This has been studied in adult male *Periplaneta americana* by Geissler and Rollo.[94] They added a drop of either lemon or kirsch flavoring to a nutritionally complete artificial diet and left the cockroaches with ample food for 3 weeks. After this time insects were provided with a choice of the diet with the familiar flavoring and the same diet with the unfamiliar flavoring; 91% of the feeding time was spent on the familiar flavored diet. However, if the same protocol was followed but with an agar and sucrose diet, rather than the nutritionally complete mix, 70% of the time spent feeding was on the diet with the novel flavor. The mechanisms of this response are unknown. If it really does involve nonassociative changes it means that the novel flavor was more phagostimulatory (or less deterrent) after feeding on the deficient than the complete diet. Alternatively (or additionally), the cockroaches may have developed a learned aversion to the flavor in the familiar but deficient diet. In the first part of the experiment, where a flavor was paired with the nutritionally complete food, the insects may have spent more time feeding on the diet with the familiar flavor because they had developed a learned preference for the flavor. Detailed studies of feeding behavior are required to distinguish these, especially of the behavior associated with the sampling of food before ingestion.[37,38,41]

Vertebrates are known to associate sensory stimuli provided by a food, especially taste and smell, with the nutritional consequences of ingesting it. The best documeted response is aversion learning, where a nutritionally inadequate food is avoided, although learned selection in favor of a nutritionally adequate food is also known.[9] At the time of writing this review, the only study on a phytophagous insect which unequivocally demonstrates food aversion learning is that of Bernays and Lee[27] on the polyphagous grasshopper, *Schistocerca americana*. Unlike other studies,[80,83] nonassociative learning and direct feedbacks were controlled for by separating the unconditioned, aversive stimulus from the food. Instead of feeding insects on a food plant which caused sickness and then testing the response to a new food, Bernays and Lee used an artificial aversive stimulus, an injection of nicotine hydrogen tartrate. Interestingly, aversion developed when the injection was paired with feeding on a marginally acceptable food plant (spinach) but not if the plant was fully acceptable (broccoli); perhaps there is a link here with the results of Geissler and Rollo, described above. Lee and Bernays[129] performed another series of experiments on *S. americana* in which it was found that insects left with spinach took progressively shorter meals until the food was completely rejected. This change in acceptability with experience did not appear to involve post-ingestive influences on the chemosensory system, since placing spinach-filled capsules in the gut of naive insects did not produce the effect. Presumably, then, the insects had associated the aversive consequences of eating spinach with prior experience of its sensory properties. Further studies are necessary to demonstrate whether aversion-learning also occurs where the unconditioned stimulus is not acute toxicity-induced sickness, but the consequence of eating a nutritionally deficient or imbalanced food.

A major problem in proving that aversion learning is involved in reduced consumption of a nutritionally unsuitable food is that of establishing what the unconditioned stimulus actually is.[76] As long as an insect is unable to compensate by altering consumption rate or utilization efficiencies, feeding on a nutritionally inadequate diet will eventually result in some sort of physiological stress, sickness, or "malaise".[65] This, whatever it is, could provide the unconditioned stimulus. One theory which is prevalent in the literature on vertebrates is that imbalances in brain neurotransmitters and neuromodulators may be involved[252] (but see next page). Recently, Cohen et al.[64] have shown that levels of 5-HT (serotonin) in the brain of *H. zea* larvae vary with diet. Insects were given a diet with casein but no digestible carbohydrate for 48 h, after which the food was replaced with either more of the same, or else a diet lacking protein but containing sucrose. Brain serotonin levels after 2 h were highest in larvae given the sucrose diets, while larvae fed the protein diet did not differ from control insects which were fed throughout on the "optimal" mix of 80:20 protein to carbohydrate. Sucrose fed larvae did not differ from insects which had been deprived of food altogether throughout the experiment. Artificially manipulating levels of brain serotonin by feeding larvae diets containing serotonin agonists and antagonists results in altered ratios of carbohydrate to protein ingestion in selection tests. These results are suggestive and warrant further investigation. Of course, such a mechanism could itself provide a direct control of selection behavior without requiring learning.

The other associatively learned response is that in which an animal develops a preference for the sensory properties of a particular food with experience of having eaten that food. Associatively learned preferences for odors and visual stimuli associated with food have been demonstrated in several insects, notably the honey bee, *Apis mellifera*.[166] There is only one example which shows that such an effect is due to the post-ingestive consequences of eating a nutritionally adequate food. Simpson and White[205] have described what students of vertebrate feeding would term a "learned specific appetite" or "learned hunger" for protein in nymphs of *Locusta migratoria*. Insects were trained for 2 days with free access to two artificial foods, which were similar in all respects except that one lacked digestible carbohydrate (the P diet) and the other lacked protein (the C diet) and each was paired with one of two distinctive secondary plant odors, carvone and linalool. The locusts were then fed for 4 h on either the P diet or the C diet, with neither odor present, after which time the insects were deficient in digestible carbohydrate or protein, respectively. The nymphs were then observed in a test chamber with the two odors present. Those insects deprived of protein showed a marked difference in their response to the two odors, approaching the source of the odor previously paired with the P diet significantly more often than the source of the odor previously paired with the C diet. Locusts deprived of digestible carbohydrate, on the other hand, did not show any discrimination between the two odors. Further experiments showed that the odors at the concentrations used were repellent to naive insects, and that the strong tendency, shown by protein-deprived nymphs, to approach the odor previously paired with the P diet was due to a reduction in the repellency of that odor, combined with maintained repellency of the odor previously paired with the C diet. Hence, locusts are able to associate an odor paired with a diet and the protein content of that diet, and the behavioral response shown to the odor depends specifically on the insect's state of protein deprivation. Such a response is likely, along with more direct feedbacks (see below), to explain the dietary selection behavior exhibited by locusts in earlier experiments.[202,217,218]

Further work is needed to show whether associatively learned preferences for nutritionally suitable foods and specific nutrients are widespread among phytophagous insects, and to establish the relationship between host-plant induction,[121] neophobia, and associative learning.[81]

d. Direct Feedbacks

There are several possible ways in which feeding on a nutritionally inadequate food could result in compensatory selection behavior without involving learning. Cohen et al.[64] suggest a simple, general mechanism whereby an insect continues to feed until perturbations in brain transmitter levels trigger locomotion and location of another food source. If the new food contains whatever it was that was lacking or in short supply in the previous diet, then eating it will rectify the neurotransmitter perturbation. The insect will continue feeding until another nutrient becomes limiting, causing new transmitter imbalances, and renewed locomotion. It is well known in invertebrates that 5-HT and related compounds are involved in behavioral arousal, sensitizing neural pathways by acting both pre- and post-synaptically.[131,162] It is also known that levels of another biogenic amine, octopamine, increase in the blood with food deprivation in locusts, as well as with other arousing or stressful situations.[72,73] Perhaps, then, feeding on a nutritionally inadequate food is the same in its effects as being deprived of food altogether. It is important to establish whether changes in levels of neuroactive compounds cause decreased feeding and increased locomotion or are simply a result of the insect having already stopped eating the deficient diet. The data of Cohen et al.[64] suggest a causal role, since larvae fed the diet containing sucrose but lacking casein had elevated levels of 5-HT yet exhibited higher consumption rates than control larvae.

A mechanism involving altered levels of transmitters could be general for all nutrient deficiencies, although it is also possible that intake of particular nutrients is regulated via specific brain transmitters.[64] How this might be mediated is unknown. It could be that deficiencies of specific nutrients affect the synthesis of particular neurotransmitters directly, as a result of shortages in precursors, for example. Alternatively, levels of nutrient reserves might be monitored by internal receptors and input from these to the CNS could cause changes in transmitter titers. It must be said, however, that the standing of the neurotransmitter hypothesis as an explanation of dietary selection in the vertebrate literature is far from firm. One of its founders, Fernstrom, has recently stated that the "current body of pharmacologic data connecting brain 5-HT release to the appetite for specific macronutrients is far from convincing,"[92] while Booth[42] goes as far as to say that it is "not even a scientific hypothesis."

The work of Abisgold and Simpson[2,3] (see Section I.A.2.b) provides another possible explanation for compensatory selection behavior involving direct feedbacks. As an insect feeds on a deficient diet, levels of limiting nutrients in storage tissues such as the blood and fat body will fall, while levels of those nutrients present in excess will rise. Blood concentrations determine how long the insect remains quiescent between meals and also selectively influence the sensitivity of taste receptors on the mouthparts. The phagostimulatory power of the diet will fall as the receptors responding to the nutrients present in excess become less sensitive. The insect consequently remains quiescent for a shorter time after feeding, rejects the deficient food more often on contact, and spends a greater proportion of its time locomoting. If the insect happens upon a food containing higher levels of the previously limiting nutrients it will accept the food and take a large meal, since the responsiveness of the taste receptors to the limiting nutrients will be high. So far this scenario could at least partly explain selective feeding for protein and perhaps digestible carbohydrate in *L. migratoria* nymphs, and probably also larvae of *S. littoralis.*[217]

It is clear that the direct control of specific nutrient intake is limited by the monitoring abilities of the insect. There are too many essential nutrients for them all to be regulated independently using specific peripheral receptor or central mechanisms. Clearly, the only way to effect compensatory feeding for many of the micronutrients would be via some general "malaise" mechanism, whether that be mediated via learned aversions or preferences or more directly, as proposed by Cohen et al. Work on vertebrates, notably the white rat, has demonstrated that certain "appetites", such as that for sodium, are directly and specifically controlled, while others such as for protein and thiamine are learned.

C. POST-INGESTIVE COMPENSATION

Intimately linked to feeding behavior are the post-ingestive consequences of eating foods varying in nutritional quality. The question arises as to whether an insect is able to regulate the efficiency with which it digests, absorbs, and incorporates absorbed nutrients into growth as part of a compensatory response to altered dietary nutrients, or whether changes in these parameters occur passively with the nature of the food and the pattern of feeding behavior. It is well known that efficiencies vary with food consumption, which in turn varies with the nutritional quality of the food. For instance, there are numerous reports of plant nitrogen levels correlating negatively with consumption rate, which in turn correlates negatively with the overall efficiency at which ingested food is utilized for growth.[149,198,227] The problem is how to decide whether a compensatory increase in consumption causes a decline in utilization efficiency, or whether a compensatory increase in utilization efficiency causes a decline in feeding.

As discussed in previous sections, it seems that blood composition (nutrient concentrations, osmolality, and hormone titers) plays an important role in determining feeding behavior. Blood nutrient composition is a function of the quantity and accessibility to digestion of nutrients in the food, the efficiency at which digested nutrients are absorbed, and the rate at which nutrients are removed from the blood for growth, respiration, storage, or excretion. The approximate indices used in most studies to measure the efficiencies of digestion and absorption of ingested food (AD), incorporation of nutrients absorbed from the gut into growth (ECD), and overall conversion of the food eaten into growth (ECI) are those derived from gravimetric balance sheets.[247] Since these are usually calculated simply on a total dry weight basis, it is likely that the inverse correlation found between consumption rates on plants varying in nitrogen content and ECI often does not involve any compensatory changes in efficiencies but just reflects constraints on digestion imposed by the food. For example, when an artificial diet is diluted twofold with cellulose, larvae of *Manduca sexta* increase their consumption rate and total dry weight ECI falls. The latter is simply a result of the inevitable decline in AD which accompanies having a higher proportion of indigestible material in the food.[228] Diluting the food by the same amount with water rather than cellulose does not affect dry weight ECI. Instances in which ECI declines with lowered plant nitrogen are mainly due to a decreased ECD rather than AD,[149] and may result from all manner of metabolic pressures consequent upon eating those low-nitrogen plants, or result from the increased consumption rate per se (see Section IV).

It is possible that compensation by altered efficiencies is occurring when AD and ECD are found, as they often are, to vary inversely when a species feeds on different plants.[198,227] For example, Volney et al.[246] found that Californian oakworms, *Phryganidia californica,* "compensated" for the reduced digestibility of old vs. young foliage by increasing ECD but not altering consumption rate. Although suggestive, there is still the possibility of other confounding effects due to uncontrollable differences between the two types of foliage. AD and ECD have also been found to vary inversely on the same food under different regimes of deprivation, but these results are confounded by differences in consumption rate during feeding periods.[102,103,193] Additionally, there is a methodological problem to be considered when interpreting the relationship between AD and ECD. The measurement of dry weight consumption in most studies involves providing insects with a known fresh weight of food and removing uneaten food at the end of the experiment. The initial dry weight of food provided is estimated from sample aliquots which are weighed fresh then dried to constant weight, while the dry weight of food left uneaten is measured directly. If the water content of the diet is underestimated, even by only a fraction of a percent (which is almost unavoidable), then dry weight consumption will be overestimated, AD will be overestimated, and ECD will be underestimated, as will ECI. The magnitude of the overestimate of consumption, and the consequent effects on the utilization indices, will depend on the proportion of the quantity of diet provided which is eaten by the insect, being greater the smaller the proportion eaten.[187] Any experimental treatment which causes one group of insects to eat a greater proportion of their food than another will result in an artefactual variation in utilization indices, with groups eating a greater fraction of their food apparently having a lower AD and a higher ECD. If the water content of the diet is overestimated the trends will be in the opposite direction. Since gravimetric balance sheets seldom tally precisely,[243] such spurious relationships are almost certainly common.

More convincing evidence of "active" post-ingestive compensation comes from the few studies in which budgets have been based on particular nutrient groups rather than total dry weights. Work such as that of Slansky and Feeny[224] on larval *Pieris rapae* and Ohmart et al.[163] on larvae of the eucalypt feeding chrysomelid, *Paropsis atomaria,* shows that the ECI for nitrogen (usually termed the nitrogen utilization efficiency, NUE) increases with increased consumption of lower nitrogen food (unlike total dry weight ECI which decreases), although at low levels of plant nitrogen the reverse may be true[104,246] Unfortunately, these studies do not report AD or ECD for nitrogen, and there is again the problem of possible confounding effects from uncontrolled factors correlated with plant nitrogen. The latter were overcome to an extent by Schroeder[194] who coated leaves of *Tilia americana* with a protein mix and fed them to larval *Datana ministra.* Larvae given coated leaves ingested less and had a lower ECI for nitrogen than did insects fed uncoated leaves. Lii et al.[132] fed larvae of the lepidopteran, *Argyrotaenia velutinana,* from hatching until 18 days old on one of several isocaloric artificial diets containing varying amounts of a balanced amino acid mix. Consumption rate was low at amino acid concentrations of 2.4 or 3% by dry weight, maximal at 3.6%, and declined at higher concentrations (4.8 and 5.4%). The response to concentrations greater than or equal to 3.6% probably reflects compensatory feeding, while concentrations of less than 3.6% were too low to sustain increased feeding. ECI for nitrogen was highest at 2.4% amino acids in the food and declined with increased concentration, suggesting a post-ingestive compensatory response, with amino acids excess to requirements at higher dietary concentrations either not being digested and absorbed as efficiently as at low concentrations, or else being excreted in a greater proportion once absorbed.

The only detailed study published to date investigating variation in utilization efficiencies for specific amino acids is that of van Loon[243] on caterpillars, *Pieris brassicae* and *P. rapae.* He fed fifth-instar larvae either on artificial diets containing 22.5% casein, 15% casein, or 15% casein supplemented with 0.41% phenylalanine, or else on young or mature foliage of *Brassica oleracea.* Budgets were derived for 17 amino acids. Consumption was not estimated in the standard manner but by the technique of summing fecal production, dry weight growth, and a precise measure of respiration (derived from undisturbed larvae feeding normally in a specially designed flow-through respirometer[244]). There are still potential problems of amino acids being absorbed then excreted (thus causing inaccuracies in measurement of AD for those compounds), and nonessential amino acids being interconverted, but the results are most instructive. In both species the ADs for individual essential amino acids as well as tyrosine varied inversely with food consumption, whereas the relationship was either absent or even positive for other amino acids. The ECD for total amino acids bore an inverse relationship with concentration in the diet, indicating increased oxidation of amino acids when dietary levels were high. This trend was exhibited for several essential amino acids. Values in excess of 100% for ECD were found for some nonessential amino acids, especially tyrosine, indicating net synthesis. Extensive conversion of phenylalanine to tyrosine apparently occurred. For several amino acids there was an inverse relationship between AD and ECD. In the case of nonessential amino acids it seems that low absorption was compensated for by net synthesis. Such results strongly imply that levels of individual amino acids are regulated independently, and that feedbacks between the mechanisms of digestion, absorption, and incorporation of nutrients to growth do occur. Other work in which apparently limiting amino acids have been supplemented in the diets of phytophagous insects and resulted in improved conversion efficiencies suggest that this is a general phenomenon. Horie and Watanabe[111] supplemented diets containing either the proteins zein or gluten with their respective limiting amino acids and caused a marked increase in growth, a reduction in uric acid production, and an increase in blood protein levels in *Bombyx mori* caterpillars. Horie and Nakasone[110] also found that levels of fatty acids and carbohydrates in the diet of silkworms affected the rate of fatty acid synthesis. Bernays and Woodhead[31] increased dry weight ECI in larval *Schistocerca gregaria* by providing extra phenylalanine in the diet.

The physiological mechanisms by which an insect might specifically regulate digestion, absorption, and metabolism of nutrients are not well known. Much more needs to be understood about the regulation of digestive enzyme secretion, absorption processes (both active and passive), and the control of nutrient metabolism.[6,15,56,57,86,155,156,231,242] Broadway and Duffey[48,49] showed that trypsin activity in the gut of larval *H. zea* and *Spodoptera exigua* was higher when the artificial diet contained more protein, and that this effect was apparently controlled by a secretogogue mechanism rather than a stretch-induced hormonal one, since the volume of food ingested was inversely related to dietary protein levels. The effect of dietary protein on enzyme secretion was more marked in *S. exigua* than in *H. zea,* particularly when tryptic activity was measured in insects after chronic exposure to the different diets. Interestingly, it seems from data on fecal production that behavioral compensation was more pronounced with changing dietary protein in the latter species.

Selective digestion of nutrients presumably requires a secretogogue control mechanism, since it is hard to imagine how a stretch-induced hormonal control could provide independent regulation of the different digestive enzyme systems. It is difficult to see, however, how a secretogogue mechanism could enable larvae feeding on food with a low protein content to increase nitrogen utilization efficiency as a compensatory response (see above).

Carlisle et al.[52] isolated a low molecular weight fraction in the hemolymph which appeared within 5 min after feeding in *L. migratoria* and caused an increase in protein synthesis. What it is about feeding which elicits release of the factor is not known. Tryptophan or isoleucine injection into the hemolymph did not stimulate its release, although it would be particularly interesting to see whether an injection of the mix of eight amino acids implicated in compensatory protein feeding in this species by Abisgold and Simpson[1,3,205] is effective.

III. COMPENSATING FOR CHANGES IN NUTRITIONAL REQUIREMENTS

Nutritional requirements vary throughout the life of an insect with, for instance, the demands of growth and development, reproduction, diapause, and migration.[71,227] Additionally, parasites and symbiotes may modify the nutritional status of an insect, while the abiotic environment imposes further changes.[62,109,113,223,227,238,245] Alteration in feeding behavior and utilization efficiencies accompany variations in nutritional requirements and are usually considered to be compensatory responses, although, strictly speaking, this is only the case, in the proximate sense, if it can be demonstrated that they are caused by the change in requirements rather than vice versa. Unfortunately, there are usually insufficient data to enable this distinction between cause and effect to be made. It is likely, however, that many of the same mechanisms involved in compensating for changes in the nutritional quality of the food are involved in the responses exhibited to changing requirements of the insect. In this section we will consider the evidence for compensation accompanying growth within and between larval stadia and with reproductive development. These are the only areas in which there is sufficient published data to allow more than an elaboration of the excellent review by Slansky and Scriber.[227]

A. COMPENSATING FOR CHANGES WITHIN A LARVAL STADIUM

During each stadium the larva must grow large enough to moult to the next stage and acquire sufficient stores of nutrients to survive the non-feeding period which precedes and follows ecdysis. The process of ecdysis and the production of new cuticle represent considerable outlays of energy and nutrients, although some savings are made by resorption of the inner layers and, perhaps, consumption of the outer layers once shed. Many insects eat increasing amounts up to about mid-stadium or apolysis, after which feeding declines.[29,198,247] Accompanying this is an increase in the mass of the insect, with specific tissues making up different proportions of growth on the different days within a stadium. Hence, changes in requirements during a stadium are both quantitative (larger insects need more food), and qualitative (demands for the relative proportions of nutrients may alter).

The behavioral mechanisms by which quantitative increases in nutrient demands are satisfied are best understood in fifth-instar *L. migratoria* nymphs and in fifth-instar *M. sexta* larvae. Reynolds et al.[180,181] recorded the feeding patterns of *Manduca* larvae fed either artificial diet or tobacco leaves. Caterpillars fed tobacco eat more on day 1 than on day 0 (when ecdysis occurs) as a result of taking meals of the same average size more frequently. The increment in consumption from day 1 to day 2 occurs because meals are both larger and more frequent, while the further rise during day 3 results solely from an increase in meal size. Insects fed artificial diet take more, larger meals on day 1 than during day 0 and subsequently only meal size increases. Differences in nutrient and water content and the physical properties of tobacco and artificial diet are likely to cause these somewhat different patterns of change in feeding parameters over the stadium. The tendency for caterpillars fed either of the two foods to eat meals of a greater size as they grow has been suggested to be a consequence of the gut volume increasing with age — more food being accommodated during a meal for the same degree of stretch of the gut wall. It was pointed out by Reynolds et al.[180] that as the gut grows the surface area to volume ratio will fall and the efficiency with which materials are digested and absorbed (AD) might be expected to decline. In fact, AD remains remarkably constant. This apparently results from the fact that food is retained in the gut for longer as consumption rises, and might represent compensation for the effects of increased size. The decline in relative growth rate which occurs during the stadium could be interpreted as evidence that this compensation is not complete. Although AD does not change, ECD (the efficiency at which nutrients absorbed from the gut are converted to body tissue) declines somewhat, possibly as a consequence of the insect spending an increasing proportion of its time feeding.

Simpson[207] observed an increase in food intake up until the 5th day of the 10-day-long fifth stadium in *L. migratoria,* and then a subsequent decrease until feeding stopped on, or around, the 8th day. When seedling wheat was the food source, the increase in consumption up to mid-stadium was primarily due to an increase in meal size.[208] This contrasted with results from *Locusta* nymphs fed mature *Agropyron* which increased intake by decreasing the time between meals.[39] The difference appears to be at least partly due to the physical properties of the two plants. The more fibrous mature grass does not pass back from the crop during a meal as readily as the softer seedling wheat. This provides less scope for increasing meal size on *Agropyron,* since stretch receptors on the crop provide inhibition of feeding.[210] In a subsequent study,[209] variations in the rate of crop emptying were found to occur after a meal on different days during the stadium, being slowest at the beginning and end of the stadium. Although the crop emptied more quickly during days 1 to 5 there was a secondary slowing on day 4. These changes in crop emptying enable nymphs fed seedling wheat to take larger meals up to mid-stadium without having to wait longer between them, and insects fed *Agropyron* to eat more often without having to take smaller meals. The variation in rate of crop emptying was shown to be independent of the size of the previous meal and is likely to be due to changes in blood composition. Around the middle of a stadium, hemolymph osmolality is low due to a rise in hemolymph volume and the rapid removal of nutrients for growth.[180,207]

Total dry weight ECI (the efficiency at which ingested food is converted to body tissues), as well as AD for protein and lipid, correlated negatively with the rate of crop emptying, being highest at the start and end of the stadium with a secondary peak at mid-stadium.[207] The AD for digestible carbohydrates was much higher and remained constant throughout the stadium, probably because these nutrients are digested and absorbed from the food much more readily than are proteins and lipids and, as a result, are less affected by variation in the retention time of food in the gut. Chapman,[58] working on *Schistocerca americana,* a grasshopper of comparable size, proposed another reason for the secondary peak in absorption of lipid and protein on day 4. Simpson[207] had reported that the dry weight of the gut increased up to mid-stadium then declined. Chapman showed that in *S. americana* a large proportion of the increase could be attributed to growth of the anterior arms of the midgut ceca, and the subsequent fall could be entirely due to their reduction. Since the anterior cecal arms are important sites of nutrient absorption, their size may well correlate with absorption efficiencies.

Changes in levels of digestive enzymes throughout a stadium have been reported in a few cases.[57,58] Khan[124] found changes in invertase activity during the fifth stadium in *L. migratoria* which followed the variations in consumption. However, Clarke and Gillott[61] did not find a similar pattern of activity for proteolytic enzymes in the same insect. Broadway and Duffey[48] showed that trypsin activity increased up to mid-stadium in fifth-instar larvae of *Spodoptera exigua* and *H. zea,* with the trend being more pronounced in the former (see also Section II.C).

That qualitative requirements vary during a stadium is suggested by the fact that different tissues grow on different days. For instance, in fifth-instar *L. migratoria* nymphs growth during the first half of the stadium is comprised mainly of an increase in protein and lipid, in the form of fat body, cuticle, and muscle, while later growth is largely an increase in body carbohydrates, especially glycogen stores in the fat body.[108,207] Horie and Watanabe[111] found that supplementation of artificial diets fed to final larval instar

B. mori with limiting amino acids had a more marked effect on growth of silk gland, fat body, and on hemolymph proteins than on growth of other tissues, suggesting that "quantitative requirements for essential amino acids vary according to the stage of tissue growth."

Whether insects alter their food preference during the course of a stadium in order to meet qualitative changes in nutritional requirements has been investigated in larval *L. migratoria* and *S. littoralis* by Simpson et al.,[217] and in larval *Supella longipalpa* and *H. zea* by Cohen et al.[65,67] In each study insects were given a choice between two artificial diets, one containing digestible carbohydrate but no protein, the other protein but no digestible carbohydrate. Selection patterns differed between the species. Locusts ate an equal quantity of protein and carbohydrate at the start and end of the fifth stadium but selected carbohydrate over protein at mid-stadium. *Spodoptera* larvae also chose equally at the start and end of the sixth stadium but ate more protein during other days. First and final instar cockroaches maintained a constant intake of protein during the stadium but ingested largest quantities of carbohydrate at the start, with intake subsequently falling. Final instar *Heliothis* larvae chose predominantly protein, with this choice being most marked during the first half of the stadium. As discussed in Section II.B.2, such selection patterns need not indicate that the insect is selectively feeding in response to changes in nutritional requirements, although experiments on *Locusta* and *Spodoptera* larvae suggest that this may be the case.[217] Work on adult *L. migratoria* by Chyb and Simpson[59] provides experimental evidence for compensatory selection during the somatic growth phase which follows adult ecdysis. Male and female locusts were provided with a choice of two diets, one containing 10, 20, or 28% carbohydrate but no protein, and the other 10, 20, or 28% protein but no digestible carbohydrate. Insects ate relatively more protein than carbohydrate diet during the first 6 days following ecdysis, after which this preference was reversed. During the first 4 days after ecdysis there was no evidence of compensatory feeding for the variation in nutritional levels in the two diets, but after this there was a strong compensatory response, with more being eaten of the more dilute diets. In other words, insects defended the ratio of protein to carbohydrate selected after the first 4 days, indicating that selection was in response to nutritional requirements. It was suggested that larval reserves obviated the need to compensate earlier in the stadium. The age-related change in preference for protein and carbohydrate corresponded with the pattern of protein and carbohydrate growth during somatic development. A parallel study by C. L. Simpson et al.,[204] using adult locusts fed on seedling wheat, found correlated changes in the relative responsiveness of chemoreceptors on the maxillary palps to stimulation with sucrose and amino acids, suggesting that selection behavior could have been modulated in part peripherally. Although levels of nutrients in the blood were not measured in this investigation, published data in studies by Hill and Goldsworthy[108] support the idea that levels of free amino acids in the hemolymph may influence sensilla responsiveness, as found for nymphs (Section II.A.2.b).

It is not known whether changes in the responsiveness of taste receptors contribute to changes in food selection during the stadium in any of the other cases described above. Schoonhoven[190] reported an increase in the responsiveness of the inositol-sensitive cell in the medial sensillum styloconicum of *M. sexta,* and also in the salicin-sensitive cell in the lateral sensillum, during the middle of the fifth stadium. In contrast, Clark[60] found that the changes in feeding rate over the final stadium of *S. littoralis* were not correlated with variations in the sensitivity of receptors of the sensillum styloconica to the four chemicals tested. Blaney and Simmonds,[36] however, working with the same species and instar, found that sensitivity to a 0.05 *M* sucrose solution (as compared to a 0.01 *M* solution as used by Clark) declines, and might therefore contribute to the above mentioned reduction in sucrose intake relative to protein over the first half of the stadium.[217] There are a number of possible explanations for these conflicting results. First, it may be that age-related reductions in sucrose sensitivity are evident only at higher concentrations, although Simmonds[200] believes this not to be the case. Secondly, there are a number of uncertainties about the methodology adopted by Clark which may affect the interpretation of the data. The author does not state when, in the course of feeding, individuals were selected for electrophysiology. This is critical in view of evidence that receptor sensitivity alters in this insect following a meal[192] (see Section II.A.4). Furthermore, it is not mentioned whether intact larvae or severed heads were used in the investigation. Blaney and Simmonds[36] have compared the responses obtained from each of these preparations and concluded that, in the intact animal, the initial firing rate on stimulation is higher, but that responses to mechanical stimulation form a significant part of this. Since Clark assessed the impulse frequency from the short interval 20 to 100 ms after the onset of stimulation, variations due to age may have been obscured. However, if age-related changes are mediated via alterations in blood composition,[3] it is perhaps better not to work with severed heads.

B. COMPENSATING FOR CHANGES BETWEEN LARVAL STADIA

Quantitative and qualitative changes in nutrient requirements also occur between stadia. The quantitative changes are self-evident. In the extreme case of some Lepidoptera, final instar larvae are up to 10,000 times heavier than newly emerged first-instar insects. The detailed behavioral mechanisms of changing feeding activity across stadia have not been reported for any insect. Generally speaking, however, it seems that the relative consumption rate actually declines as larvae progress through larval development, although this is by no means always the case.[227] Associated with changes in relative consumption rate are variations in both AD, which usually declines, and ECD, which usually rises.[227]

Again, it is not possible to distinguish cause and effect from most studies, although the basis of observed changes in consumption rate and utilization indices has been speculated upon by a number of workers.[7,10,125,126,144] One feature which seems common is that, relative to younger larvae, later instars possess a greater proportion of lipid and reserve materials. As a result, the proportion of metabolically active tissues falls in successive instars, leaving more nutrients available for conversion into biomass rather than to fuel metabolic needs. This could account for a rise in ECD as the larva develops.[227] Increases in body mass could lead to a reduction in AD as the surface area to volume ratio of the gut declines, unless accompanied by changes in digestive physiology and/or the relative sizes of different parts of the alimentary canal (see previous section). The net outcome of the interaction between AD and ECD might determine the pattern of change in relative consumption rate, although the reverse is also possible (see Section II.C).

It is not uncommon to find that the suitability of a plant or plant part depends on the age of the insect and relates to changes in selection behavior.[227,249] Differences in the foods selected between instars could indicate compensatory feeding for changes in qualitative nutritional requirements.[249] There are other possible explanations, however, including changing plant availability, as well as environmental and physical constraints on the insect.[227,249] Young instars may be unable to feed on the same food as older insects because their mouthparts are too small.[55] Equally, being small may enable an insect to feed more selectively, since the nutritional heterogeneity of the plant is greater than it is for bigger larvae, and also to chew food into smaller and more digestible fragments.

Studies using artificial diets are needed in order to separate nutritional from other confounding factors, but even then there are still likely to be problems in distinguishing cause from effect. For example, fourth-instar western spruce budworms, *Choristoneura occidentalis,* fed an artificial diet had a lower ECI than fifth- and sixth-instar larvae, and yet fourth and fifth instars had similar relative growth rates.[34] Blake and Wagner proposed that the higher relative consumption rate of younger larvae represented a compensatory response to low ECI. However, the possibility that variation in relative consumption caused rather than followed the trend in ECI still cannot be excluded.

In addition to general trends in nutritional requirements, some insects have specific needs during certain stadia. Delvi and Pandian,[77] in a comparative analysis of data from several species of Orthoptera and the silkworm, showed that fifth-instar larval *B. mori,* unlike the grasshoppers, increase their feeding rate considerably relative to earlier instars. This was related to the production of the pupal cocoon. AD remained constant despite the higher ingestion rate. Horie and Watanabe[111,112] found that about 45% of ingested and 70% of digested nitrogen was used to produce silk for the cocoon.

C. COMPENSATING FOR CHANGES ASSOCIATED WITH REPRODUCTION

Sex differences in food consumption and utilization are to be expected. Most attention has been focused on the requirements for ovarian development in adult females, although the nutritional and energetic needs of males for finding females, patrolling territories, spermatogenesis, production of accessory fluids, and spermatophores can be considerable.[227] Much depends on the particular reproductive strategy employed and whether or not the insect feeds as an adult.[227]

Sex differences may or may not be apparent in larvae or adults, depending upon differences in the magnitude and timing of the accumulation of reserves. Male and female larvae are likely to be most different where the adult does not feed.[227] The female is often the larger of the two sexes and may have a longer larval life, in terms of stadium duration or number, in order to achieve this greater mass. Differences in consumption indices and utilization efficiencies also occur between male and female larvae in some species, and are summarized in Slansky and Scriber's review.[227] In those cases where feeding is required as an adult in order to stimulate the initiation or maintenance of reproductive development there are also reports of variation in the consumption, selection, and utilization of food between the sexes, with stage in the reproductive cycle, and with mating status. Once again, these are discussed in some detail in the review of Slansky and Scriber and do not warrant further treatment here.

The only case in which the underlying mechanisms of compensatory feeding has been investigated is the control of the protein "hunger" which accompanies vitellogenesis in females of various species of blowflies. Female blowflies exhibit temporary periods of specific dietary selection during which they are attracted by odors emanating from decaying proteinaceous material and preferentially ingest protein over carbohydrate in a choice test.[29,79,213] They appear to recognize the taste of protein sources by responding to salts and certain amino acids.[14,50,172] Belzer,[16-20] working on the black blowfly, *Phormia regina,* proposed that the levels of some reserve material related to protein synthesis fall as vitellogenesis begins and so stimulate protein hunger. As protein is ingested, the eggs develop and the abdomen becomes distended. This then inhibits further protein feeding until oviposition occurs when the abdominal distension is removed, allowing renewed feeding. Because carbohydrate intake is regulated by a different combination of stretch receptor inputs, a gravid fly continues to ingest some sugar, despite having a distended abdomen.

Rachman[173] proposed that Belzer's "reserve material" was related to the carbohydrate stores of the fly. This conclusion was based on the finding that reducing carbohydrate reserves by starvation, flight, or prefeeding on a weak sugar solution inhibited protein feeding, while increasing the fly's carbohydrate reserves by prefeeding on a concentrated sugar solution promoted protein feeding. Rachman suggested that the drain on carbohydrate reserves associated with egg production causes the inhibition of protein feeding in gravid flies and the preference for sugar which they exhibit. She also maintained that the effects of other treatments such as allatectomy, removal of brain neurosecretory cells, chemosterilization, ovariectomy, and treatment with inhibitors of protein metabolism, all of which inhibit protein feeding, are explicable in terms of interference with carbohydrate metabolism.

While the experiments of Rachman indicate that carbohydrate reserves modulate the response to protein, they do not, in fact, prove that carbohydrate reserves cause protein "hunger". This is an important distinction and requires further investigation. That carbohydrate reserves should influence protein feeding is plainly sensible, since the fly cannot survive on protein alone and there is no point seeking out and ingesting protein when energy reserves are below a critical level. The separate control of protein from carbohydrate feeding also seems sensible. What is needed are dietary selection studies in which flies are offered a threeway choice of water, protein, and sugar, with the latter two being provided in one of several concentrations. This is an elaboration of the experimental design used by Roberts and Kitching[183] for the Australian sheep blowfly, *Lucilia cuprina*. If *ad libitum* intake is recorded over an extended period it should be possible to note, not only whether flies select (this is already well known), but whether they "defend" consumption of either nutrient by altering the amounts of the solutions eaten in inverse proportion to their concentration. If protein intake is dependent on the concentration of the solution, and this occurs independently of sugar intake, then there would be strong evidence for a separate control of protein hunger.

IV. CONSTRAINTS ON COMPENSATION

A combination of behavioral and post-ingestive compensation is often not sufficient to make up for a change in dietary quality or nutritional need, if indeed any compensatory response is exhibited at all. The efficiency of compensation exhibited depends on the nature and magnitude of dietary change, the species of insect, its physiological state, its previous dietary experience, and environmental conditions.[198,222,227] Some of this variability almost certainly reflects ecological constraints on compensatory abilities, for instance the increased risks of predation or parasitism associated with feeding for longer might outweight the deleterious effects on survivorship and reproductive success of not compensating completely.[227] There are no studies directly testing such theories, however. Other constraints on compensation are more obvious in terms of the physiology of the insect. These are generally discussed as the metabolic "costs" of compensation. In this review we have purposefully avoided the term, and have instead used the more general "consequences". The point is that not all of the physiological constraints on compensatory feeding need be metabolic costs. Some undoubtedly are, however.

Martin and Van't Hof[147] have proposed that metabolic costs should be classified as being either "processing costs" or "acquisition costs". The former is "a metabolic cost imposed by the operation of physiological and biochemical processes involved in extracting useful nutrients from ingested food, eliminating waste products generated during the processing of the food, and countering the potential harmful effects of toxins or other non-nutrient chemicals present in the food." The latter is "a cost associated with foraging or feeding." There are numerous possible examples of both sorts of metabolic cost associated with compensatory feeding and food utilization. These are listed by Slansky and Scriber,[227] Martin and Van't Hof,[147] and others, but there has been little work attempting to measure their relative effects. Recent advances in respirometry which allow oxygen consumption and metabolic rates to be measured in insects feeding under more normal conditions than those in a Gilsen flask ought to see an improvement in the quantification of metabolic costs.[244]

As mentioned above, not all physiological constraints need be metabolic costs. Three examples spring to mind. The commonly reported phenomenon whereby continued dilution of the entire nutrient complement of the diet results in incomplete compensation by increased feeding (see Section II.A.1) is just as likely to reflect the limitations of gustatory receptors as it is the metabolic costs of increased consumption, increased production of peritrophic membrane, or whatever. This is suggested by the fact that the degree of compensation exhibited by blowflies and cockroaches to dilution of a carbohydrae diet depends on the phagostimulatory power of the sugar used.[98,99] Since phagostimulatory input is essential for the initiation of feeding, an insect will not eat large amounts of a food that is so dilute that it cannot provide sufficient excitation to cross even the lowest of behavioral thresholds.

Another consequence of increased feeding is an increased rate of passage of each meal through the gut, although this need not necessarily be the case[2] (see Section II.A.2.b). This could lead to reduced digestive efficiencies[180,199] but studies on the efficiency at which individual nutrients are digested and absorbed are needed to establish whether this is a significant constraint. Slansky and Wheeler, in their study on the effect of dietary dilution on larval *Anticarsia gemmatalis*,[228] found that AD for total utilizable nutrients was hardly affected by increased passage of food through the gut of compensating larvae. They also found little evidence of increased metabolic costs associated with compensatory feeding.

The third, and completely overlooked consequence of compensatory feeding (although see comments in the Discussion section of Schroeder[194]), is the possibility that the control mechanisms regulating intake of different nutrient groups interfere with each other. It is usually considered that the poor performance of insects fed a nutritionally imbalanced diet is the result of their respiring or excreting nutrients which are excess to requirements in order to gain sufficient of those which are limiting. Although excess nutrients are eliminated in this way,[133,237] this does not mean that the metabolic costs involved are the primary factors limiting compensation and growth. In cases where compensation entails eating more protein than is needed it is more likely that high levels of free amino acids in the blood inhibit feeding and prevent the insect from compensating fully for other limiting nutrients. The increased metabolic costs of removing the unwanted amino acids from the blood may be secondary to the "jamming" of the feeding control system. Once again, more needs to be known about the mechanisms by which intake of specific nutrients is regulated before this idea can be tested.

Another constraint on compensation would appear to be the previous dietary experience of an insect. Food plant induction provides a behavioral impediment to compensatory selection by inhibiting the acceptance of alternative foods (see review by Jermy[221]), although, as stated earlier, it is not known how the strength of induction is influenced by nutritional quality of the food. There is some evidence that insects may also become "physiologically adapted" to a food plant and impaired in their ability to utilize alternative foods once ingested.[102,103,191,195,196,236,237] This is an interesting idea which requires detailed studies on artificial diets to elucidate the effect of nutrients, rather than allelochemicals and other factors, in causing the response. Care must also be taken to separate pre- from post-ingestive effects.[103]

A further issue which has been virtually unstudied is the relative advantages and disadvantages of different compensatory strategies. Is it best for an insect in any particular situation to increase consumption, select from alternative foods, or alter utilization efficiencies? Griswold and Trumble[104] found that penultimate and final instar larvae of *Spodoptera exigua* in a no-choice test compensated, presumably for protein, by eating nearly three times as much celery petiole as did those given leaves. However, if given a choice, late instar larvae feed preferentially on the nutritionally less favorable petioles on a complete plant. This could be because of direct responses to differences in microclimate, or represent an evolved response to avoid predation or parasitism. Studies of the relative performance of insects under choice and no-choice situations have been carried out by Waldbauer, Cohen, Friedman and colleagues and are discussed in Section II.B.2. Presumably, factors such as the mobility, growth rate, type of post-embryonic development, and degree of diet specialization exhibited by an insect are important in determining which compensatory responses are shown.

V. COMPENSATION AND ALLELOCHEMICALS

Secondary plant compounds influence phytophagous insects in numerous ways. They may be sequestered, act as phagostimulants or deterrents, and may have detrimental effects once eaten. Insects sequester allelochemicals for many reasons, including for use as pheromone precursors, defensive compounds, and as nutrients in their own right.[30,87] Insects might be expected to exhibit compensatory responses to ensure these compounds are acquired in sufficient quantities from the diet, but we could find no evidence which demonstrated this for compounds derived from normal food plants. *Danaus chrysippus* retains glycoside poisons for protection against birds, for which the plant *Calotropis gigantea* is a better source than *Asclepias curassavica*. Larvae provided only with the latter failed to compensate.[148] Some species, however, are known to gather pyrrolyzidine alkaloids independently of true feeding behavior, an activity termed pharmacophagy.[43] How this specialized form of dietary selection is regulated is unknown.

Although there have been many investigations into the growth and performance of insects fed diets containing allelochemicals, very few have considered how these compounds interact with compensatory responses for dietary nutrients. This is of particular importance to compensation by increased consumption, since eating more in order to gain sufficient limiting nutrients entails ingesting increased quantities of allelochemicals. It is virtually impossible to alter concentrations of nutrients and allelochemicals independently in plant tissue because of the complex interdependency between them, but a small number of studies have investigated feeding, utilization, and growth on artificial diets varying in nutrient and allelochemical content.

Lincoln et al.[134] studied the effects of varied protein and resin contents in an artificial diet on larvae of the checkerspot butterfly, *Euphydryas chalcedona*. The resin used was extracted from the larval food plant, *Diplacus aurantiacus*. Phenolic compounds, such as this resin, are known to precipitate proteins and it has been suggested that they might reduce plant digestibility by binding to digestive enzymes or to plant proteins in the gut.[90,182] However, while growth and survivorship were adversely affected by high resin levels, food digestibility (AD) and nitrogen utilization efficiency were not, suggesting that the resin had not interfered with protein digestion (although it appears that in deriving these indices no correction was made for the dry weight of resin in food or feces). Larvae increased relative consumption of low protein diets independently of resin level, and there was no significant interaction term between resin and protein levels in their influence on relative growth rates. This indicates that, although the resin was detrimental to growth and survival, eating more of it as a consequence of compensatory feeding for protein did not increase its effects. There is an important methodological problem in this experiment which deserves a mention. In the preparation of diets the dry ingredients were mixed with water, thus providing a medium in which the phenolic resins could form chemical complexes before being ingested by the insect. The resins occur as a surface layer on the intact leaves of the plant, so they would not normally contact soluble leaf proteins until after the leaf is bitten.

Martin et al.[146] showed that another phenolic, tannic acid, did not precipitate digestive enzymes or dietary protein *in vitro* because of the action of surfactants in the gut fluids of larval *M. sexta* and *Schistocerca gregaria.* Maintenance of a high pH in the gut lumen of lepidopteran larvae might also prevent phenolics from reducing protein digestibility, while some acridids possess gut enzymes which degrade hydrolyzable tannins, as well as peritrophic membranes which preferentially adsorb tannins.[21]

Although AD and relative consumption rate for *E. chalcedona* larvae were not influenced by resin in the diet, ECD was reduced. This is a commonly reported effect of ingesting allelochemicals in the food. There are several possible reasons, including increased metabolic costs of digestion, selective excretion, and the production of detoxifying enzymes such as mixed function oxidases.[47,227,253] Neal,[160] however, found little metabolic cost involved in the induction of detoxification enzymes in *H. zea.*

Johnson and Bentley[122] investigated the interaction between dietary protein and host plant alkaloids in larvae of the southern armyworm, *Spodoptera eridania.* Two types of protein diet were prepared: C diets comprised a minimal amount of wheatgerm, with high or low levels of casein added to give a total dietary protein content of 36 and 20%, respectively; Wg diets contained 16.7% protein in the form of wheatgerm (low protein) to which was added 16.7% casein to yield a high protein diet. Various levels of sparteine and lupinine were also incorporated. The concentrations used were designed to mimic those occurring in plants of the genus *Lupinus.*

Larvae compensated by eating more of the low protein than the high protein diets. The effect of adding sparteine depended on the nature of the diet. For the Wg diets there was no significant interaction between protein and alkaloid content (in terms of growth and survivorship), but with C diets there was an interaction, growth being significantly reduced at low protein levels. This difference was probably due to the wheatgerm diets being nutritionally superior to those containing mainly casein, either because they provided a better balance of amino acids or because they contained more of some other limiting nutrient. The alkaloids sparteine and lupinine were additive rather than "synergistic" in their effects.

In those larvae that survived to the fifth instar on wheatgerm diets, sparteine had no significant effect on any of the growth and utilization indices measured over a single 24-h period. Similarly, dietary protein did not affect relative growth rate or final larval weight, but AD, ECD, and ECI were reduced at low protein levels. Hence, larvae that managed to survive the toxic effects of sparteine with continued exposure from hatching managed to compensate completely for a reduction in dietary protein without any deleterious effects of the resultant increase in alkaloid intake.

The toxic effects of sparteine were greatest in the first larval stadium (as measured by mortality and growth) but then declined. This implies that "physiological adaptation" had occurred. The basis of a reduced sensitivity to allelochemicals with continued exposure is poorly understood, but could involve induction of detoxification enzymes,[47] which may in turn facilitate habituation to the deterrent sensory properties of the compound, thus overcoming inhibition of feeding.[69,234] Continued exposure may also cause a change in the sensitivity of gustatory receptors responding to the deterrent[36,192] (see Section II.A.2.b).

Hare[106] raised larvae of the potato beetle, *Leptinotarsa decemlineata,* on artificial diets containing one of three concentrations of protein (25, 50, and 100 mg/ml) and one of five concentrations of five glycoalkaloids (0, 2, 4, 8, and 16 mg/ml). For four of the alkaloids there was no significant interaction with protein in larval weight gain, but for the most toxic, tomatine, there was. Unfortunately, consumption was not measured.

Slansky and Wheeler,[229] in a study on the velvetbean caterpillar, *Anticarsia gemmatalis,* also found a significant interaction between the presence of caffeine and the degree of dietary dilution in their effect on growth and survivorship. Increased consumption to compensate for dietary dilution resulted in the ingestion of deleterious quantities of the allelochemical. There was no significant effect of caffeine on relative consumption rate, suggesting that reduced total consumption was a secondary consequence of reduced growth, occasioned by the toxicity of the compound, rather than due to caffeine acting as a feeding deterrent.

The interaction between dietary nutrients and allelochemicals has also been studied in nymphs of the locusts *L. migratoria* (an oligophage) and *Schistocerca gregaria* (a polyphage) by Raubenheimer and Simpson.[177] This is the only study to investigate detailed behavioral mechanisms and to specifically tease apart the pre- and post-ingestive effects of an allelochemical in its interactions with nutrients. Insects were provided with artificial diets varying in protein (14 or 28%), digestible carbohydrate (14 or 28%), and tannic acid (0 or 10%) content. Total consumption and detailed feeding behavior were recorded over a 12-h period in both choice and no-choice experiments. There were no interactive effects of nutrients and tannic acid, despite the fact that both species compensated for dilution of dietary protein by increasing consumption. Tannic acid did influence feeding as a main effect, however. It caused an increase in amounts eaten by *Schistocerca* in both choice and no-choice experiments. This increased consumption was due to an increase in the number of meals taken. A shorter latency period before and a longer duration of the first meal by naive insects indicated a phagostimulatory rather than a post-ingestive effect of tannic acid. Long-term experiments showed that the stimulatory effect was only apparent for the first 24 h of exposure, but this temporary enhancement nonetheless resulted in the insects being heavier at adult ecdysis and moulting somewhat sooner. Unlike for *Schistocerca* no effect of tannic acid was found on the feeding behavior of *Locusta* in the 12-h no-choice test. When given a choice, however, this species took significantly more meals on the diet lacking tannic acid, these being of similar average size to meals taken on the tannic acid diet. More insects took their first meals on the tannic acid-free diet in the choice test, indicating a deterrent effect of tannic acid for *Locusta.* Long-term studies[176] in no-choice situations have since shown only subtle deleterious effect on *Locusta* of tannic acid.

It might be concluded from the limited amount of data available that allelochemicals do not interfere with compensatory feeding as significantly as might be expected, except perhaps in cases where the compounds are highly toxic and are not normally encountered by the insect in question or when food quality is particularly poor (see also Reference 49). Phytophagous insects have presumably evolved means of either avoiding ingesting toxins or reducing the problems associated with increasing intake in order to compensate for changing dietary nutrients or nutritional needs.

VI. CONCLUDING REMARKS

It it apparent that, although there are major gaps in our understanding of the mechanisms controlling nutritional compensation in insects, much progress has been made in recent years. In fact, the subject has become one of the most actively pursued in the field of insect feeding behavior. Not surprisingly, the best understood species are those for which most is known about the basic mechanisms regulating food intake. We would go as far as to say that knowledge of the control of feeding and, especially, detailed analyses of feeding behavior are essential prerequisites to an investigation of the mechanisms of compensation.

While much more is known about vertebrates in some areas of the subject, in others, studies on phytophagous insects have demonstrated novel mechanisms. The finding in locusts that specific nutrient feedbacks for amino acids influence behavior by regulating gustatory sensitivity, as well as by acting more centrally, is a case in point.

Although there are no general principles apparent as yet, a likely candidate is the central role of the hemolymph in regulating intake of macronutrients. As stated by Abisgold and Simpson,[2] "It is not surprising that levels of blood nutrients [and osmolality] are involved in the regulation of feeding behavior. The blood serves as a repository for stored nutrients, as well as those newly arrived from digested food in the gut. Hence levels of blood nutrients provide an ideal indicator not only of long-term nutrient status, but also of time since, and quality of, the previous meal". It now remains to be shown whether or not there is generally a "blood-related control" by investigating other species and other nutrient groups. More effort is also required to further elucidate the basis and nature of regulation via specific sensitivity changes in taste receptors, as well as to establish the role of central neural and endocrine control.

The extent to which phytophagous insects can regulate intake of the more than 30 essential macro- and micronutrients is not known. It is hard to imagine that each is controlled by specific, direct feedback mechanisms.[249] The "malaise hypothesis" discussed by Cohen et al.[64] provides a possible general feedback which could act directly or, more likely, provide an unconditioned stimulus for associatively learned responses. Both associative and nonassociative learning are clearly implicated in compensatory feeding by phytophagous insects, but there is still much to be done.

A key area in which we are severely limited is post-ingestive compensation. To date much reliance has been placed on the use of standard gravimetric techniques. Such budgets have been of value to the study of nutritional ecology, but are very crude and of only limited usefulness when considering physiological control mechanisms, particularly the nature of the relationship between feeding behavior and post-ingestive physiology.

Additionally, while it is true that artificial diets are of restricted value to ecologists, their use is vital to the student of nutritional compensation, at least in the early stages of an investigation. The judicious choice of study insects (oligo- and polyphagous representatives of both endo- and exopterygotes, for instance) could conceivably provide insights into the evolution of feeding and compensatory strategies which would then direct studies on plants.

Finally, an awareness and understanding of the nature of compensatory responses by phytophagous insects is necessary if the manipulation of pest-crop systems is to be maximally effective.[78,154] Otherwise, there is the very real danger that compensatory feeding in response to altered plant or insect characteristics will actually exacerbate problems of pest control.

ACKNOWLEDGMENTS

We would like to thank Reg Chapman and Frank Slansky for critically reviewing the manuscript and Angela Douglas, Dave Raubenheimer, Peter White, and Dawn Bazeley for their helpful comments. Many thanks also to Wally Blaney, Monique Simmonds, David Rollo, Terrence Geissler, Frank Slansky, Gregory Wheeler, Dave Raubenheimer, and Julie Abisgold for allowing us to discuss unpublished data. Finally, thanks to Lindsay Barton Browne for use of his database on insect feeding.

REFERENCES

1. **Abisgold, J. D.,** Compensation by Locusts for Changes in Dietary Nutrients: Behavioural and Physiological Mechanisms, Doctoral thesis, University of Oxford, 1988.
2. **Abisgold, J. D. and Simpson, S. J.,** The physiology of compensation by locusts for changes in dietary protein, *J. Exp. Biol.,* 129, 329, 1987.
3. **Abisgold, J. D. and Simpson, S. J.,** The effect of dietary protein levels and haemolymph composition on the sensitivity of the maxillary palp chemoreceptors of locusts, *J. Exp. Biol.,* 135, 215, 1988.
4. **Al-Zubaidi, F. S. and Capinera, J. L.,** Application of different nitrogen levels to the host plant and cannibalistic behaviour of beet armyworms, *Spodoptera exigua* (Hubner) (Lepidoptera: Noctuidae), *Environ. Entomol.,* 11, 1687, 1983.
5. **Al-Zubaidi, F. S. and Capinera, J. L.,** Utilization of food and nitrogen by the beet armyworm *Spodoptera exigua* (Hubner) (Lepidoptera: Noctuidae), in relation to food type and dietary nitrogen levels, *Environ. Entomol.,* 13, 1604, 1984.
6. **Appelbaum, S. W.,** Biochemistry of digestion, in *Comprehensive Insect Physiology, Biochemistry and Pharmacology,* Vol. 4, Kerkut, G. A. and Gilbert, L. I., Eds., Pergamon Press, Oxford, 1985, 279.
7. **Bailey, C. G.,** A quantitative study of consumption and utilization of various diets in the bertha armyworm, *Mamestra configurata* (Lepidoptera: Noctuidae), *Can. Entomol.,* 108, 1319, 1976.
8. **Baines, D. M.,** Studies of weight changes and movements of dyes in the caeca and midgut of fifth-instar *Locusta migratoria migratorioides* (R. and F.), in relation to feeding and food deprivation, *Acrida,* 8, 95, 1979.
9. **Baker, B. J., Booth, D. A., Duggan, J. P., and Gibson, E. L.,** Protein appetite demonstrated. Learned specificity of protein-cue preference to protein need in adult rats, *Nutr. Res.,* 7, 481, 1987.
10. **Bannerjee, T. C. and Haque, N.,** Dry-matter budgets for *Diacrisia casignetum* larvae fed on sunflower leaves, *J. Insect Physiol.,* 30, 861, 1984.
11. **Barbosa, P., Martinat, P., and Waldvogel, W.,** Development, fecundity and survival of the herbivore *Lymantria dispar* and the number of plant species in its diet, *Ecol. Entomol.,* 11, 1, 1986.
12. **Barnes, O. L.,** Further tests of the effect of food plants on the migratory grasshopper, *J. Econ. Entomol.,* 56, 396, 1965.
13. **Barton Browne, L.,** Regulatory mechanisms in insect feeding, *Adv. Insect Physiol.,* 11, 1, 1975.
14. **Barton Browne, L. and Kerr, R. W.,** Influence of sex and prior protein feeding on preferences by the housefly, *Musca domestica,* between sucrose solutions and solutions of L-leucine or sodium phosphate buffers, *Entomol. Exp. Appl.,* 41, 135, 1986.
15. **Beenakkers, A. M. T.,** Regulation of lipid metabolism, in *Endocrinology of Insects,* Downer, R. G. H. and Laufer, H., Eds., Alan R. Liss, New York, 1983, 441.
16. **Belzer, W. R.,** *The Control of Protein Ingestion in the Black Blowfly, Phormia regina (Meigen),* Ph.D. thesis, University of Pennsylvania, Philadelphia, 1970.
17. **Belzer, W. R.,** Patterns of selective protein ingestion by the blowfly *Phormia regina, Physiol. Entomol.,* 3, 169, 1978.
18. **Belzer, W. R.,** Factors conducive to increased protein feeding by the blowfly *Phormia regina, Physiol. Entomol.,* 3, 251, 1978.
19. **Belzer, W. R.,** Recurrent nerve inhibition of protein feeding in the blowfly *Phormia regina, Physiol. Entomol.,* 3, 259, 1978.
20. **Belzer, W. R.,** Abdominal stretch in the regulation of protein ingestion by the black blowfly, *Phormia regina, Physiol. Entomol.,* 4, 7, 1979.
21. **Bernays, E. A.,** Plant tannins and insect herbivores: an appraisal, *Ecol. Entomol.,* 6, 353, 1981.
22. **Bernays, E. A.,** Regulation of feeding behaviour, in *Comprehensive Insect Physiology, Biochemistry and Pharmacology,* Vol. 4, Kerkut, G. A. and Gilbert, L. I., Eds., Pergamon Press, Oxford, 1985, 1.
23. **Bernays, E. A.,** Evolutionary contrasts in insects: nutritional advantages of holometabolous development, *Physiol. Entomol.,* 11, 377, 1986.
24. **Bernays, E. A. and Chapman, R. F.,** The control of changes in peripheral sensilla associated with feeding in *Locusta migratoria, J. Exp. Biol.,* 57, 755, 1972.
25. **Bernays, E. A. and Chapman, R. F.,** The regulation of food intake by acridids, in *Experimental Analysis of Insect Behaviour,* Barton Browne, L., Ed., Springer-Verlag, Berlin, 1974, 48.
26. **Bernays, E. A. and Chapman, R. F.,** The effect of haemolymph osmotic pressure on the meal size of nymphs of *Locusta migratoria,* L., *J. Exp. Biol.,* 61, 473, 1974.
27. **Bernays, E. A. and Lee, J. C.,** Food aversion learning in the polyphagous grasshopper, *Schistocerca americana, Physiol. Entomol.,* 13, 131, 1988.
28. **Bernays, E. A. and Mordue, A. J.,** Changes in palp tip sensilla of *Locusta migratoria* in relation to feeding: the effects of different levels of hormone, *Comp. Biochem. Phys-*

iol., 45A, 451, 1973.

29. **Bernays, E. A. and Simpson, S. J.,** Control of food intake, *Adv. Insect Physiol.*, 16, 59, 1982.

30. **Bernays, E. A. and Woodhead, S.,** Incorporation of dietary phenols into the cuticle in the tree locust *Anacridium melanorhodon*, *J. Insect. Physiol.*, 28, 601, 1982.

31. **Bernays, E. A. and Woodhead, S.,** The need for high levels of phenylalanine in the diet of *Schistocerca gregaria* nymphs, *J. Insect Physiol.*, 30, 489, 1984.

32. **Bernays, E. A., Blaney, W. M., and Chapman, R. F.,** Changes in chemoreceptor sensilla on the maxillary palps of *Locusta migratoria* in relation to feeding, *J. Exp. Biol.*, 57, 745, 1972.

33. **Bignell, D. E.,** Effects of cellulose in the diets of cockroaches, *Entomol. Exp. Appl.*, 24, 54, 1978.

34. **Blake, E. A. and Wagner, M. R.,** Effect of sex and instar on food consumption, nutritional indices, and foliage wasting by the western spruce budworm, *Choristoneura occidentalis*, *Environ. Entomol.*, 13, 1634, 1984.

35. **Blaney, W. M.,** Chemoreception and food selection in locusts, *Trends Neurosci.*, 4, 35, 1981.

36. **Blaney, W. M. and Simmonds, M. S. J.,** Electrophysiological activity in insects in response to antifeedants, *TDRI*, Misc. Pub. Lond., 1983.

37. **Blaney, W. M. and Simmonds, M. S. J.,** Food selection by locusts: the role of learning in rejection behaviour, *Entomol. Exp. Appl.*, 39, 273, 1985.

38. **Blaney, W. M. and Simmonds, M. S. J.,** Experience: a modifier of neural and behavioural sensitivity, in *Insects-Plants*, Labeyrie, V., Fabres, G., and Lachaise, D., Eds., W. Junk, Dordrecht, 1987, 237.

39. **Blaney, W. M., Chapman, R. F., and Wilson, A.,** The pattern of feeding of *Locusta migratoria* (L.) (Orthoptera, Acrididae), *Acrida*, 2, 119, 1973.

40. **Blaney, W. M., Schoonhoven, L. M., and Simmonds, M. S. J.,** Sensitivity variations in insect chemoreceptors; a review, *Experientia*, 42, 13, 1986.

41. **Blaney, W. M., Winstanley, C., and Simmonds, M. S. J.,** Food selection by locusts: an analysis of rejection behaviour, *Entomol. Exp. Appl.*, 38, 35, 1985.

42. **Booth, D. A.,** Central dietary "feedback onto nutrient selection": not even a scientific hypothesis, *Appetite*, 8, 195, 1987.

43. **Boppre, M.,** Insects pharmacophagously utilizing defensive plant chemicals (Pyrrolizidine alkaloids), *Naturwissenschaften*, 73, 17, 1986.

44. **Bounias, M.,** Effects of ascorbic acid and dehydroascorbic acid *per os* on the larval

glycemia and amino-acidemia of artificially fed *Laspereysia pomonella* (Lepidoptera), *Insect Biochem.*, 10, 521, 1980.

45. **Bowdan, E.,** The effect of deprivation on the microstructure of feeding by the tobacco hornworm caterpillar, *J. Insect Behav.*, 1, 31, 1988.

46. **Bowdan, E. and Dethier, V. G.,** Coordination of a dual inhibitory system regulating feeding behaviour in the blowfly, *J. Comp. Physiol. A*, 158, 713, 1986.

47. **Brattsten, L. B. and Ahmad, S.,** *Molecular Aspects of Insect-Plant Associations*, Plenum Press, New York, 1986.

48. **Broadway, R. M. and Duffey, S. S.,** The effect of dietary protein on the growth and digestive physiology of larval *Heliothis zea* and *Spodoptera exigua*, *J. Insect Physiol.*, 32, 673, 1986.

49. **Broadway, R. M. and Duffey, S. S.,** The effect of plant protein quality on insect digestive physiology and the toxicity of plant proteinase inhibitors, *J. Insect Physiol.*, 34, 1111, 1988.

50. **Busse, F. K., Jr. and Barth, R. K., Jr.,** The physiology of feeding-preference patterns in female black blowflies *(Phormia regina* Meigen): modifications in responsiveness to salts subsequent to salt feeding, *J. Insect Physiol.*, 31, 23, 1985.

51. **Calow, P.,** Ecology, evolution and energetics: a study in metabolic adaptation, *Adv. Ecol. Res.*, 10, 1, 1977.

52. **Carlisle, J., Loughton, B., and Ampelford, E.,** Feeding causes the appearance of a factor in the haemolymph that stimulates protein synthesis, *J. Insect Physiol.*, 33, 493, 1987.

53. **Cartier, J. J.,** Factors of host plant specificity and artificial diets, *Bull. Entomol. Soc. Am.*, 14, 18, 1968.

54. **Chandra, S. and Williams, G.,** Frequency-dependent selection in the grazing behaviour of the desert locust *Schistocerca gregaria*, *Ecol. Entomol.*, 8, 13, 1983.

55. **Chaplin, S. B. and Chaplin, S. J.,** Comparative growth energetics of a migratory and nonmigratory insect: the milkweed bugs, *J. Anim. Ecol.*, 50, 407, 1981.

56. **Chapman, R. F.,** Structure of the digestive system, in *Comprehensive Insect Physiology, Biochemistry and Pharmacology*, Vol. 4, Kerkut, G. A. and Gilbert, L. I., Eds., Pergamon Press, Oxford, 1985, 165.

57. **Chapman, R. F.,** Coordination of digestion, in *Comprehensive Insect Physiology, Biochemistry and Pharmacology*, Vol. 4., Kerkut, G. A. and Gilbert, L. I., Eds., Pergamon Press, Oxford, 1985, 213.

58. **Chapman, R. F.,** Variations in the size of the midgut caeca during the fifth instar of the grasshopper, *Schistocerca americana* (Drury), *J. Insect Physiol.*, 34, 329, 1988.

59. **Chyb, S. and Simpson, S. J.,** Dietary selection in adult *Locusta migratoria, Entomol. Exp. Appl.,* in press, 1990.

60. **Clark, J. V.,** Changes in the feeding rate and receptor sensitivity over the last instar of the african armyworm, *Spodoptera exempta, Entomol. Exp. Appl.,* 27, 144, 1980.

61. **Clarke, K. U. and Gillott, C.,** Studies on the effects of the removal of the frontal ganglia in *Locusta migratoria* (L.). I. The effect on protein metabolism, *J. Exp. Biol.,* 46, 13, 1967.

62. **Cloutier, C.,** Amino acid utilization in the aphid *Acyrthosiphon pisum* infected by the parasitoid *Aphidius smithi, J. Insect Physiol.,* 32, 263, 1986.

63. **Cohen, A. C. and Pantana, R.,** Efficiency of food utilization by *Heliothis zea* (Lepidoptera: Noctuidae) fed artificial diets on green beans, *Can. Entomol.,* 116, 139, 1984.

64. **Cohen, R. W., Friedman, S., and Waldbauer, G. P.,** Physiological control of nutrient self-selection in *Heliothis zea* larvae: the role of serotonin, *J. Insect Physiol.,* 34, 935, 1988.

65. **Cohen, R. W., Heydon, S. L., Waldbauer, G. P., and Friedman, S.,** Nutrient self-selection by the omnivorous cockroach *Supella longipalpa, J. Insect Physiol.,* 33, 77, 1987.

66. **Cohen, R. W., Waldbauer, G. P., and Friedman, S.,** Natural diets and self-selection: *Heliothis zea* larvae and maize, *Entomol. Exp. Appl.,* 46, 161, 1988.

67. **Cohen, R. W., Waldbauer, G. P., Friedman, S., and Schiff, N. M.,** Nutrient self-selection by *Heliothis zea* larvae: a time-lapse film study, *Entomol. Exp. Appl.,* 44, 65, 1987.

68. **Cook, A. G.,** Nutrient chemicals as phagostimulants for *Locusta migratoria, Ecol. Entomol.,* 2, 113, 1977.

69. **Cottee, P. K., Bernays, E. A., and Mordue, A. J.,** Comparisons of deterrency and toxicity of selected secondary plant compounds to an oligophagous and polyphagous acridid, *Entomol. Exp. Appl.,* 46, 241, 1988.

70. **Dadd, R. H.,** Observations on the palatability and utilization of food by locusts, with particular reference to the interpretation of performance in growth trials using synthetic diets, *Entomol. Exp. Appl.,* 3, 283, 1960.

71. **Dadd, R. H.,** Nutrition: organisms, in *Comprehensive Insect Physiology, Biochemistry and Pharmacology,* Vol. 4, Kerkut, G. A. and Gilbert, L. I., Eds., Pergamon Press, Oxford, 1985, 313.

72. **Davenport, A. P. and Evans, P. D.,** Stress-induced changes in the octopamine levels of insect haemolymph, *Insect Biochem.,* 154, 135, 1984.

73. **Davenport, A. P. and Evans, P. D.,** Changes in haemolymph octopamine levels associated with food deprivation in the locust *Schistocerca gregaria, Physiol. Entomol.,* 9, 269, 1984.

74. **Davey, K. G. and Treherne, J. E.,** Studies on crop function in the cockroach *(Periplaneta americana L.).* I. The mechanism of crop emptying, *J. Exp. Biol.,* 40, 763, 1963.

75. **David, W. A. L. and Gardiner, B. O. C.,** Feeding behaviour of adults of *Pieris brassicae* (L.) in a laboratory culture, *Bull. Entomol. Res.,* 52, 741, 1961.

76. **Delaney, K. and Gelperin, A.,** Post-ingestive food-aversion learning to amino acid deficient diets by the terrestrial slug *Limax maximus, J. Comp. Physiol. A,* 159, 281, 1986.

77. **Delvi, M. R. and Pandian, T. J.,** Rates of feeding and assimilation in the grasshopper *Poecilocerus pictus, J. Insect Physiol.,* 18, 1829, 1972.

78. **Denno, R. F. and McClure, M. S., Eds.,** *Variable Plants and Herbivores in Natural and Managed Systems,* Academic Press, New York, 1983.

79. **Dethier, V. G.,** *The Hungry Fly,* Harvard University Press, Cambridge, MA, 1976.

80. **Dethier, V. G.,** Food aversion learning in two polyphagous caterpillars, *Diacrisia virginica* and *Estigmene congrua, Physiol. Entomol.,* 5, 321, 1980.

81. **Dethier, V. G.,** Induction and aversion-learning in polyphagous arctiid larvae (Lepidoptera) in an ecological setting, *Can. Entomol.,* 120, 125, 1988.

82. **Dethier, V. G. and Bowdan, E.,** Relations between differential threshold and sugar receptor mechanisms in the blowfly, *Behav. Neurosci.,* 98, 791, 1984.

83. **Dethier, V. G. and Yost, M. T.,** Oligophagy and the absence of food-aversion learning in tobacco hornworms, *Manduca sexta, Physiol. Entomol.,* 4, 125, 1979.

84. **Dethier, V. G., Evans, D. R., and Rhoades, M. V.,** Some factors controlling the ingestion of carbohydrates by the blowfly, *Biol. Bull.,* 111, 204, 1956.

85. **Dicke, M., Sabelis, M. N., and Groeneveld, A.,** Vitamin A deficiency modifies response of predatory mite *Amblyseius potentillae* to volatile kairomone of two-spotted spider mite *Tetranychus urticae, J. Chem. Ecol.,* 12, 1389, 1986.

86. **Dow, J. A. T.,** Insect midgut function, *Adv. Insect Physiol.,* 19, 187, 1987.

87. **Duffey, S. S.,** Sequestration of plant natural products by insects, *Annu. Rev. Entomol.,* 25, 447, 1980.

88. **Dwivedy, A. K.,** Dietary cholesterol utilization by the housefly, *Musca domestica, Insect Biochem.,* 15, 137, 1985.

89. **Edgecomb, R. S., Murdock, L. L., Smith,**

A. B., and Stephen, M. D., Regulation of tarsal taste threshold in the blowfly, *Phormia regina, J. Exp. Biol.,* 127, 79, 1987.

90. **Feeny, P.,** Plant apparency and chemical defense, *Recent Adv. Phytochem.,* 10, 1, 1976.

91. **Fernando-Warnakulasuriya, G. J. P., Tsuchida, K., and Wells, M. A.,** Effects of dietary lipid content on lipid transport and storage during larval development of *Manduca sexta, Insect Biochem.,* 18, 211, 1988.

92. **Fernstrom, J. D.,** Food-induced changes in brain serotonin synthesis: is there a relationship to appetite for specific nutrients?, *Appetite,* 8, 163, 1987.

93. **Fox, L. R. and Macauley, B. J.,** Insect grazing on *Eucalyptus* in response to variation in leaf tannins and nitrogen, *Oecologia,* 29, 145, 1977.

94. **Geissler, T. G. and Rollo, C. D.,** The influence of nutritional history on the response to novel food by the cockroach, *Periplaneta americana* (L.), *Anim. Behav.,* 35, 1905, 1988.

95. **Geissler, T. G. and Rollo, C. D.,** Unpublished data, 1988.

96. **Gelperin, A.,** Control of crop emptying in the blowfly, *J. Insect Physiol.,* 12, 331, 1966.

97. **Gelperin, A.,** Neural control systems underlying insect feeding behaviour, *Am. Zool.,* 12, 489, 1972.

98. **Gelperin, A. and Dethier, V. G.,** Long term regulation of sugar intake by the blowfly, *Physiol. Zool.,* 40, 218, 1967.

99. **Gordon, H. T.,** Intake rates of various solid carbohydrates by male German cockroaches, *J. Insect Physiol.,* 14, 41, 1968.

100. **Gordon, H. T.,** Quantitative aspects of insect nutrition, *Am. Zool.,* 8, 131, 1968.

101. **Gordon, H. T.,** Interpretations of insect quantitative nutrition, in *Insect and Mite Nutrition,* Rodriguez, J. G., Ed., North-Holland, Amsterdam, 1972, 73.

102. **Grabstein, E. M. and Scriber, J. M.,** Host-plant utilization by *Hyalophora cercropia* as affected by prior feeding experience, *Entomol. Exp. Appl.,* 32, 262, 1982.

103. **Grabstein, E. M. and Scriber, J. M.,** The relationship between restriction of host plant consumption, and post-ingestive utilization of biomass and nitrogen in *Hyalophora cecropia, Entomol. Exp. Appl.,* 31, 202, 1982.

104. **Griswold, M. J. and Trumble, J. T.,** Consumption and utilization of celery, *Apium graveolens,* by the beet armyworm *Spodoptera exigua, Entomol. Exp. Appl.,* 38, 73, 1985.

105. **Hall, M. J. R.,** Central control of tarsal threshold for proboscis extension in the blowfly, *Physiol. Entomol.,* 5, 17, 1980.

106. **Hare, J. D.,** Growth of *Leptinotarsa decemlineata* larvae in response to simultaneous variation in protein and glycoalkaloid ecology, *J. Chem. Ecology.,* 13, 39, 1987.

107. **van Herrewege, C.,** Régulation de la prise de nourriture, après un jeune, chez les males de la blatte Germanique dans différentes conditions alimentaires, *Entomol. Exp. Appl.,* 17, 234, 1974.

108. **Hill, L. and Goldsworthy, G. J.,** Growth, feeding activity and the utilization of reserves in larvae of *Locusta, J. Insect Physiol.,* 14, 1085, 1968.

109. **Hodge, C.,** Growth and nutrition of *Melanoplus differentialis* Thomas (Orthoptera: Acrididae). I. Growth on a satisfactory mixed diet and on diets of single food plants, *Physiol. Zool.,* 6, 306, 1933.

110. **Horie, Y. and Nakasone, S.,** Effects of the levels of fatty acids and carbohydrates in a diet on the biosynthesis of fatty acids in larvae of the silkworm, *Bombyx mori, J. Insect Physiol.,* 17, 1441, 1971.

111. **Horie, Y. and Watanabe, K.,** Effects of various kinds of dietary protein and supplementation with limiting amino acids on growth, haemolymph components and uric acid excretion on the silkworm, *Bombyx mori, J. Insect Physiol.,* 29, 187, 1983.

112. **Horie, Y. and Watanabe, K.,** Daily utilization of nitrogen in food by the silkworm, *Bombyx mori* (Lepidoptera: Bombycidae), *Appl. Entomol. Zool.,* 21, 289, 1986.

113. **Houk, E. J. and Griffiths, G. W.,** Intracellular symbiotes of the Homoptera, *Annu. Rev. Entomol.,* 25, 161, 1980.

114. **House, H. L.,** Effects of low levels of the nutrient content of a food and nutrient imbalance on the feeding and the nutrition of a phytophagous larva, *Celerio euphorbia* (Linnaeus) (Lepidoptera, Sphingidae), *Can. Entomol.,* 97, 62, 1965.

115. **House, H. L.,** Effects of varying the ratio between amino acids and other nutrients in conjunction with a salt mixture on the fly *Agria affinis* (Fall.), *J. Insect Physiol.,* 12, 299, 1966.

116. **House, H. L.,** The role of nutritional factors in food selection and preference as related to larval nutrition of an insect *Psuedo-sarcophaga affinis* (Diptera, Sarcophagidae), on synthetic diets, *Can. Entomol.,* 99, 1310, 1967.

117. **House, H. L.,** Effects of different proportions of nutrients on insects, *Entomol. Exp. Appl.,* 12, 651, 1969.

118. **House, H. L.,** Choice of food by larvae of the fly, *Agria affinis,* related to dietary proportions of nutrients, *J. Insect Physiol.,* 16, 2041, 1970.

119. **House, H. L.,** Changes from initial food choice in a fly larva, *Agria affinis,* as related to dietary proportions of nutrients, *J. Insect Physiol.,* 17, 1051, 1971.

120. **House, H. L.,** Relations between dietary proportions of nutrients, growth rate, and choice of food in the fly larva, *Agria affinis, J. Insect Physiol.,* 17, 1225, 1971.

121. **Jermy, T.,** The role of experience in the host selection of phytophagous insects, in *Perspectives in Chemoreception and Behaviour,* Chapman, R. F., Bernays, E. A., and Stoffolano, J., Jr., Eds., Springer-Verlag, New York, 1987, 143.

122. **Johnson, N. D. and Bentley, B. L.,** Effects of dietary protein and lupine alkaloids on growth and survivorship of *Spodoptera eridania, J. Chem. Ecol.,* 14, 1391, 1988.

123. **Jones, C. G., Hoggard, M. P., and Blum, M. S.,** Pattern and process in insect feeding behaviour: a quantitative analysis of the Mexican bean beetle, *Epilachna varivestis, Entomol. Exp. Appl.,* 30, 254, 1981.

124. **Khan, M. A.,** Studies on the secretion of digestive enzymes in *Locusta migratoria* (L.). II. Invertase activity, *Entomol. Exp. Appl.,* 7, 125, 1964.

125. **Koller, C. N. and Leonard, D. E.,** Comparison of energy budgets for the spruce budworm *Choristoneura fumiferana* (Clemens) on balsam fir and whitespruce, *Oecologia,* 49, 14, 1981.

126. **Latheef, H. A. and Harcourt, D. G.,** A quantitative study of food consumption, assimilation, and growth in *Leptinotarsa decemlineata* (Coleoptera: Chrysomelidae) on two host plants, *Can. Entomol.,* 104, 1271, 1972.

127. **Le Berre, J. R. and Mainguet, A. M.,** Nutrition du criquet migrateur *Locusta migratoria* L. I. Compartement alimentaire, survie et variation ponderale en fonction de quelques substances, *Rev. Comp. Anim.,* 7, 203, 1973.

128. **Leckstein, P. M. and Llewellyn, M.,** The role of amino acids in diet intake and selection and the utilization of dipeptides by *Aphis fabae, J. Insect Physiol.,* 20, 877, 1974.

129. **Lee, J. C. and Bernays, E. A.,** Declining acceptability of a food plant for the phytophagous grasshopper, *Schistocerca americana:* the role of food aversion learning, *Physiol. Entomol.,* 13, 291, 1988.

130. **Le Magnen, J.,** *Hunger,* Cambridge University Press, Cambridge, 1985.

131. **Lent, C. M. and Dickinson, M. H.,** The neurobiology of feeding in leeches, *Sci. Am.,* 78, 77, 1988.

132. **Lii, G. Y., Garlich, J. D., and Rock, G. C.,** Protein and energy utilization by the insect *Argyrotaenia velutinana* (Walker), fed diets containing graded levels of an amino acid mixture, *Comp. Biochem. Physiol. A,* 52, 615, 1975.

133. **Lincoln, D. E., Couvert, D., and Sionit, N.,** Response of an insect herbivore to host plants grown in carbon dioxide and enriched atmospheres, *Oecologia,* 69, 556, 1986.

134. **Lincoln, D. E., Newton, T. S., Ehrlich, P. R., and Williams, K. S.,** Coevolution of the checkerspot butterfly *Euphydryas chalcedona* and its larval food plant *Diplacus aurantiacus:* larval response to protein and leaf resin, *Oecologia,* 52, 216, 1982.

135. **Loughton, B. G. and Tobe, S. S.,** Blood volume in the African migratory locust, *Can. J. Zool.,* 47, 1333, 1969.

136. **Louveaux, A.,** Regulation of meal size and growth of fifth instar nymphs of *Locusta migratoria* (Orthoptera, Acrididae) in different conditions of starvation and temperature, *Ann. Nutr. Aliment.,* 31, 85, 1977.

137. **McClain, E. and Feir, D.,** Role of amino acid concentration in the feeding of *Oncopeltus fasciatus* on a liquid and a solid diet, *J. Insect Physiol.,* 19, 287, 1973.

138. **McFarland, J. H. and Thorsteinson, A. J.,** Development and survival of the two striped grasshopper, *Melanoplus bivittatus* (Say) (Orthoptera: Acrididae), on various single and multiple plant diets, *Acrida,* 9, 63, 1980.

139. **McGinnis, A. J. and Kasting, R.,** Comparison of tissues from solid- and hollowstemmed spring wheats during growth. IV. Apparent dry matter utilization and nitrogen balance in the two-striped grasshopper, *Melanoplus bivittatus* (Say), *J. Insect Physiol.,* 12, 671, 1966.

140. **McGinnis, A. J. and Kasting, R.,** Dietary cellulose: effect on food consumption and growth of a grasshopper, *Can. J. Zool.,* 45, 365, 1967.

141. **McLean, P. L. and Kinsey, M. G.,** Probing behaviour of the pea aphid, *Acyrthosiphon pisum.* IV. Effects of starvation on certain probing activities, *Ann. Entomol. Soc. Am.,* 62, 987, 1969.

142. **McNeill, S. and Southwood, T. R. E.,** The role of nitrogen in the development of insect/plant relationships, in *Biochemical Aspects of Plant and Animal Coevolution,* Harborne, J., Ed., Academic Press, London, 1978, 71.

143. **Ma, W-C.,** Dynamics of feeding responses in *Pieris brassicae* Linn. as a function of chemosensory input: a behavioural, ultrastructural and electrophysiological study, *Meded. Landbourwhoges, Wageningen,* 72, 1972.

144. **Mackey, A. P.,** Growth and bioenergetics of the moth *Cyclophragma leucosticta* Grunberg, *Oecologia,* 32, 367, 1978.

145. **Manoukas, A. G.,** Effect of excess levels of individual amino acids upon survival, growth and pupal yield of *Dacus oleae* (Gmel.) larvae, *Z. Ang. Entomol.,* 91, 309, 1981.

146. **Martin, J. S., Martin, M. M., and Bernays, E. A.,** Failure of tannic acid to inhibit diges-

tion or reduce digestibility of plant protein in gut fluids of insect herbivores, *J. Chem. Ecol.,* 13, 605, 1988.

147. **Martin, M. M. and Van't Hof, H. M.,** The cause of reduced growth of *Manduca sexta* larvae on a low-water diet: increased metabolic processing costs or nutrient limitation? *J. Insect Physiol.,* 34, 515, 1987.

148. **Mathavan, S. and Bhaskaran, R.,** Food selection and utilization in a danid butterfly, *Oecologia,* 18, 55, 1975.

149. **Mattson, W. J., Jr.,** Herbivory in relation to plant nitrogen content, *Annu. Rev. Ecol. Syst.,* 11, 119, 1980.

150. **Mei, N.,** Intestinal chemosensitivity, *Physiol. Rev.,* 65, 211, 1985.

151. **Mittler, T. E.,** Effects of dietary amino acids on the feeding rate of the aphid *Myzus persicae, Entomol. Exp. Appl.,* 13, 432, 1970.

152. **Mittler, T. E. and Dadd, R. H.,** Gustatory discrimination between liquids by the aphid *Myzus persicae* (Sulzer), *Entomol. Exp. Appl.,* 7, 315, 1964.

153. **Montgomery, M. E.,** Life-cycle nitrogen budget for the gypsy moth, *Lymantria dispar,* reared on artificial diet, *J. Insect Physiol.,* 28, 437, 1982.

154. **Moran, N. and Hamilton, W. D.,** Low nutritive quality as a defense against herbivores, *J. Theor. Biol.,* 86, 247, 1980.

155. **Mullins, D. H.,** Haemolymph composition, in *Comprehensive Insect Physiology, Biochemistry and Pharmacology,* Vol. 3, Kerkut, G. A. and Gilbert, L. I., Eds., Pergamon Press, Oxford, 1985, 355.

156. **Mullins, D. E. and Cochran, D. G.,** Nitrogen metabolism, in *Endocrinology of Insects,* Downer, R. G. H. and Laufer, H., Eds., Alan R. Liss, New York, 1983, 451.

157. **Muthhukrishnan, J. and Delvi, M. R.,** Effect of ration levels on food utilization in the grasshopper *Poecilocerus pictus, Oecologia,* 16, 227, 1974.

158. **Navon, A., Nesbitt, J., Henzel, W. M., and Lipke, H.,** Effects of ascorbic acid deficiency on growth and cuticle composition of *Manduca sexta* and *Spodoptera littoralis, Insect Biochem.,* 15, 285, 1985.

159. **Nayer, J. K. and Sauerman, D. M.,** Long term regulation of sucrose intake by the female mosquito *Aedes taeniorhynchus, J. Insect Physiol.,* 20, 1203, 1974.

160. **Neal, J. J.,** Metabolic costs of mixed function oxidase induction in *Heliothis zea, Entomol. Exp. Appl.,* 435, 175, 1987.

161. **Hestel, D., Galun, R., and Friedman, S.,** Long-term regulation of sucrose intake by the adult Mediterranean fruit fly *Ceratitis capitata* (Wiedermann), *J. Insect Physiol.,* 31, 533, 1985.

162. **Nusbaum, M. P. and Kristan, W. B., Jr.,** Swim initiation in the leech by serotonin containing interneurones, cells 21 and 61, *J. Exp. Biol.,* 122, 277, 1986.

163. **Ohmart, C. P., Stewart, L. G., and Thomas, J. R.,** Effects of food quality, particularly nitrogen concentrations of *Eucalyptus blakelyi* foliage on the growth of *Paropsis atomaria* larvae (Coleoptera: Chrysomelidae), *Oecologia,* 65, 543, 1985.

164. **Omand, E.,** A peripheral sensory basis for behavioural regulation, *Comp. Biochem. Physiol., A.,* 38, 265, 1971.

165. **Omand, E. and Zabara, J.,** Response reduction in dipteran chemoreceptors after sustained feeding or darkness, *Comp. Biochem. Physiol. A,* 70, 469, 1981.

166. **Papaj, D. R. and Prokopy, J.,** Ecological and evolutionary aspects of learning in phytophagous insects, *Annu. Rev. Entomol.,* 34, 315, 1989.

167. **Peterson, S. S., Scriber, J. M., and Coors, J. G.,** Silica, cellulose and their interactive effects on the feeding performance of the southern armyworm, *Spodoptera eridania* (Cramer) (Lepidoptera: Noctuidae), *J. Kansas Entomol. Soc.,* 61, 169, 1988.

168. **Phillips, N. H.,** Compensatory intake can be consistent with an optimal foraging model, *Am. Nat.,* 123, 867, 1984.

169. **Pickford, R.,** Development, reproduction and survival of *Melanoplus bilituratus* (Wilk.) (Orthoptera: Acrididae) reared on various food plants, *Can. Entomol.,* 94, 859, 1962.

170. **Pulliam, H. R.,** Diet optimization with nutrient constraints, *Am. Nat.,* 109, 765, 1975.

171. **Pullin, A. S.,** Influence of the food plant, *Urtica dioica,* on larval development feeding efficiencies, and voltinism of a specialist insect, *Inachis io, Hol. Ecol.,* 9, 72, 1986.

172. **Rachman, N. J.,** The sensitivity of the labellar sugar receptors of *Phormia regina* in relation to feeding, *J. Insect Physiol.,* 25, 733, 1979.

173. **Rachman, N. J.,** Physiology of feeding preference patterns of female black blowflies *(Phormia regina* Meigen). I. The role of carbohydrate reserves, *J. Comp. Physiol.,* 139, 59, 1980.

174. **Rachman, N. J., Busse, F. K., Jr., and Barth, R. H., Jr.,** Physiology of feeding preference patterns of female black blowflies *(Phormia regina):* alterations in responsiveness to salts, *J. Insect Physiol.,* 28, 625, 1982.

175. **Ramesh, N. and Lipke, H.,** Acid ribonuclease activity in insect scurvy, in *Proc. Joint Meeting of the Entomology Societies of America and Canada,* Toronto, 1982, 64.

176. **Raubenheimer, D.,** unpublished data.

177. **Raubenheimer, D. and Simpson, S. J.,** The effects of simultaneous variation in protein, carbohydrate and tannic acid on the feeding behaviour of larval *Locusta migratoria* (L.)

and *Schistocerca gregaria* (Forskal). I. Short term studies, *Physiol. Entomol.,* in press.

178. **Reese, J. C. and Beck, S. D.,** Interrelationships of nutritional indices and dietary moisture in the black cutworm *(Agrotis ipsilon)* digestive efficiency, *J. Insect Physiol.,* 24, 473, 1978.

179. **Reynolds, S. E. and Bellward, K.,** Water balance in *Manduca sexta* caterpillars: water recycling from the rectum, *J. Exp. Biol.,* 141, 33, 1989.

180. **Reynolds, S. E., Nottingham, S. F., and Stephens, A. E.,** Food and water economy and its relations to growth in fifth-instar larvae of the tobacco hornworm, *Manduca sexta, J. Insect Physiol.,* 31, 119, 1985.

181. **Reynolds, S. E., Yeomans, M. R., and Timmins, W. A.,** The feeding behaviour of caterpillars *(Manduca sexta)* on tobacco and on artificial diet, *Physiol. Entomol.,* 11, 39, 1986.

182. **Rhoades, D. F. and Cates, R. G.,** A general theory of plant antiherbivore chemistry, *Recent Adv. Phytochem.,* 10, 168, 1976.

183. **Roberts, J. A. and Kitching, R. L.,** Ingestion of sugar, protein and water by adult *Lucilia cuprina* (Wied.) (Diptera, Calliphoridae), *Bull. Entomol. Res.,* 64, 81, 1974.

184. **Roessingh, P. and Simpson, S. J.,** Volumetric feedback and the control of meal size in *Schistocerca gregaria, Entomol. Exp. Appl.,* 36, 279, 1984.

185. **Rollo, C. D.,** Resource allocation and time budgeting in adults of the cockroach, *Periplaneta americana:* the interaction of behaviour and metabolic reserves, *Res. Pop. Ecol.,* 26, 150, 1984.

186. **Schiff, N. M., Waldbauer, G. P., and Friedman, S.,** Dietary self-selection for vitamins and lipid by larvae of the corn earworm, *Heliothis zea, Entomol. Exp. Appl.,* 46, 240, 1988.

187. **Schmidt, D. J. and Reese, J. C.,** Sources of error in nutritional index studies of insects on artificial diet, *J. Insect Physiol.,* 32, 193, 1986.

188. **Schoonhoven, L. M.,** Loss of hostplant specificity by *Manduca sexta* after rearing on an artificial diet, *Entomol. Exp. Appl.,* 10, 270, 1967.

189. **Schoonhoven, L. M.,** Sensitivity changes in some insect chemoreceptors and their effect on food selection behaviour, *Proc. Koninkl. Ned. Akad. Wetensch. (C.),* 72, 491, 1969.

190. **Schoonhoven, L. M.,** On the variability of chemosensory information, *Symp. Biol. Hung.,* 16, 261, 1976.

191. **Schoonhoven, L. M. and Meerman, J.,** Metabolic costs of changes in diet and neutralization of allelochemicals, *Entomol. Exp. Appl.,* 24, 689, 1978.

192. **Schoonhoven, L. M., Blaney, W. M., and Simmonds, M. S. J.,** Inconstancies of chemoreceptor sensitivities, in *Insect-Plants,* Labeyrie, V., Fabres, G., and Lachaise, D., Eds., W. Junk, Dordrecht, 1987, 141.

193. **Schroeder, L. A.,** Effect of food deprivation on the efficiency of utilization of dry matter, energy and nitrogen by larva of the cherry scallop moth, *Calocalpe undulata, Ann. Entomol. Soc. Am.,* 69, 55, 1976.

194. **Schroeder, L. A.,** Protein limitation of a tree leaf feeding lepidopteran, *Entomol. Exp. Appl.,* 41, 115, 1986.

195. **Scriber, J. M.,** The effects of sequentially switching foodplants upon biomass and nitrogen utilization by polyphagous and stenophagous *Papilio* larvae, *Entomol. Exp. Appl.,* 25, 203, 1979.

196. **Scriber, J. M.,** The behaviour and nutritional physiology of southern armyworm larvae as a function of plant species consumed in earlier instars, *Entomol. Exp. Appl.,* 31, 359, 1982.

197. **Scriber, J. M.,** Host-plant suitability, in *Chemical Ecology of Insects,* Bell, W. J. and Carde, R. T., Eds., Chapman and Hall, London, 1984, 160.

198. **Scriber, J. M. and Slansky, F., Jr.,** The nutritional ecology of immature insects, *Annu. Rev. Entomol.,* 26, 183, 1981.

199. **Sibly, R. M.,** Strategies of digestion and defecation, in *Physiological Ecology,* Townsend, C. R. and Calow, P., Eds., Blackwell Scientific, Oxford, 1981.

200. **Simmonds, M. S. J.,** Personal communication, 1988.

201. **Simmonds, M. S. J. and Blaney, W. M.,** unpublished data, 1988.

202. **Simmonds, M. S. J., Simpson, S. J., and Blaney, W. M.,** Detailed behavioural analysis of selection behaviour of larval *Locusta migratoria* and *Spodoptera littoralis,* submitted.

203. **Simpson, C. L.,** Unpublished data, 1988.

204. **Simpson, C. L., Chyb, S., and Simpson, S. J.,** Chemoreceptor sensitivity changes in relation to dietary selection behaviour of adult *Locusta migratoria, Entomol. Exp. Appl.,* in press, 1990.

205. **Simpson, C. L., Simpson, S. J., and Abisgold, J. D.,** The role of various amino acids in the protein compensatory response of *Locusta migratoria,* in *Proc. 7th Int. Symp. on Insect-Plant Relations,* in press.

206. **Simpson, S. J.,** An oscillation underlying feeding and a number of other behaviours in fifth-instar *Locusta migratoria* nymphs, *Physiol. Entomol.,* 6, 315, 1981.

207. **Simpson, S. J.,** Changes in the efficiency of utilization of food throughout the fifth instar of *Locusta migratoria* nymphs, *Entomol. Exp. Appl.,* 31, 265, 1982.

208. **Simpson, S. J.,** Patterns in feeding: a behavioural analysis using *Locusta migratoria* nymphs, *Physiol. Entomol.,* 7, 325, 1982.

209. **Simpson, S. J.,** Changes in the rate of crop emptying during the fifth instar of *Locusta migratoria* and their relationship to feeding and utilization, *Entomol. Exp. Appl.,* 33, 235, 1983.

210. **Simpson, S. J.,** The role of volumetric feedback from the midgut in the regulation of meal size in fifth-instar *Locusta migratoria* nymphs, *Physiol. Entomol.,* 8, 451, 1983.

211. **Simpson, S. J.,** The pattern of feeding, in *A Biology of Grasshoppers,* Chapman, R. F. and Joern, T., Eds., John Wiley & Sons, New York, 1990.

212. **Simpson, S. J. and Abisgold, J. D.,** Compensation by locusts for changes in dietary nutrients: behavioural mechanisms, *Physiol. Entomol.,* 10, 443, 1985.

213. **Simpson, S. J. and Bernays, E. A.,** The regulation of feeding: locusts and blowflies are not so different from mammals. *Appetite,* 4, 313, 1983.

214. **Simpson, S. J. and Ludlow, A. R.,** Why locusts start to feed: a comparison of causal factors, *Anim. Behav.,* 34, 480, 1986.

215. **Simpson, S. J. and White, P. R.,** Associative learning and locust feeding: evidence for a "learned hunger" for protein, *Anim. Behav.,* in press.

216. **Simpson, S. J., Barton Browne, L., and van Gerwen, A. C. M.,** The patterning of compensatory sugar feeding in the Australian sheep blowfly, *Physiol. Entomol.,* 14, 91, 1989.

217. **Simpson, S. J., Simmonds, M. S. J., and Blaney, W. M.,** A comparison of dietary selection behaviour in larval *Locusta migratoria* and *Spodoptera littoralis, Physiol. Entomol.,* 13, 225, 1988.

218. **Simpson, S. J., Simmonds, M. S. J., Blaney, W. M., and Jones, J. P.,** Compensatory dietary selection occurs in larval *Locusta migratoria* but not *Spodoptera littoralis* after a single deficient meal during *ad libitum* feeding, *Physiol. Entomol.,* in press.

219. **Simpson, S. J., Simmonds, M. S. J., and Blaney, W. M.,** unpublished data.

220. **Simpson, S. J., Simmonds, M. S. J., Wheatley, A. R., and Bernays, E. A.,** The control of meal termination in the locust, *Anim. Behav.,* 36, 1216, 1988.

221. **Sinoir, Y.,** Étude de quelques facteurs conditionnant la prise de nourriture chez les larves du criquet migrateur, *Locusta migratoria migratorioides* (Orthoptera, Acrididae). II. Facteurs internes, *Entomol. Exp. Appl.,* 11, 443, 1968.

222. **Slansky, F., Jr.,** Toward a nutritional ecology of insects, in *Proc. 5th Int. Symp. on Insect-Plant Relationships,* Visser, J. H., and Minks, A. K., Eds., Pudoc, Wageningen, 1982, 253.

223. **Slansky, F., Jr.,** Nutritional ecology of endoparasitic insects and their hosts: an overview, *J. Insect Physiol.,* 32, 255, 1986.

224. **Slansky, F., Jr. and Feeny, P.,** Stabilization of the rate of nitrogen accumulation by larvae of the cabbage butterfly on wild and cultivated food plants, *Ecol. Monogr.,* 47, 207, 1977.

225. **Slansky, F., Jr. and Panizzi, J.,** Seed eaters life-history strategies, in *Nutritional Ecology of Insects, Mites, Spiders and Related Invertebrates,* Slansky, F., Jr. and Rodriguez, J. G., Eds., John Wiley & Sons, New York, 1987.

226. **Slansky, F., Jr. and Rodriguez, J. G., Eds.,** *Nutritional Ecology of Insects, Mites, Spiders and Related Invertebrates,* John Wiley & Sons, New York, 1987.

227. **Slansky, F., Jr. and Scriber, J. M.,** Compensatory increases in food consumption and utilization efficiencies by velvetbean caterpillars mitigate impact of diluted diets on growth, *Entomol. Exp. Appl.,* 51, 175, 1989.

228. **Slansky, F., Jr. and Wheeler, G. S.,** Compensatory increases in food consumption and utilization efficiencies by velvetbean caterpillars mitigate impact of diluted diets on growth, *Entomol. Exp. Appl.,* 51, 175, 1989.

229. **Slansky, F., Jr. and Wheeler, G. S.,** Diet dilution-induced compensatory feeding: ingestion of lethal allelochemical dose, in press, 1989.

230. **Stadler, E. and Hanson, F. E.,** Influence of induction of host preference on chemoreception of *Manduca sexta:* behavioural and electrophysiological studies, *Symp. Biol. Hung.,* 16, 267, 1976.

231. **Steele, J. E.,** Endocrine control of carbohydrate metabolism in insects, in *Endocrinology of Insects,* Downer, R. G. H. and Lanfer, H., Eds., Alan R. Liss, New York, 1983, 427.

232. **Strong, D. R., Lawton, J. H., and Southwood, T. R. E.,** *Insects on Plants — Community Patterns and Mechanisms,* Blackwell Scientific, Oxford 1984.

233. **Sudlow, L.C., Edgecomb, R. S., and Murdock, L. L.,** Regulation of labellar and tarsal taste thresholds in the black blowfly, *Phormia regina, J. Exp. Biol.,* 130, 219, 1987.

234. **Szentesi, A. and Bernays, E. A.,** A study of behavioural habituation to a feeding deterrent in nymphs of *Schistocerca gregaria, Physiol. Entomol.,* 9, 329, 1984.

236. **Taghon, G. L.,** Beyond selection: optimal ingestion rate as a function of food value, *Am. Nat.,* 118, 202, 1981.

236. **Taylor, M. F. J.,** The dependence of development and fecundity of *Samea multiplicatus* on early larval nitrogen intake, *J.*

Insect Physiol., 30, 779, 1984.

237. **Taylor, M. F. J.,** Early diet affects growth, and food nitrogen use in the final instar of the salvinia moth, unpublished, 1988.

238. **Thompson, S. N.,** Biochemical and physiological effects of metazoan endoparasites on their host species, *Comp. Biochem. Physiol.*, 74B, 183, 1983.

239. **Timmins, W. A.,** Control of feeding in *Manduca sexta* larvae, Ph.D. thesis, University of Bath, 1988.

240. **Timmons, W. A., Bellward, K., Stamp, A. J., and Reynolds, S. E.,** Food intake, conversion efficiency, and feeding behaviour of tobacco hornworm caterpillars given artificial diets of varying nutrient and water content, *Physiol. Entomol.*, 13, 303, 1988.

241. **Treherne, J. E.,** The absorption of glucose from the alimentary canal of the locust *Schistocerca gregaria* (Forsk.), *J. Exp. Biol.*, 35, 297, 1957.

242. **Turunen, S.,** Absorption, in *Comprehensive Insect Physiology, Biochemistry and Pharmacology, Vol. 4,* Kerkut, G. A. and Gilbert, L. I., Eds., Pergamon Press, Oxford, 1985.

243. **van Loon, J. J. A.,** Sensory and Nutritional Effects of Amino Acids and Phenolic Plant Compounds on the Caterpillars of two Pieris Species, Doctoral thesis, Landbouw Universiteit, Wageningen, 1988, 103.

244. **van Loon, J. J. A.,** A flow-through respirometer for leaf chewing insects, *Entomol. Exp. Appl.*, 49, 265, 1988.

245. **Vinson, S. B. and Iwantsch, G. F.,** Host regulation by insect parasitoids, *Q. Rev. Biol.*, 55, 143, 1980.

246. **Volney, W. J. A., Milstead, J. E., and Lewis, V. R.,** Effect of food quality, larval density, and photoperiod on the feeding rate of the Californian oakworm (Lepidoptera: Dioptidae), *Environ. Entomol.*, 12, 792, 1983.

247. **Waldbauer, G. P.,** The consumption and utilization of food by insects, *Adv. Insect Physiol.*, 5, 229, 1968.

248. **Waldbauer, G. P. and Bhattacharya, A. K.,** Self-selection of an optimum diet from a mixture of wheat fractions by the larvae of *Tribolium confusum, J. Insect Physiol.*, 19, 407, 1973.

249. **Waldbauer, G. P. and Friedman, S.,** Dietary self-selection by insects, in *Endocrinological Frontiers in Physiological Insect Ecology,* Sehnal, F., Zabza, A.,and Denlinger, D. L., Eds., Wroclaw Technical University Press, Wroclaw, 1988, 403.

250. **Waldbauer, G. P., Cohen, R. W., and Friedman, S.,** Self-selection of an optimal nutrient mix from defined diets by larvae of the corn earworm *Heliothis zea* (Boddie), *Physiol. Zool.*, 57, 590, 1984.

251. **White, T. C. R.,** The importance of a relative shortage of food in animal ecology, *Oecologia*, 33, 71, 1978.

252. **Wurtman, R. J. and Wurtman, J. J., Eds.,** *Nutrition and the Brain,* Vols. 1-5, Raven New York, 1977, 1979.

253. **Yu, S. J.,** Consequences of induction of foreign compound-metabolizing enzymes in insects, in *Molecular Aspects of Insect-Plant Associations,* Brattsten, L. B. and Ahmad, S., Eds., Plenum Press, New York, 1986, 153.

254. **Zucoloto, F. S.,** Feeding habits of *Ceratitis capitata* (Diptera: Tephritidae): Can larvae recognise a nutritionally effective diet?, *J. Insect Physiol.*, 33, 349, 1987.

6

Mother Doesn't Know Best: Selection of Hosts by Ovipositing Insects

Steven P. Courtney and Travis T. Kibota
Biology Department
University of Oregon
Eugene, Oregon

TABLE OF CONTENTS

I. Introduction: How Theory Fails Us 162
    A. Proximate Factors 163
        1. Plant Chemistry 163
        2. Plant Morphological Characteristics 164
        3. Environmental Variation 164
        4. Biotic Interactions 165
        5. Insect Physiological Factors 165
    B. Ontogenic Factors 165
    C. Phylogenetic Factors 166
    D. Ultimate Factors — The Effects of Natural Selection 168
II. Toward a Theory of Host Selection 169
    A. Proximate 174
    B. Ontogenetic and Experiential 174
    C. Phylogeny/Genetic 176
    D. Ultimate Factors 179
III. Chemistry Is Important; Chemistry Is Unimportant 180
IV. Unanswered Questions 181
    A. Mechanistic 181
    B. Experience and Learning 181
    C. Genetics/Evolutionary 182
References 182

I. INTRODUCTION: HOW THEORY FAILS US

Over the past 40 years there has been an extraordinary effort to understand the relationships of plant-feeding insects and their hosts. Anyone who is studying such problems knows the mountain of literature which must be scaled. Perhaps 10% of the papers in journals such as *Ecology* consider phytophagous insects; in more specialized journals, such as *Journal of Chemical Ecology* or *Journal of Economic Entomology,* the figure is much higher. Moreover, the production of such research shows no sign of deceleration. Despite all this effort, we have collectively developed only a skimpy theoretical framework with which to understand insect diet breadth or plant defense strategies (for example). We will argue that new developments are certainly needed. If theory is the mainspring for our research programs, then we might indeed consider a drought alert.

It is instructive to compare progress in this field with the recent advances made in behavioral ecology, where theory has played a preeminent role in shaping the development of the science. For example, there is now a wealth of detailed observations and experiments which reveal the general principles underlying diet choice by predators. We cannot say the same for herbivores.

Several factors probably explain such differences; for instance, much work on plant-feeding insects is necessarily directed to applied goals. Insects are also more complex research organisms than, say, birds. Foraging by birds can usually be understood as a simple optimality problem, where individual learning can solve the problem. Insects can be viewed as more "hard-wired", where the role of sensory physiology is critical to an understanding of host choice. This means that to understand insect diet we must additionally know about the physiology, the genetics, and the phylogeny of host selection. Instead of an ecological approach, we are forced to adopt an evolutionary one.

This complexity also leads to a final problem with theory; it hasn't really proven very useful.

Consider the classic paper by Ehrlich and Raven,[36] which stimulated much research into the chemistry of plant-insect interactions. This paper developed an evolutionary theory concerning reciprocal effects of one organism on another. However, because of the very generality of the model Ehrlich and Raven proposed, it has been of little use in making predictions at the level of specific taxa; here phylogenetic constraints may determine why particular host shifts have occurred and others have not.

If the influential theories of plant-insect interactions can be counted on the fingers of one hand, then the evolutionary approach pioneered by Ehrlich and Raven deserves to be recognized as the opposable thumb. Next to it, index and middle finger, can be ranked theories of plant defense which predict plant chemistry from likelihood of insect attack[41,129] or from resource availability.[16,17] These theories have generality, and may explain large scale patterns in defense chemistry. Once again, however, they may fail at the level of specific taxa if other evolutionary influences (such as phylogeny) are more important.

These three theories each have considerable power to explain observed patterns in nature; they have each stimulated important research, and have frequently been invoked to explain the results of particular studies. However they have not proven rich in generating new testable hypotheses. Is this an inevitable consequence of any work in an evolutionary, hence historical, framework? We believe that it is not — that more predictive theories are both possible and desirable. In this paper we will examine one aspect of insect diet — the choice of hosts by ovipositing females — and discuss what an ideal theory would tell us at different scales of questions, from the mechanistic to the evolutionary. We will then examine how existing theory measures up to these demands. A simple, mechanistic model will then be explored in more detail. We expect that this model will be superceded by more sophisticated developments. Nevertheless, this simple approach is a first step to a comprehensive theory.

Our inspiration for this approach has been a forceful restatement by Charnov and Skinner[13] of ideas originally put forward by Tinbergen.[155] These authors argue that there are four "causes" for any behavioral pattern:

1. Proximate causes such as the interaction of physiology with sensory cues;
2. Ontogenetic causes including all developmental changes such as learning;
3. Ultimate causes reflecting the action of natural selection;
4. Phylogenetic causes determined by evolutionary history.

We agree with Charnov and Skinner that these four levels of causation should be studied together, since they yield complementary information. The problem now is to develop models which will allow this. What should go into them?

A. PROXIMATE FACTORS

Plant characteristics such as color, shape, size, and chemistry have long been recognized as important influences on insect host selection.[32,47,83,170] Physiological factors of the insects themselves, such as motivational state or receptor sensitivity, also play important roles, as do the interactions between the various physiological factors, which may be under additional environmental influence. Together these may all be considered as determining the proximate mechanisms of host choice.

1. Plant Chemistry

Plant secondary chemistry has attained a preeminent position as an explanation for patterns of host utilization. Thus Renwick and Radke:[128] "the host ranges of phytophagous insects are determined to a large extent by the presence or absence of specific chemicals in potential host plants"; and Pereyra and Bowers:[106] "Although other factors are certainly involved in the decision of whether or not to oviposit on an individual host plant or leaf (see References 139, 140), leaf chemicals have been suggested . . . to be critical as oviposition cues."

Recently this preeminence has begun to fade somewhat. For instance, Bernays and Graham[6] argue that biotic interactions are more important than host chemistry in determining host range (see section below) in spite of the preeminent role of chemistry at the behavioral level. The responses to Bernays and Graham are instructive. Some[35,46,134] reject the notion that plant chemistry is unimportant; others[4,25,74,76,124,134,147,154] wish to seek an ultimate approach which incorporates plant chemistry as one of several possible influences on diet.

Clearly host chemistry may have large effects on the performance of individual feeding insects,[7] but it is by no means clear that we should expect any clear correlation between chemically mediated suitability and actual host use. Table 1 lists 12 reasons why host chemistry may be unrelated to host choice; doubtless others can be found.

Even if host chemical suitability does not explain insect diet particularly well, chemistry may provide proximate cues. Chemistry is clearly important in oviposition site selection[42,100,117,133] and particular chemicals may be recognized by insects as deterrents or as specific positive stimulants, even if the chemicals are entirely neutral in effect;[5,20,95,127] in such cases, host chemistry is being used as a correlate of host identity. An ideal theory of host selection will then incorporate responses of insects to host-specific cues, such as chemistry, without making any necessary assumptions about the relationship of such cues to host suitability.

Table 1
TWELVE REASONS WHY HOST CHEMISTRY MAY BE UNIMPORTANT IN DETERMINING HOST CHOICE

A. Other factors determine host suitability
    1. Mechanical factors[21,99]
    2. Predation, competition, other biological interactions[5,72,107,142]
    3. Host phenology[168,171]
    4. Microorganisms[158]
    5. Micro-climatic conditions[73]

B.   6. Insects have mixed diets (i.e., are grazers)[152]

C. Suitability for larvae is unimportant
    7. Availability or predictability is more important[22,23,121,167]
    8. Adults select hosts where they can themselves feed (e.g., on pollen)[158]
    9. Adults select hosts where mate-encounter frequency is high[19]

D. 10. Suitable and unsuitable hosts are indistinguishable to females[15]

E. 11. Evolution has not yet eliminated maladaptive host use because insufficient time has passed or genetic variance is absent[14]

F. 12. Females select habitats where optimal hosts do not occur[10,24]

A number of authors have recently reemphasized that the insect searching for its host plant typically uses an array of senses.[33,57,59,95,100,116] In many cases these are used in sequence; in other cases, different sense organs may be employed simultaneously, and the insect may respond to a host "gestalt", which is only partly determined by chemical make-up.[34,100,144] Our ideal theory should incorporate any aspect of host recognition, central and peripheral, neural phenomena, physical and chemical factors.

2. Plant Morphological Characters

Searching insects often use general plant morphological traits in locating hosts. The color, shape, and size of a plant affects host selection in a wide variety of insect herbivores.[3,42,56,85,86,93,112] Discrimination may be fine-tuned or rather unspecific. For instance, Cappucino and Kareiva[10] note that the butterfly *Pieris virginiensis* is attracted to general upright forms. Another butterfly, *Battus philenor,* by contrast, not only discriminates between the leaf shapes of two congeneric host plants[104] but may also be able to distinguish between high and low quality conspecifics.[105] Jones et al.[78] have argued that intraspecific discrimination is important and may explain observed interspecific patterns across hosts of *Pieris*. Forsberg[45] has shown that two pierines do discriminate against large host individuals of several species. Similarly Thomas[151] has found that leaf age is more important than host species for two beetles associated with *Passiflora* spp. We expect that such patterns are widespread. As leaves age, they change in secondary chemistry, as well as nutritional value (water and nitrogen concentrations)[118,136] and toughness, as reinforcing lignins become deposited in the leaves.[16,40,135]

3. Environmental Variation

Variation among hosts may be caused by environmental factors. For example, previous herbivore damage is thought to induce changes in a number of plants, resulting in decreased acceptability to subsequent herbivores[60,119,160] (see also Reference 169). Conversely, some plants which are subject to drought or other environmental stress become more susceptible to insect attack[63,87,89] (see also Reference 110).

Environmental influences may also operate directly on the insects. For instance, many insects, being exothermic, are inactive during cold spells. They may then fail to utilize hosts which are available during such periods.[10,52] Plants that occur in shaded areas have a different herbivore fauna than those that inhabit sunny spots.[18,97]

4. Biotic Interactions

Evolutionary responses to biotic effects are discussed later. It is sufficient here to note that other organisms may make hosts more or less attractive to ovipositing females. For instance, the cabbage root fly, *Delia radicum,* avoids hosts which have frass from the caterpillar *Evergestis forficalis.*[79] Conversely, lycaenid butterflies which have mutualistic relations with ants may preferentially oviposit on those host plants bearing ants.[107-109]

The presence of other neighboring host plants may also have strong effects, both on host location and on host acceptance by females. The "resource concentration hypothesis"[131] suggests that specialized herbivores should be found in the densest patches of their hosts. Several phytophages[2,9,31,81,138,145,164,165] as well as mycophagous flies[172] do seem to show this pattern. Generalist species on the other hand may not utilize high density patches so intensely. Inverse density-dependent relations have been described in a number of taxa.[28,80,136]

The presence of non-host plant species may also have effects, for instance by masking host odors.[1] The well-known association between monocultural agriculture and insect outbreaks is sometimes explained by the absence of such an interference effect on the insect.[131,146]

5. Insect Physiological Factors

Internal factors also determine an insect's response to its host. Peripheral or central processes may impinge on both receptor sensitivity and on the response to cues. The number of eggs that a female insect carries may be one important parameter which affects females' responses. Jones[77] describes how females of *Pieris rapae* normally lay only one egg per host (the larvae are cannibalistic). However, when especially gravid, these females will return to the host and lay further eggs. Age[52] and number of matings[163] may impinge on eggload if the female continues to mature eggs. Fitt[44] has described how tephritid flies may broaden host range if a female begins to accumulate eggs. Another fly, *Delia antiqua,* accepts unusual hosts most frequently when highly fecund,[57] while Singer[140] has shown experimentally that butterflies broaden their range of acceptable hosts as the time since last oviposition increases. Roininen and Tahvanainen[130] show that, in a specialist sawfly, females which accept low-ranking hosts lay larger clutches, implying a larger eggload. In the field, Courtney and Forsberg[29] have found that host density may affect host breadth. Two pierids are monophagous on high-ranking hosts when these are abundant, but accept other plants if the highest ranked hosts are rare. Prokopy et al.[115] show a similar effect of host density. Other physiological parameters, such as food deprivation in *Drosophila* species,[155] may alter responses to potential hosts for oviposition.

Just as host chemistry cannot be seen as a single-factor explanation of host choice, so insect physiological status is only one facet of host acceptance. To be successful, any theory of host choice must incorporate all potential proximate effects on the recognition and acceptance of a plant by its insect herbivore.

B. ONTOGENETIC FACTORS

Ontogeny, the second of Tinbergen's four areas of study, deals with the lifetime dynamics of an individual. While the mechanical development of an organism undoubtedly plays an important role in determining host use (an extreme example being the trajectory toward fertility for one ontogenetic path vs. sterility along another), this branch of ontogenetic research has not yet been linked to host selection in phytophagous insects. Studies have, however, begun to be published on the effects of experience, here defined as an aspect of ontogeny.

The experience that an individual accumulates over a lifetime is unique to it and may therefore result in development of unique responses. For instance, past encounters with potential host plants are known to affect detection and selection of hosts in future encounters. In their excellent and extensive review of insect learning, Papaj and Prokopy[103] give many examples of such changes. The influence of larval experience on adult behavior (Hopkin's host selection principle) is now largely discounted in favor of the influence of early adult experience, which may indeed be of the larval host.[62,65,67,68,104] Adult experience of hosts may be one of the most important factors causing differences between individual insects in the same population.[121] Single events may also have large influences upon the behavior of females.[156] An ideal theory of host choice must therefore accommodate such effects of experience on an underlying template of decision processes.

Singer[140] and Thompson[153] have emphasized the need to distinguish between the ranking of potential hosts by an insect, and the specificity of the insect; two insects may rank hosts similarly but show very different specificity of responses. Papaj and Rausher[104] have pointed out that experience may in any one situation modify ranking or specificity or both together. At present the evidence suggests that modification in specificity is most common — in phasmids,[12] Lepidoptera,[120,140] and Diptera.[61,64,65,111]

The effects of experience on host specificity appear to result in more than just a simple change in host range. Jaenike[65] documented cross-induction of oviposition in *Drosophila melanogaster* — females which had been previously exposed to a host were more responsive to it (induced acceptance) as well as other hosts which they had not experienced (cross-induction). Papaj and Prokopy[102] found similar effects in the tephritid *Rhagoletis pomonella* on exposure to two potential hosts. In both these insects, cross-induction appears asymmetrical; exposure to the low-ranking host results in acceptance of the higher ranked host, but not vice versa. Some of the results of Hanula[54] on experience in a weevil suggest similar behavior. In *Dacus tryoni*,[114] and *Ceratitis capitata*,[115] asymmetrical cross-induction occurs but here it appears to be mainly aversion to an unfamiliar fruit. Chemical similarity might explain cross-induction to alternative hosts,[65] but it is hard to see that this would explain Jaenike's own results where the hosts were apples, oranges, tomatoes, grapes, and onions. Induction due to similar host chemistry also fails to explain the asymmetry of the responses seen. Papaj et al. (cited in Reference 114) suggest that similarity in size of alternative hosts may explain cross-induction in Tephritidae. Is asymmetrical induction of acceptance adaptive? This seems highly unlikely. We suspect instead that it is an epiphenomenon of changes in specificity. These results will therefore only be understood when we have mechanistic knowledge of the physiology of acceptance, as well as a thorough understanding of the evolution of specificity changes (Tinbergen's first and fourth questions).

Perhaps the greatest challenge to a comprehensive theory of host selection is to incorporate experiential effects. Learning has a big influence on host choice. It is relatively easy to demonstrate; its effects have begun to be modeled in isolation[68,123] but relatively little attempt has been made so far to assimilate it into the wider corpus of theory.

C. PHYLOGENETIC FACTORS

Evolutionary history has been seen as an important influence on host plant use since Ehrlich and Raven's[36] synthesis of host and butterfly phylogeny. We include in this section all historical effects including the presence of genetic variability for host use; in the following section we will consider how natural selection and other forces may act on such variability.

A central issue is the evolution of novel host associations and of host shifts. Ehrlich and Raven[36] argued that a majority of host shifts are to plants which share similar chemistry with the ancestral host — the insects which make shifts are then "preadapted" to their new hosts. Jermy[75] and Strong et al.[148] have shown that some host shifts do fit this pattern. Chew[14] gives some clear-cut examples for pierid butterflies which readily adopt introduced hosts in Cruciferae and related families. However, such cases are certainly not a majority. Strong et al.[148] show that in many cases host shifts occur between taxa which show no obvious phytochemical similarity. In the *Drosophila quinaria* species complex, the ancestral hosts are almost certainly fungi but sibling species may use araceous, solanaceous or curcubitaceous hosts which do not appear chemically close. Other shifts to unrelated host taxa may be equally dramatic. In the pierid genus *Colias*, the presumed ancestral hosts are legumes but some species utilize *Salic, Vaccinium*, and other novel plants. Watt (personal communication) has suggested that such shifts may come about due to close ecological proximity (e.g., similar growth form in the same habitats) rather than by chemical similarity. If this is the case, then we need to understand the mechanics of host selection in order to understand which routes to host shifts are more likely to occur.

Host shifts apparently occur with equal ease in polyphagous and stenophagous species.[11] Many of the actual host shifts recorded by Strong et al.[148] were range expansions by generalist feeders but examples were also given where rather strict specialists shifted to use novel hosts. If the phylogeny of host shifts is examined it is seen that taxa which make radical changes in hosts are often specialized on the new host, but not infallibly so. Hence *Papilio oregonius* appears specialized on *Artemesia* relative to "normal" umbellifer feeding *Papilio*;[153] *Colias interior* feeds only on *Vaccinium*, not on legumes; *Drosophila magnaquinaria* feeds exclusively on skunk cabbage (*Araceae*), not on mushrooms, while *D. quinaria* uses both substrates as well as others. In many species which do become specialized to use novel hosts, the larvae remain able to feed on the ancestral host. We suspect therefore that much specialization is behavioral, affecting only the adult, and comes about after the host shift has occurred. An ideal model will therefore allow host shifts to occur in specialists as well as generalists, by means of changes in specificity.

An explicitly phylogenetic approach has only rarely been applied to host use of any insect group, despite the obvious power of such techniques for studying the history of associations.[8,41a,94a,96] Exceptions include sibling taxa in *Rhagoletis* which use alternative hosts, populations of *Drosophila mojavensis* which use different cactus species, and three derived populations of the Colorado potato beetle which use three different novel hosts. In all cases, ancestral and derived females alike respond more to (prefer) the ancestral host over the novel hosts.[38,43,58,113] In *Heliconius* spp., all taxa feed well on recently evolved *Passiflora*, but only some are able to use "primitive" hosts.[143] Unfortunately we do not know which was the first group of hosts used by the butterflies. We badly need further examples where the directions of host shifts can be assured, and females can be assessed for their propensity to adopt hosts used in other populations. Such analyses allow us to assess the effects of "preadaptation", for instance, in the host recognition mechanisms of the insects. A growing consensus is that host shifts are primarily behavioral,[49,50,75] and that the physiological ability to feed on a host evolves independently of the shift itself. Phylogenetic analysis should allow us to determine whether these abilities arise before or after host colonization.

Large-scale changes in host use occupy one end of an evolutionary spectrum, at the other end of which are small-scale, within population effects. Genetic variance in host use has long been suspected from geographic differentiation, but has only recently been studied in any detail. Futuyma and Peterson[50] review the literature; more recent studies include those by Jaenike,[70] James et al.,[71] Rossiter,[132] and Thompson.[153] An important finding from these studies is that genetic variance for host use is widespread, even the rule. The full gamut of genetic effects can be demonstrated from single gene effects,[11] to sex-linked loci,[152] to polygenic control,[26,88] to complex epistatic interactions[69] and correlation between host use characters.[98,141,159] Oviposition may also be affected by cytoplasmically inherited factors.[162] Clearly no one pattern is emerging as most common, although polygenic variation does seem to be present when it is looked for.

The results of Singer and his co-workers on the butterfly *Euphydryas editha* seem to suggest that genetic variance may exist for specificity of host associations. Other data, however, suggest that ranking of hosts may vary.[141] One emergent result is that genetic variance may exist for use of hosts which have never been experienced by the population concerned (nor by their ancestors). Hence *Drosophila mojavensis* populations show significant heritability for acceptance of novel hosts (cacti) by females.[88] This pattern is repeatedly found in other *Drosophila* such as the stenophagous *D. suboccidentalis,*[26] and the distantly related, polyphagous *D. busckii,*[30] both of which show variance for use of commercial mushrooms and cucumbers.

Our ideal model of host selection must therefore allow differences between individuals which reflect genetic variance. Specificity and host ranking should be explicitly included as variables, preferably independent of each other. The model should allow this variance to exist regardless of whether the host has even been encountered by the insect lineage, which implies that the traits need not be subject to direct selection.

D. ULTIMATE FACTORS — THE EFFECTS OF NATURAL SELECTION

The evolution of host choice under natural selection has been a major research focus, and several excellent reviews are already available.[5,49,51,123,152] Without reiterating the results of these authors, it is worth emphasizing a single point: the expectations of a naive adaptationist viewpoint are not met. For instance, many hosts contain compounds which are deterrent to insects but which have few post-ingestive effects,[5] implying that many insects are avoiding foods which are perfectly suitable to be eaten. Again, in many cases the adult oviposition range is restricted to a few hosts, although larvae will develop as well or even better on other hosts.[130,152,166] *Drosophila magnaquinaria* is a strictly monophagous species, whose range matches that of its host, the western skunk cabbage *Lysichitum americanum.* Females have never been found to lay eggs on alternative hosts outside of host patches. They will, on the other hand, readily oviposit on alternative hosts placed within skunk cabbage patches and in the lab. In addition, the larvae perform better on the alternative hosts, which include mushrooms, the ancestral hosts.[84] A similar result may be found in cactophilic flies (W. Starmer personal communication). Again, such results suggest that evolution of host choice operates largely through female oviposition behavior, as several authors have already argued.[11,49,75] They also show that an ideal model of host selection should focus on such behavior and not on the fitness of larvae on alternative substrates. Of course it will be interesting to incorporate larval fitness later, to see if this alters the results obtained.

If intrinsic suitability of a host as larval food is neither the only nor even the major selective agent favoring specialization, why are so many insects so specialized? Bernays and Graham[6] have reopened a debate on the role of biotic interactions in determining host choice. They focus on the effects of predators, although competitors and parasites are also likely to play a role. Jaenike[66] for instance has shown that amanitin containing fungi are rarely infested with nematodes which adversely affect *Drosophila* larvae; *Drosophila* which oviposit on such hosts are therefore utilizing "enemy-free space". Smiley[142] similarly argued that *Heliconius* spp. could become specialized on hosts as a consequence of competition. Alternative routes to specialization may lie through mating rendezvous sites, which will restrict host affiliations,[19] or through constraints on perception, so that specialists find hosts more readily than generalists.[23] We agree with Futuyma and Moreno (Reference 151; and see Table 1) that all these factors "are likely sources of selection for specialization [but that] the evidence on their role is circumstantial at best."

II. TOWARD A THEORY OF HOST SELECTION

Our survey of the literature shows that each level of investigation has problems that can only be solved by input from other levels. A recurring theme is that behavioral evolution is the rule, and that this needs to be understood from a mechanistic basis. Macro- and micro-evolutionary patterns alike show that host acceptance must be understood in terms of female responses to stimuli. Some behaviors appear to require explanations that invoke epiphenomena of other traits rather than selection on the characteristics themselves: cross-induction of acceptance, and the existence of genetic variance for use of hosts which are not encountered in the wild. On the other hand, in order to understand the action of natural selection, it seems essential to study and incorporate these epiphenomena into our models.

Several authors have discussed mechanistic approaches to the search and acceptance of hosts. Rausher[123] for instance has developed a simple model of learning which allows individuals to switch between alternative hosts. Females alighting on a leaf of a particular shape, which proves to be a suitable host, are more likely to respond to that leaf shape in future. Females will then receive a positive feedback effect from successful host encounters. Host switching ocurs when females accidentally land on a suitable host of a different shape. This model is essentially a restatement of a rule formulated by behaviorists interested in predator foraging: the relative payoff sum (RPS) learning rule.[55] Extensive literature on this learning rule is already available.[90,91,126] The essential features of it are shown in Figure 1.

In the RPS learning rule animals have a tendency to perform one of two or more alternative behaviors dependent upon past success (payoff) with that behavior, as well as the length of memory for such successes, and an innate capacity for switching. Switching could be seen as errors or mistakes, but may also be seen as sampling in a fluctuating environment which may allow changing conditions to be tracked.

Battus philenor females face a relatively simple choice: which of two leaf types to approach. Success is then easy to measure — it is the number of encounters with a particular leaf type which prove to be hosts. In other insects more complex decisions are required: whether or not to accept a particular host from an array of potential food plants. A particular problem here is the notion of "payoff". Unlike predators, ovipositing insects receive no direct calorific reward for an action but their fitness is affected through survival of their offspring. In this case, payoff may best be understood in terms of oviposition rate when different hosts are accepted, discounted by a constant which represents host acceptability (we make no assumption here that acceptability is related to suitability). Ward[161] has modeled this sort of situation using a dynamic programming approach, and has produced some satisfyingly intuitive results: (1) A discrimination phase should exist during which only high-ranking hosts are accepted (cf. Singer[139]); (2) specificity depends on factors potentially affecting searching efficiency including time

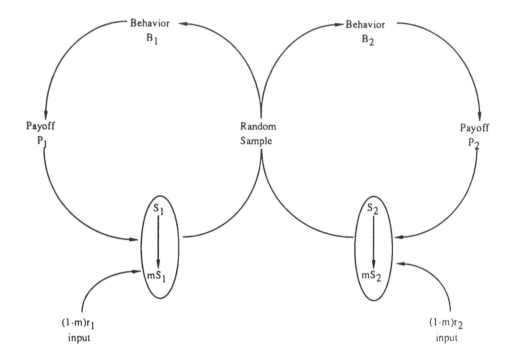

FIGURE 1. The Relative Payoff Sum (RPS) learning rule. A hypothetical animal is faced with a choice between two alternative behaviors (such as acceptance of two alternative hosts). Which of these two alternatives is performed is probabilistic. *Two* state variables (S) affect this probability; a high value for S_1 relative to S_2 increases the probability that the animal will perform behavior 1. Whichever behavior is chosen, the animal receives a payoff (in the case of feeding animals this is presumed proportional to nutritional value). This payoff increases the relevant value for S: hence an animal receiving a large payoff will have a higher S for that behavior, and will have a higher probability of repeating it. Two other factors affect the values of S: m (a decay function, equivalent to the effect of memory) and r (a residual input, which reflects an inbuilt bias to performing a particular behavior). To see how these also affect host choice, consider the case where m is low and r is large. Values of S will then largely depend on r, and the effects of memory or payoffs will be small. By contrast, when r is small and m large, inherent biases of the animal will have little effect, and the probability that an animal will choose one behavior or another will largely depend on past successes (relative payoffs) with either strategy. (Redrawn from Harley, C. B., in *J. Theor. Biol.*, 89, 611, 1981. With permission.)

available for search, abundance of high-ranking hosts, relative differences among hosts, and mortality during search; (3) individual differences in host selection should lead to population structures which approximate the "ideal free distribution" of Fretwell and Lucas.[48]

An alternative approach has been followed by Miller and Strickler[94] and others who have emphasized the plethora of factors, both external and internal to the insect, which may affect an oviposition decision. Miller and Strickler formulated the "rolling fulcrum" model of host choice, shown in Figure 2, which bears some similarities to the studies of Wiklund,[167] Singer,[139,140] Singer et al.,[141] Ng,[98] and Thompson.[153] Miller and Strickler envisage that at each host encounter the insect's decision is determined by its response to its own physiological condition as well as to external stimuli. This approach allows any effect of past oviposition to be of any kind or even entirely absent (if past oviposition affects only some physiological state such as eggload in the abdomen). Some may find this "black-box" approach unappealing; we find it refreshing in that it retains the flexibility to build in only those complexities which are important in each specific system.

Miller and Strickler leave unspecified the ways in which alternative hosts stimulate

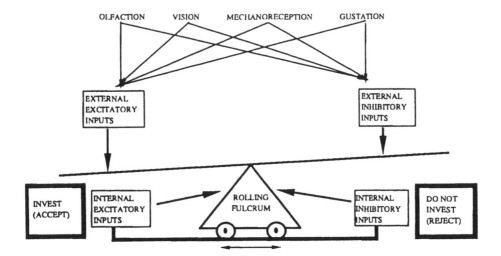

FIGURE 2. The rolling fulcrum model of Miller and Strickler (1984). Sensory input, from whatever source, has both excitatory and inhibitory effects. Similarly, internal factors affect acceptance. The internal factors determine the ways in which the external factors will influence acceptance. If the balance of stimulation favors excitation, then the host is accepted. (Redrawn from Miller, J. R. and Strickler, K., in *Chemical Ecology of Insects*, Bell, W. J. and Carde, R., Eds., Sinauer, Sunderland, MA, 1984, 127. With permission.)

insects, and the mechanism by which information is processed by the individual. Their model also is uninformative about the phylogeny or genetics controlling host selection. Singer[140] and Thompson[153] have developed an approach which may point the way to a more general model. They emphasize that females encounter hosts one at a time, and that females may change diet breadth during their lives, such that specificity may alter. Females may then change from monophagy to a more generalist-type habit, or vice versa. Interestingly, rank-order does not change, in that a host which is usually accepted early in life as the sole host will never be accepted less than other hosts in later, polyphagous life. Singer[140] therefore argues that females change specificity of responses but not rank-order.

This inflexibility of rank-order of hosts is central to the hierarchy-threshold model of Courtney et al.,[27] which is described below and in Figure 3. This is a model of individual oviposition which is mechanistic, incorporating information about peripheral sense-organs as well as central physiological processes, but which makes predictions at all four of Tinbergen's levels. It is a conceptually simple model which makes only the one simplifying assumption about rank-order invariability. This assumption is certainly incorrect in some circumstances (e.g., classical host-specific conditioning), and the model may be elaborated in this and many other ways. However it is instructive to explore a simple model, to see which phenomena may be explained. Our ultimate goal is a theory of host choice which has both power and generality — just as foraging theory is for predators. By beginning with the simplest possible model, we should be able to determine where further complexities are essential and need to be added. One important point is that the model does not distinguish between stimulants and deterrents of oviposition. An unacceptable plant therefore may be either deterrent or not recognized as acceptable.[33,95]

The model treats specificity and rank-order of hosts as separate issues. Specificity is a central process which is affected by internal factors: physiological condition, age, nutritional status, travel time since last host, etc. Just as in the rolling fulcrum model, all these are integrated to determine the motivational state toward oviposition at a

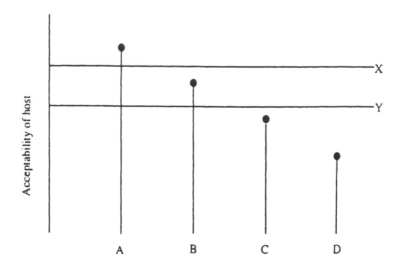

FIGURE 3.   The hierarchy-threshold model of Courtney et al. (1989). An insect encounters individual host plants one at a time, and makes a decision as to whether or not to accept each one. Sensory input from the host (A to D) determines its 'acceptability'. The balance of attractants and deterrents yields a net positive stimulation: acceptability. All senses can be involved in determining this balance. Acceptability thus concerns external stimuli. Rank-order of hosts is determined solely by their relative stimulatory effects. Thus host A is highly stimulatory, while host D is less so; host A is therefore higher-ranked than host D. Whether or not either host is actually accepted depends however on internal factors, which set a motivational threshold for acceptance (X or Y in the Figure). If the acceptability of the particular plant is above this critical value, it will be accepted, and oviposition will occur. If the acceptability is too low, the plant will be rejected. As internal factors change, this critical threshold will also change. For instance, in the Figure, the insect is initially highly selective of hosts (condition X) and accepts only host A. But as conditions change (for instance, the female ages, increases her eggload, or flies a long way without finding a host), she becomes more motivated to oviposit, the critical threshold decreases (to condition Y), and hosts of type B are now accepted. Note that changes in diet therefore come about without any changes in the responses of the sense-organs to the hosts. Under this formulaton, rank-order is a function of acceptability (external stimulation), while specificity is a function of the motivational threshold (internal condition). (Redrawn from Courtney, S. P., Chen, G. K., and Gardener, A., in *Oikos*, 55, 55, 1989. With permission.)

particular host encounter. Whether or not this motivation is sufficient to cause oviposition depends on the "acceptability" of the host, which is determined by the stimuli received at the periphery. Oviposition will occur if the stimulus differential is positive (i.e., attractants outweigh deterrents) and exceeds the current motivational threshold. Ranking of various alternative hosts is then a consequence of their differing acceptabilities and is therefore purely a result of peripheral inputs from different hosts. Specificity is the interaction of motivational state with acceptabilities of the various hosts.

This approach immediately focuses attention on important characteristics which warrant investigation. For instance, some authors have identified insect populations which are polymorphic for host use strategies, where some individuals are more generalist than others.[26,98,104,141] In the model, generalists may take three different forms (Figure 4). First, all hosts could have similar acceptabilities, so that sense organs respond similarly to all of them. Secondly, females could have very high motivation to oviposit, such that the thresholds for acceptance are low. Thirdly, females could show

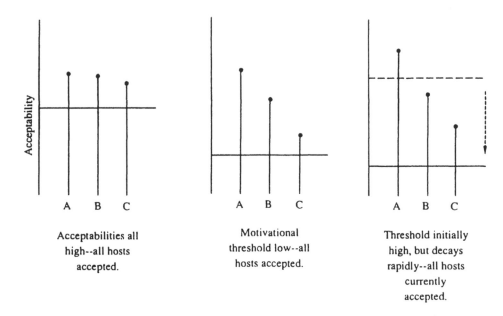

FIGURE 4. Three ways in which an individual insect could be a generalist under the hierarchy-threshold model. The sense-organs of the animal may perceive each host similarly, so that each gives sufficient stimulation to fall above the critical threshold for acceptance. This would be the case in a mutant individual whose receptors showed no specificity in response. Alternatively, internal factors may result in lowered specificity of response (middle figure). For instance, a highly fecund female might begin adult life as a generalist, because her threshold to oviposit is very low. Lastly, a female may exhibit a very rapid response to stimuli, so that the critical threshold decays swiftly. Acceptability, response to internal factors (setting the threshold), and the rate of decay of the threshold are all presumably evolved characters. It is therefore possible that, within a single population, individuals which share a generalist phenotype may nevertheless have quite different genotypes for host-ranking and specificity.

high sensitivity to the internal factors responsible for changes in motivation, so that the acceptance threshold decays rapidly. All three mechanisms will produce phenotypes with low specificity, but the underlying genotypes are very different. Evolution will similarly proceed in very different directions when selection acts on these phenotypes.

Currently we have little information on whether generalist individuals do fit any or all of these phenotypes. Association of low specificity with high fecundity[27,44,77] (see below) might indicate that the second mode (individuals with high motivation) is common. However, results with *Euphydryas editha*,[98] showing complex larval-adult genotype associations, implies that the first mode (lack of sensory discrimination between hosts) is also a real possibility. Ng's fascinating study shows that a single population of insects may consist of both generalist and specialist genotypes. He argues that failure to recognize such diversity will lead to an improper understanding of how evolution proceeds. We agree with this view and add that even more variation may exist in the groups that Ng defines. Figure 4(C) shows an individual whose oviposition threshold is sensitive to internal factors. As physiological states change, the insect becomes more or less specialized. Broad classifications — e.g., generalists vs. specialists[23,51] — will not adequately describe this type of individual and a more dynamic approach (e.g., Reference 161) is needed.

Courtney et al.[27] give a list of 18 specific testable predictions that arise from the model; we have not repeated these here. We focus instead on each of Tinbergen's levels of enquiry and consider how the hierarchy-threshold model fares in explaining the problems recognized in the section above and in setting directions for future research.

A. PROXIMATE

The proximate causes of host choice encompass a wide range of external stimuli and internal responses. The hierarchy-threshold model is successful at incorporating these, as it makes no differentiation between different sensory cues that might be used in host detection and acceptance. Any of the many senses of insects might be incorporated into the "acceptability" of a particular host; this also allows the insect to respond to a gestalt of the host. The model can be applied to any stage of the host location sequence — the "catenary process" of host selection.[82] Thus, it may be useful to distinguish between the acceptabilities of different hosts at different points in that sequence. A useful distinction in studying insect oviposition is between pre-alighting and post-alighting discrimination. The hierarchy-threshold model suggests that acceptabilities of different hosts may well change along the host location sequence. Hence resultant changes in ranking and specificity may be expected. Papaj and Rausher[104] have already argued from a different (evolutionary) standpoint that pre- and post-alighting preferences may not be identical. Observed patterns of host selection certainly conform to these predictions. *Drosophila magnaquinaria* is a strictly monophagous species in nature, and yet will readily accept other hosts in the field, but only where the normal host is nearby.[84] Similar results are found in the lab in this and other insects.[130] The papilionid *Zerynthia rumina* similarly will occasionally accept non-hosts when these are in the vicinity of the normal host (Courtney, personal observations). In these and many other cases, insects appear to be specialized at an early point of the host location sequence, but less so later in the same sequence. This strongly implies the operation of different, noninteracting sensory modalities in the different stages. The model suggests that it would be fruitful to understand how acceptabilities vary between hosts along the whole sequence. Papaj and Rausher suggest that selection favors strong discrimination early in the sequence, as this minimizes lost time spent investigating inappropriate hosts. If this is true, a major facet of host selection is habitat selection at the broad scale, yet this has received relatively little attention from students of insect-plant relations.[24,150]

The hierarchy-threshold model does satisfy the criteria demanded at the proximate level. It is a theory of behavior, which does not make any necessary assumptions about the relationship between host acceptability and host suitability; these may be positively related or negatively related, as is found in nature.[152] It also allows external stimuli to interact with internal factors, such as egg number, which is known to affect insect oviposition.[27,44,101,163] It does not, however, allow central processes to change the acceptability of individual hosts. For instance, if increasing egg number were found to increase detection of one phytochemical but not others, this would violate the assumptions of the theory. Such patterns have not yet been described, but it would be instructive to carry out a more systematic search for them. Violation of this assumption would imply that acceptability of each host might then change independently of the alternatives. It would also imply that the very pattern of changes in specificity and ranking was itself hard-wired, under genetic control, and hence presumably an evolved trait. We do not believe this will prove to be the case.

B. ONTOGENETIC AND EXPERIENTIAL

One of the successes of the model is that it provides a framework for incorporating learning into the more general theory of host selection. It suggests that we should investigate learning effects on ranking and upon specificity rather differently. It also allows learning to take several forms. Thus an integration of information over a lengthy time period, incorporating memory effects, is possible, and would be incorporated into changes of the motivational state. This would then wed the approach taken by students of foraging theory with that taken more normally by those studying phytophagous insects.

Alternatively, single events of large effect are also understandable (e.g., Reference 156) if they cause irreversible changes in, say, the threshold for host acceptance. In both cases we can make predictions about how learning will change the animal's diet: we predict that effects on specificity are quite independent of effects on host ranking. Note that the majority of cases which have been investigated do seem to support this prediction.

The theory is also robust in dealing directly with the higher-order experiential effects of cross-inductions. The reason for this is that induction of acceptance of a low-ranking host should involve lowered specificity (see Figure 3); hence other hosts are now accepted, even though there has been no direct induction from them. Note that this is true, even if these alternative hosts are recognized by different sensory receptors. Hence, in Jaenike's[65] study, apples, oranges, onions, etc. may each cause different receptors to fire, but cross-induction still takes place because it is the central change in specificity which has occurred rather than a change in the sensory processes of detection. The model therefore suggests that we should routinely see cross-induction of acceptance of hosts which bear no chemical similarity. We find this more appealing as an explanation of Jaenike's results than the suggestion that these hosts have some hidden chemical constituent in common. Alternative models should however also be examined. Changes in receptor sensitivity do correlate with induced changes in preference, so that any generalized response would produce cross-induction. E. Bernays has pointed out to us that such changes could occur, for instance, in receptors responding to all chemicals affecting sodium channels. Cross-induction would then occur for all hosts containing such chemicals.

By the same reasoning, the theory explains the asymmetry of cross-induction; experience of a low-ranking host which causes large changes in specificity will result in acceptance of all higher-ranking hosts. Experience of a host may not however cause acceptance of hosts ranked below it. Hence *D. busckii* exposed to avocado (a very low ranked host) show little specificity on subsequent tests, and willingly accept banana and other potential hosts. Siblings exposed to banana, conversely, do not show increased acceptance of avocado (Courtney, Gardner, Clements, in preparation). The theory therefore makes some rather bold predictions about the ways that induction of acceptance occurs; it will be instructive to see how well these predictions are met by data.

A different sort of learning is not allowed in the model: changes in the responses of specific sensory neurons independently of other neurons. Such changes would make acceptability change independently of propensity to oviposit. Clearly induced responses at the receptor level do occur. The success of the theory in general application then becomes a test of how often this assumption is violated. If specific receptor changes are frequent, this could be readily incorporated into the model. It should be possible, for instance, to develop a theory which accounts for simultaneous changes at the receptors (affecting acceptability) and in the CNS (affecting propensity to oviposit and hence specificity). We may then consider the relative contribution of different forms of learning to the changes in oviposition behavior of a single individual.

Although such an elaboration of theory is beyond the scope of this paper, it is a conceptually easy step. For instance, it is easy to see that the effects of one learning process will depend on the rate at which it proceeds relative to another. Hence if changes in neurons are rapid relative to changes in specificity, then we may expect big changes in rank-order of hosts. If, alternatively, the processes are more comparable in rate, we should see relatively little change in rank-order. Our intuition is that this last situation is the case. For example, the classical induction seen in *D. melanogaster*[65] and *D. busckii* (Courtney, Gardner, Clements, in preparation) is indeed accompanied by changes in acceptability, but these changes are small and do not affect rank-order greatly.

C. PHYLOGENY/GENETICS

We have stressed repeatedly that many factors affect acceptance, and that several different receptors and receptor modalities may be involved in recognition of a host. It should therefore come as no surprise to us that no one factor can be invoked as the sole cause for shifts in host affiliation. The hierarchy-threshold model is hence entirely compatible with the view that similarities in host chemistry may facilitate, but are not necessary, and are certainly not vital to the process of host shifts. The theory does suggest that similarities may occur prior to the incorporation of a new host into the diet, but these need not be chemical in nature. We have already emphasized the catenary nature of host selection, where different sense organs may be used at different points in the host location sequence. It is possible that host shifts may occur due to similarities in the stimuli received at any one of these stages. Thus female insects which search visually might lay eggs on a novel host which resembled the normal host in shape and color, in the absence of other cues. Insects parasitic on fruit, whose hosts have evolved recognizable common patterns to attract dispersers, may be one profitable group in which to study such host shifts. D. H. Janzen (in preparation) has recorded one host shift in Costa Rican beetles which appears to be best explained by similarities in visual image. Similarly, shifts to hosts which share little chemical similarity (such as the hosts of *Colias*) but close ecological proximity are explicable by changes in acceptability at an early stage in the host location sequence.

One of the problems of working with host shifts is that it is not often possible to predict when such shifts are likely. There is, however, considerable interest in doing so, because of the importance of host shifts in the origin of pest insects. The scenario derived from Ehrlich and Raven's classic paper, where most but not all host shifts are dependent upon chemical similarity, clearly fails to explain observed patterns.[148] Does the hierarchy-threshold model fare any better? We believe that it has the potential to do so, particularly when used within a phylogenetic framework. For instance, it should prove possible to observe a population's potential for use of [novel] hosts, such as those used by other members of its taxonomic lineage. It may then be possible to determine that certain host shifts are most likely, which stimuli will be most important, and at which position in the host location sequence the shift would be likely to occur. For instance, we are assured that steps late in the host location sequence are no barrier to the use of novel hosts by members of the *D. quinaria* complex; our studies suggest that early phases in host selection are associated with host shifts. Moreover, certain host shifts appear to have occurred several times (such as that to use of skunk cabbage) suggesting that there is indeed some sensory basis for such shifts.

Though larval fitness will undoubtedly place some limits on which shifts are successful, the model allows a dissection of larval performance and adult behavior. This suggests the possibility of preadaptation for host shifts at the behavioral level, as well as at the level of host suitability. Those studies which have accumulated to date have concentrated on larval preadaptation to use of novel hosts (e.g., Reference 149). The hierarchy-threshold model suggests that behavioral shifts may be a more profitable area for study.

The model does not regard specialization as an evolutionary dead-end (see Reference 51). Hence specialists are as likely to evolve the use of new hosts as are generalists. This matches well with observed host colonization events. The theory does make the suggestion, however, that all host shifts, regardless of the original host use strategy of the insect, should involve changes in specificity rather than changes in host ranking. Studies are needed which observe actual host shifts in the present. Comparing different taxa which shifted in the past will not provide useful information, since such taxa will have undergone additional adaptation to the host (following the shift) resulting in new rankings (see below).

It is becoming commonplace to observe genetic variance for host acceptance, even when the hosts are not used by the population concerned, and even if there is no evolutionary history of such use. It is difficult to invoke direct natural selection to explain such results, but they are readily understandable from the hierarchy-threshold model, which suggests that hosts may serendipitously evoke similar responses from insects (see Reference 14 for some examples). Chemical similarity is just one example, where two different hosts may be perceived similarly by an insect's chemoreceptors. The theory focuses attention on just what those similarities are. It also argues that observed genetic variance may take several forms: for instance, variants may have different responses to different hosts (different acceptabilities, and hence different host rankings), or they may differ in specificity. Again, a number of studies suggest that variance in specificity may prove more common.

The theory is comprehensive in allowing any form of genetic control of host acceptance. Because of its formulation, the model does focus on acceptance as a threshold trait (see Reference 39 for a definition) and polygenic inheritance of such threshold characters has been confirmed.[26,30,88] However it also allows single genes of large effect (which may, for instance, cause large changes in specificity, or in detection of a particular host constituent). It also allows epistatic interaction between traits. Consider the following example (Figure 5): An insect selects among two hosts, X and Y (X is the lower ranked host). Each insect has two loci affecting the decision to accept X; character A controls acceptability of X, while character B controls the oviposition threshold. In addition, each character has two alternative phenotypes: A' and A" show low and high acceptability of X, respectively; B' and B" show low and high propensity to oviposit. As shown in Figure 5, only one of the four combinations of phenotypes (A"B") accepts host X, therefore demonstrating epistasis. Jaenike[69] has described how genetic variance for host use in *Drosophila tripunctata* appears to be polygenic, but with additional epistatic interactions. This is to be expected under the hierarchy-threshold model if there are quantitative characters affecting specificity and acceptability independently of one another.

The theory also predicts extensive pleiotropic gene action, and positive genetic correlations between use of different hosts, even when these are chemically dissimilar. This follows from the action of quantitative characters affecting specificity but not host ranking. Genes which cause a generalized lowering of specificity should result in acceptance of different hosts, even though these may be detected and recognized using different stimuli, and even if there is no evolutionary history of association with either host. This unusual prediction has been addressed in our lab, and has been corroborated for two species.[27,30] It is to be expected that this pattern will be found in other species where changes in specificity are not associated with changes in rank-order: for instance, in Tephritidae[44] and some butterflies.[29] Similarly we expect genetic correlations between use of novel hosts and the intrinsic factors affecting specificity. We have found such correlations between use of novel hosts and the number of eggs that females have in the abdomen.[27,30]

Such correlations are however only expected where genetic variance in specificity is common, but genetic variance in acceptability is rare (M. Geber, personal communication). Since we have found such correlations in two taxonomically distant members of the same genus, *Drosophila,* we suspect that variance in specificity is indeed more common than variance in preference determined by rank-order. Moreover, our results argue that this pattern is maintained for long periods in evolution (*D. suboccidentalis* and *D. busckii* are in separate subgenera, implying a long time since separation). What could explain such a pattern? One of the basic propositions of quantitative genetics is that there should be little genetic variablity for characters which are under strong di-

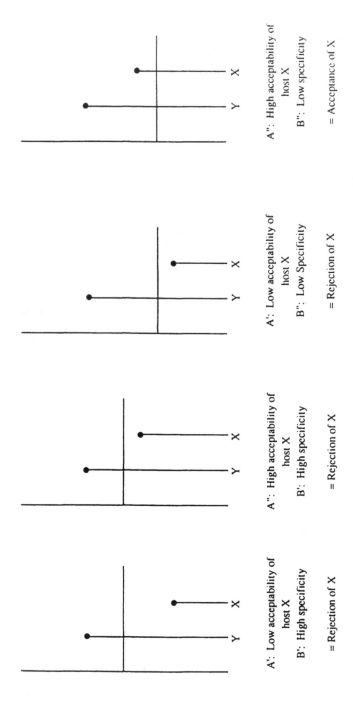

FIGURE 5. Epistasis in host choice under the hierarchy-threshold model. Two characters (A, B) each have two alternative states. A' individuals show low acceptability of host X, A" individuals are more stimulated by it. B' individuals have high specificity, B" individuals low specificity. Of the four possible combinations of phenotypes, only A"B" accepts host X. For further discussion see text.

rectional selection, but that variance may be maintained under other selection regimes, such as balancing selection or stabilizing selection. We suspect specificity is under stabilizing selection, and that mutations which affect the acceptability of low-ranking hosts are slow to accumulate. This will then result in high genetic variance for specificity, but lowered variance in rank-order, maintaining the observed correlations. A major prediction for future research is that host ranking evolves less rapidly than changes in specificity. Populations where host-ranking may vary (e.g., Reference 141) therefore are important areas for extended study.

D. ULTIMATE FACTORS

The hierarchy-threshold model does not explicitly incorporate the effects of natural selection acting through larval fitness and the suitability of alternative hosts. It would be possible to create such a theory, and to make predictions about which hosts might be used by populations. Ultimately it will be of interest to know the relative rates of evolution for adult and larval characters. It is sufficient here to note that there is no necessary correlation between host suitability and acceptance — nor is this indeed routinely found in nature.[152] Other models have been proposed which consider the joint evolution of host acceptance and host suitability (through larval adaptation to the host).[53,122,125]

We suspect that these models are considering unusual circumstances, and may largely be inappropriate, since the evolution of host choice appears to be essentially behavioral. The hierarchy-threshold model is not only much simpler than such models, it is more appropriate because it incorporates the constraints operating on behavior, which are probably of major effect.

Of more interest is the proposition that host use evolves primarily through changes in specificity. How can this result in adaptive changes in host use? Figure 6 shows one case where two populations differ in host use, one having evolved from the other. No evolution has occurred at loci affecting acceptability of the altenative hosts, but there have been changes in specificity at two stages in the host selection sequence. The ancestral population shows high specificity at an early stage of host selection, but low specificity later, in the pattern that seems typical of *Drosophila*. The derived population has lowered specificity in the early part of selection, but higher specificity later (e.g., a generalized visual attraction followed by a more specific olfactory response). For such processes to operate, it is necessary that different sets of sense organs be responsible for recognition of, and response to, each host. Messina et al.[92] have recently shown this pattern in *Callosobruchus maculatus,* where different ablations resulted in different oviposition preferences. Jaenike[67] describes how the responses of *Drosophila tripunctata* to alternative hosts change at different stages of the host location sequence. Note that we could reverse the direction of evolution: there is no apparent need for earlier or later stages to be more or less specific. The two populations are monophagous on different hosts, even though neither has changed in the strength of response to host stimuli. Since the derived population accepts all hosts early in the sequence, it may be that females spend more time on close-range identification and rejection of unacceptable hosts. In this case any additional behaviors which limit such contacts (such as area-restricted search) will be favored by selection.

Application of the theory to a sequence of behaviors therefore allows the evolution of monophagy on novel hosts due only to shifts in specificity. It will be interesting to see if empirical results demonstrate such a pattern. What is needed is an investigation of receptor sensitivity in a taxon where host shifts have been described. The geographic populations of the Colorado potato beetle would be well suited.[58] We would predict that the populations would not differ in response to host stimuli at any level. Since this

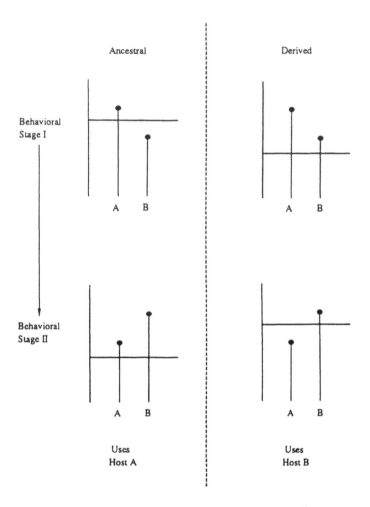

FIGURE 6. Host shifts without changes in host ranking. At least two senses are used in host choice, one following the other, and acceptability of alternative hosts differs for the two steps (cf. Messina et al. 1987). In the ancestral population, the acceptance threshold for stage 1 is high, for stage 2 low. As a consequence, the insect is specific in stage 1 and accepts only host A, which is the sole host used. In the derived population, specificity is low at stage 1, but higher at stage 2. All hosts are therefore accepted at stage 1, but only host 2 at stage 2. The derived population therefore uses a host which the ancestral population does not, even though no change in receptor sensitivity has taken place.

is the opposite of naive, intuitive theory (that insects should respond more to their natural host), it is interesting to speculate that there is a bias against reporting such results. We hope that this paper will emphasize the positive value of such "negative results".

III. CHEMISTRY IS IMPORTANT; CHEMISTRY IS UNIMPORTANT

Undoubtedly the dominant approach to understanding insect-plant interactions has been to investigate the effects of the diverse array of plant constituents on insect feeding and host selection. Recently there have been calls for a more pluralistic approach which will take account of, for instance, biotic interactions.[6] Our Table 1 is one more argument in a chorus of voices, that are collectively arguing that host chemistry is not sufficient to explain the full range of phenomena concerning host choice. Nevertheless there is a persistent trend to seek chemical explanations wherever possible. For instance, Jaenike[65] in his ground-breaking discovery of cross-induction of acceptance, sought an explanation for his results in unknown chemical similarities between apples, oranges, grapes, tomatoes, and onions. Similarly, reviewers of our manuscripts have criticized us for assuming that skunk cabbage, fungi, tomatoes, and cucumbers were chemically dissimilar. Again, the vehemence of some of the respondents and critics of Bernays and Graham indicates the high investment many have placed in the "host-chemistry/ insect diet" paradigm.

We believe that it is no longer tenable to argue that host chemistry has been the major force in the evolution of diet of plant-feeding insects. Both of us have come slowly and reluctantly to this view, having been thoroughly versed in orthodoxy. However, our experience with a number of insects, and the growing impetus given by reviews (such as Thompson's[152] demonstration of poor correlations between host choice and host suitability) have forced our reappraisal.

How can we reconcile this pluralistic view on the evolution of diet with the undoubted prominence of olfactory and taste cues in host detection and selection? We believe that the hierarchy-threshold model suggests a way, because it distinguishes acceptability and suitability. Tinbergen's separation of ultimate and proximate causes of behavior emphasizes that the evolutionary factors favoring a behavior need to be considered separately from the mechanistic stimulus of it. Our paper, as well as many other works, have repeatedly shown that cues from the host, such as chemical constituents, are central to any understanding of host selection, up to and including the phylogenetic patterns at large taxonomic levels. However, we find no necessary connection between these cues and host suitability per se. Hence host chemistry is important; it is a major, probably *the* major cue used in host selection by many insects. Host chemistry is also unimportant; it is only one facet of host suitability, and has little power in predicting the course of evolution.

IV. UNANSWERED QUESTIONS

A. MECHANISTIC

1. Is there a gestalt recognition of hosts, or do all sense organs respond independently, each separately affecting acceptance? How can we distinguish gestalt recognition from a "stimulus sieve" as in Figure 6, where several receptors act in sequence?

2. What are the effects of physiological characters on specificity? Behavioral rhythms and eggload appear to be important influences; what are the others?

3. How does hormonal control of such processes interact with host choice?

4. Do observed changes of diet in the field (e.g., Reference 29) parallel relevant physiological changes?

5. How can we study and distinguish the different processes which affect motivation and hence specificity?

6. Do multiple-choice tests give comparable results to no-choice tests? The hierarchy-threshold model implies that they should not.

7. Are "mistakes" in the early phases of host acceptance really mistakes, or are they sampling behavior?

B. EXPERIENCE AND LEARNING

8. What is the role of single learning events of large effect? How can such effects evolve?

9. Is cross-induction of acceptance the rule?

10. How is memory of host plant encounters mediated? Can learning mutants be exploited to eliminate, e.g., long-term memory, and does this affect host choice?

11. Does memory of different hosts last equally long (as in the RPS learning rule), or does it vary with host type (e.g., an encounter with a high-ranked host is remembered longer than a lower-ranked host)? If the latter, what is the relationship between the ratio of memory retention and larval fitness on the two hosts?

12. Is there an effect of travel time, independent of changes in eggload? Effect of travel time implies an influence of memory or of energy reserve levels; eggload merely implies a physiological change.

13. Is habituation important relative to other types of learning? If so, rank-order of hosts may change.

C. GENETICS/EVOLUTIONARY

14. Are there more examples of radical host shifts which are ascribable to similarities in stimuli received early in the host selection sequence?

15. Are the larvae of derived populations, using novel hosts, better suited to the ancestral hosts of the taxon? At how large a taxonomic scale does any such effect persist?

16. Is larval fitness a coevolving trait with female behavior (cf. Reference 122) or is it essentially independent? If independent, does it evolve before host shifts (preadaptation) or after them?

17. Is there indirect selection on acceptability of different hosts, due to pleiotropic effects through linked patterns of recognition?

18. Is there more than one type of generalist in a population?

19. If one part of the host recognition sequence is eliminated (e.g., in olfactory mutants), does the rest of host acceptance proceed normally?

20. When ancestral and derived populations differ in host plants, is this ascribable solely to differences in specificity, rather than to differences in the sensitivity of receptors?

21. Is rank-order of hosts subject to directional selection, for instance, through correlated effects on sensitivity to host stimuli? Is specificity subject to stabilizing selection?

REFERENCES

1. **Atsatt, P. and O'Dowd, A. J.,** Plant defense guilds, *Science,* 193, 24, 1976.
2. **Bach, C. E.,** Effects of host plant patch size on herbivore density: patterns, *Ecology,* 69, 1090, 1988.
3. **Banerjee, B.,** Can leaf aspect affect herbivory? A case study with tea, *Ecology,* 68, 839, 1987.
4. **Barbosa, P.,** Some thoughts on "the evolution of host range," *Ecology,* 69, 912, 1988.
5. **Bernays, E. A. and Chapman, R. F.,** Evolution of deterrent responses by phytophagous insects, in *Perspectives in Chemoreception and Behavior,* Chapman, R. F., Bernays, E. A., and Stoffolano, J. G., Eds., Springer-Verlag, New York, 1987, 159.
6. **Bernays, E. A. and Graham, M.,** On the evolution of host specificity in phytophagous arthropods, *Ecology,* 69, 886, 1988.
7. **Blau, P. A., Feeny, P. P., Contardo, L., and Robson, D.,** Allylglucosinolate and herbivorous caterpillars: a contrast in toxicity and tolerance, *Science,* 200, 1296, 1978.
8. **Brooks, D. R.,** Macroevolutionary comparisons of host and parasite phylogenies, *Annu. Rev. Ecol. Syst.,* 19, 235, 1988.
9. **Brussard, R. P. and Ehrlich, P. R.,** The population structure of *Erebia epipsodea* (Lepidoptera: Satyrniidae), *Ecology,* 51, 119, 1970.
10. **Cappucino, N. and Kareiva, P.,** Coping with a capricious environment: a population study of a rare pierid butterfly, *Ecology,* 66, 152, 1985.

11. **Carson, H. L. and Ohta, A. T.,** Origin of the genetic basis of colonizing ability, in *Evolution Today,* Proc. 2nd Int. Congress of Systematic and Evolutionary Biology, Scudder, G. G. E. and Reveal, J. L., Eds., Pittsburgh, Hunt Inst. for Botanical Documentation, Carnegie-Mellon University, 1981, 365.

12. **Cassidy, M. D.,** Development of an induced food plant preference in the Indian stick insect *Carausius morosus, Entomol. Exp. Appl.,* 24, 87, 1978.

13. **Charnov, E. L. and Skinner, S. W.,** Complementary approaches to the understanding of parasitoid oviposition decisions, *Environ. Entomol.,* 14, 383, 1985.

14. **Chew, F. S.,** Coevolution of pierid butterflies and their cruciferous food plants. II. The distribution of eggs on potential food plants, *Evolution,* 31, 568, 1977.

15. **Chew, F. S.,** The effects of introduced mustards (Cruciferae) on some native North American cabbage butterflies (Lep. Pieridae), *Atala,* 5, 13, 1977b.

16. **Coley, P. D.,** Herbivory and defensive characteristics of tree species in a lowland tropical forest, *Ecol. Monogr.,* 53, 209, 1983.

17. **Coley, P. D., Bryant, J. P., and Chapin, F. S., III,** Resource availability and plant antiherbivore defense, *Science,* 230, 895, 1985.

18. **Collinge, S. K. and Louda, S.,** Patterns of resource use by a drosophilid (Diptera) leaf miner on a native crucifer, *Ann. Entomol. Soc. Am.,* 81, 733, 1988.

19. **Colwell, R. K.,** Population structure and sexual selection for host fidelity in the speciation of hummingbird flower mites, in *Evolutionary Processes and Theory,* Karlin S. and Nevo, E., Eds., Academic Press, New York, 1986.

20. **Cottee, P. K., Bernays, E. A., and Mordue, A. J.,** Comparisons of deterrency and toxicity of selected secondary plant compounds to an oligophagous and a polyphagous acridid, *Entomol. Exp. Appl.,* 46, 241, 1988.

21. **Courtney, S. P.,** Coevolution of pierid butterflies and their cruciferous food plants. III. *Anthocharis cardamines* survival, development, and oviposition, *Oecologia,* 51, 91, 1981.

22. **Courtney, S. P.,** Hostplant apparency and *Anthocharis cardamines* oviposition, *Oecologia,* 52, 258, 1982.

23. **Courtney, S. P.,** Models of hostplant location by butterflies: the effect of search images and search efficiency, *Oecologia,* 59, 317, 1983.

24. **Courtney, S. P.,** Habitat versus food plant selection, in *The Biology of Butterflies,* Vane-Wright, R. I. and Ackery, P. R., Eds., Academic Press, London, 1984.

25. **Courtney, S. P.,** If it's not coevolution, it must be predation? *Ecology,* 69, 910, 1988.

26. **Courtney, S. P. and Chen, G. K.,** Genetic and environmental variation in oviposition behaviour in the mycophagous *Drosophila suboccidentalis* Spcr., *Func. Ecol.,* 2, 521, 1988.

27. **Courtney, S. P., Chen, G. K., and Gardner, A.,** A general model for individual host selection, *Oikos,* 55, 55, 1989.

28. **Courtney, S. P. and Courtney, S.,** The 'edge-effect' in butterfly oviposition: causality in *Anthocharis cardamines* and related species, *Ecol. Entomol.,* 7, 131, 1982.

29. **Courtney, S. P. and Forsberg, J.,** Host use by two pierid butterfies varies with host density, *Func. Ecol.,* 2, 67, 1988.

30. **Courtney, S. P. and Hard, J.,** Genetic correlations amongst host choice characters: a test of the hierarchy-threshold model, *Heredity,* in press, 1989.

31. **Cullenward, M. J., Ehrlich, P. R., White, R. R., and Holdren, C. E.,** The ecology and population genetics of an alpine checkerspot butterfly, *Euphydryas anicia, Oecologia,* 38, 1, 1979.

32. **Dethier, V. G.,** *Chemical Insect Attractants and Repellents,* Blakiston Co., Philadelphia, 1947.

33. **Dethier, V. G.,** Mechanism of host-plant recognition, *Entomol. Exp. Appl.,* 31, 49, 1982.

34. **Dethier, V. G. and Crnjar, R. M.,** Candidate codes in the gustatory system of caterpillars, *J. Gen. Physiol.,* 79, 549, 1982.

35. **Ehrlich, P. R. and Murphy, D. D.,** Plant chemistry and host range in insect herbivores, *Ecology,* 69, 908, 1988.

36. **Ehrlich, P. R. and Raven, P. H.,** Butterflies and plants: a study in coevolution, *Evolution,* 18, 586, 1964.

37. **Ellis, P. R., Cole, R., Crisp, P., and Hardman, J. A.,** The relationship between cabbage root fly egg-laying and volatile hydrolysis products of radish, *Ann. Appl. Biol.,* 95, 283, 1980.

38. **Etges, W. J. and Heed, W. B.,** Sensitivity to larval densities in populations of *Drosophila mojavensis:* influences of host plant variation on components of fitness, *Oecologia,* 71, 375, 1987.

39. **Falconer, D. S.,** *An introduction to Quantitative Genetics,* 2nd ed., Longman, New York, 1981.

40. **Feeny, P. P.,** Seasonal changes in oak leaf tannins and nutrients as a cause of spring feeding of winter moth caterpillars, *Ecology,* 51, 565, 1970.

41. **Feeny, P. P.,** Plant apparency and chemical defence, *Recent Adv. Phytochem.,* 10, 1, 1976.

41a. **Feeny, P. P.,** Chemical constraints on the

evolution of swallowtail butterfies, in *Herbivory: Tropical and Temperate Perspectives*, 2nd ed., Price, P. W., Lewinsohn, T. M., Benson, W. W., and Fernandes, G. W., Eds., John Wiley & Sons, New York, 1989.

42. **Feeny, P. P., Rosenberry, L., and Carter, M.,** Chemical aspects of oviposition behavior in butterflies, in *Herbivorous Insects: Host-Seeking Behavior and Mechanisms*, Ahmad, S., Ed., Academic Press, London, 1983.

43. **Fellows, D. P. and Heed, W. B.,** Factors affecting host plant selection in desert-adapted cactophilic *Drosophila, Ecology,* 53, 850, 1972.

44. **Fitt, G. P.,** The influence of a shortage of hosts on the specificity of oviposition behavior in species of *Dacus* (Diptera: Tephritidae), *Physiol. Entomol.,* 11, 133, 1986.

45. **Forsberg, J.,** Size discrimination among conspecific host plants in two pierid butterflies; *Pieris napi* L. and *Pontia daplidice* L., *Oecologia,* 72, 52, 1987.

46. **Fox, L. R.,** Diffuse coevolution within complex communities, *Ecology,* 69, 906, 1988.

47. **Fraenkel, G.,** The raison d'être of secondary plant substances, *Science,* 129, 1466, 1959.

48. **Fretwell, S. D. and Lucas, H. L.,** On territorial behavior and other factors influencing habitat distribution in birds, *Acta Biotheor.,* 19, 16, 1970.

49. **Futuyma, D. J.,** Selective factors in the evolution of host choice by phytophagous insects, in *Herbivorous Insects: Host-Seeking Behavior and Mechanisms*, Ahmad, S., Ed., Academic Press, New York, 1983.

50. **Futuyma, D. J. and Peterson, S. C.,** Genetic variation in the use of resources by insects, *Annu. Rev. Entomol.,* 30, 217, 1985.

51. **Futuyma, D. J. and Moreno, G.,** The evolution of ecological specialization, *Annu. Rev. Ecol. Syst.,* 19, 207, 1988.

52. **Gossard, T. W. and Jones, R. E.,** The effects of age and weather on egg-laying in *Pieris rapae* L. *J. Appl. Ecol.,* 14, 65, 1977.

53. **Gould, F.,** Genetics of plant-herbivore systems: interactions between applied and basic study, in *Variable Plants and Herbivores in Natural and Managed Systems*, Denno, R. F., and McClure, M. S., Eds., Academic Press, New York, 1983, 599.

54. **Hanula, J. C.,** Oviposition preference and host recognition by the black vine weevil *Otiorhynchus sulcatus* (Coleoptera: Curculionidae), *Environ. Entomol.,* 17, 694, 1988.

55. **Harley, C. B.,** Learning the evolutionary stable strategy, *J. Theor. Biol.,* 89, 611, 1981.

56. **Harris, M. O. and Miller, J. R.,** Color stimuli and oviposition behavior of the onion fly, *Delia antiqua* (Meigen), *Ann. Entomol. Soc. Am.,* 76, 766, 1983.

57. **Harris, M. O. and Miller, J. R.,** Host-acceptance behaviour in an herbivorous fly, *Delia antiqug, J. Insect Physiol.,* 34, 179, 1988.

58. **Harrison, G. D.,** Host-plant discrimination and evolution of feeding preference in the Colorado potato beetle *Leptinotarsa decemlineata, Physiol. Entomol.,* 12, 407, 1987.

59. **Hattori, M.,** Host-plant factors responsible for oviposition behaviour in the lima bean pod borer, *Etiella zickenella* Treitschke, *J. Insect Physiol.,* 34, 191, 1988.

60. **Haukioja, E. and Niemela, P.,** Birch leaves as a resource for herbivores: seasonal occurrence of increased resistance in foliage after mechanical damage of adjacent leaves, *Oecologia,* 39, 151, 1979.

61. **Hoffman, A. A.,** Effects of experience on oviposition and attraction in *Drosophila:* comparing apples and oranges, *Am. Nat.,* 126, 41, 1985.

62. **Hoffman, A. A.,** Early adult experience in *Drosophila melanogaster, J. Insect. Physiol.,* 34, 197, 1988.

63. **Hughes, P. R., Weinstein, L. H., Laurence, J. A., Sacher, R. F., Dickie, A. I., and Johnson, L. M.,** Modification of insect-plant relations by plant stress, in *Proc. of the 5th Int. Symp. on Insect-Plant Relationships*, Visser, J. H. and Minks, A. K., Eds., Centre for Agricultural Publication and Documentation, Wageningen, The Netherlands, 1982, 445.

64. **Jaenike, J.,** Environmental modification of oviposition behavior in *Drosophila, Am. Nat.,* 119, 784, 1982.

65. **Jaenike, J.,** Induction of host preference in *Drosophila melanogaster, Oecologia,* 58, 320, 1983.

66. **Jaenike, J.,** Parasite pressure and the evolution of amanitin tolerance in *Drosophila, Evolution,* 39, 1295, 1985.

67. **Jaenike, J.,** Genetic complexity of host-selection behavior in *Drosophila, Proc. Natl. Acad. Sci. U.S.A.,* 83, 2148, 1986.

68. **Jaenike, J.,** Effects of early adult experience on host selection in insects: some experimental and theoretical results, *J. Insect Behav.,* 1, 3, 1988a.

69. **Jaenike, J.,** Genetics of oviposition-site preference in *Drosophila tripunctata, Heredity,* 59, 363, 1988b.

70. **Jaenike, J.,** in press, 1989.

71. **James, A. C., Jakubczak, J., Riley, M., and Jaenike, J.,** On the causes of monophagy in *Drosophila quinaria, Evolution,* 42, 626, 1988.

72. **Janzen, D. H.,** Two ways to be a tropical big moth: Santa Rosa saturniids and sphyngids, *Oxford Surveys in Evolutionary Biology,* 1, 85, 1984.

73. **Janzen, D. H.,** A host plant is more than its chemistry, *Ill. Nat. Hist. Surv. Bull.,* 33, 141,

1985.

74. **Janzen, D. H.,** On the broadening of insect-plant research, *Ecology,* 69, 95, 1988.

75. **Jermy, T.,** Evolution of insect/hostplant relationships, *Am. Nat.,* 124, 609, 1984.

76. **Jermy, T.,** Can predation lead to narrow food specialization in phytophagous insects? *Ecology,* 69, 902, 1988.

77. **Jones, R. E.,** Movement patterns and egg distribution in cabbage butterflies, *J. Anim. Ecol.,* 46, 195, 1977.

78. **Jones, R. E., Hart, J. R., and Bull, G. D.,** Temperature, size, and egg-production in the cabbage butterfly, *Pieris rapae L., Aust. J. Ecol.,* 30, 223, 1982.

79. **Jones, T. H. and Finch, S.,** The effect of a chemical deterrent, released from the frass of a caterpillar of the garden pebble moth, on cabbage root fly oviposition, *Entomol. Exp. Appl.,* 45, 283, 1987.

80. **Juliano, S. A. and Borowicz, V. A.,** Parasitism of a frugivorous fly, *Rhagoletis cornivora,* by the wasp *Opius richmondi* — relationships to fruit and host density, *Can. J. Zool.,* 65, 1326,1987.

81. **Kareiva, P.,** Experimental and mathematical analyses of herbivore movements: quantifying the influence of plant spacing and quality on foraging discrimination, *Ecol. Monogr.,* 52, 261, 1982.

82. **Kennedy, J. S.,** Mechanisms of host plant selection, *Ann. Appl. Biol.,* 56, 317, 1965.

83. **Kennedy, J. S., Booth, C. O., and Kershaw, W. J. S.,** Host finding by aphids in the field. III. Visual attraction, *Ann. Appl. Biol.,* 49, 1, 1961.

84. **Kibota, T. T.,** Jack-of-One-Trade, Master-of-None: The Ecology of *Drosphila magnaquinaria,* M. S. thesis, University of Oregon, Eugene, 1990.

85. **Lanier, G. N.,** Integration of visual stimuli, host odorants, and pheromones by bark beetles and weevils in locating and colonizing host trees, in *Herbivorous Insects Host-Seeking Behavior and Mechanisms,* Ahmad, S., Ed., Academic Press, New York, 1983, 161.

86. **Levinson, H. Z. and Haisch, A.,** Optical and chemosensory stimuli involved in host recognition and oviposition of the cherry fruit fly *Rhagoletis cerasi L. Z. Angew. Entomol.,* 97, 85, 1984.

87. **Lewis, A. C.,** Plant-quality and grasshopper feeding — effects of sunflower condition on preference and performance in *Melanopus differentialis, Ecology,* 65, 836, 1984.

88. **Lofdahl, K.,** A genetic analysis of habitat selection in the cactophilic species *Drosophila mojavensis,* in *Evolutionary Genetics of Invertebrate Behavior,* Huettel, M. D., Ed., Plenum Press, New York, 1987, 153.

89. **Louda, S. M., Huntly, N. J., and Dixon, P. M.,** Insect herbivory across a sun shade gradient — response to experimentally-induced in situ plant stress, *Acta Oecol. Oecol. Gen.,* 8, 357, 1987.

90. **Maynard Smith, J.,** *Evolution and the Theory of Games,* Cambridge University Press, 1982.

91. **Maynard Smith, J.,** Game theory and the evolution of behavior, *Brain Behav. Sci.,* 7, 95, 1984.

92. **Messina, F. J., Barmore, J. L., and Renwick, J. A. A.,** Host selection by ovipositing cowpea weevils: patterning of input from separate sense organs, *Entomol. Exp. Appl.,* 43, 169, 1987.

93. **Miller, J. R. and Harris, M. O.,** Host acceptance behavior in an herbivorous fly, *Delia antiqua, J. Insect Physiol.,* 34, 179, 1988.

94. **Miller, J. R. and Strickler, K.,** Finding and accepting host plants, in *Chemical Ecology of Insects,* Bell, W. J. and Carde, R., Eds., Sinauer, Sunderland, Mass., 1984, 127.

94a. **Miller, J. S.,** Hostplant relationships in the Papilionidae (Lepidoptera): parallel cladogenesis or colonization? *Cladistics,* 3, 105, 1987.

95. **Mitchell, B. K.,** Adult leaf beetles as models for exploring the chemical basis of host-plant recognition, *J. Insect Physiol.,* 34, 213, 1988.

96. **Mitter, C. E. and Brooks, D. R.,** Phylogenetic aspects of coevolution, in *Coevolution,* Futuyma, D. J. and Slatkin, M., Eds., Sinauer, Sunderland, Mass., 1983, 65.

97. **Moore, L. V., Myers, J. H., and Eng, R.,** Western tent caterpillars prefer the sunny side of the tree, but why? *Oikos,* 51, 321, 1988.

98. **Ng, D.,** A novel level of interactions in plant-insect systems, *Nature,* 344, 611, 1988.

99. **Nielsen, J. K.,** Hostplant descrimination within Cruciferae: feeding responses of four leaf beetles (Col. Chrys.) to glucosinolates, curcubitacins, and cardenolides, *Entomol. Exp. Appl.,* 24, 41, 1978.

100. **Nottingham, S. F.,** Host-plant finding for oviposition by adult cabbage root fly, *Delia radicum, J. Insect Physiol.,* 34, 227, 1988.

101. **Odendaal, F. J.,** Mature egg number influences the behavior of female *Battus philenor* butterflies, *J. Insect Behav.,* 2, 15, 1989.

102. **Papaj, D. R. and Prokopy, R. J.,** The effect of prior adult experience on components of habitat preference in the apple maggot fly (*Rhagoletis pomonella*), *Oecologia,* 76, 538, 1988.

103. **Papaj, D. R. and Prokopy, R. J.,** Ecological and evolutionary aspects of learning in phytophagous insects, *Annu. Rev. Entomol.,* 34, 315, 1989.

104. **Papaj, D. R. and Rausher, M. D.,** Individual variation in host location by phytophagous insects, in *Herbivorous Insects: Host-Seeking Behavior and Mechanisms,* Ahmad, S., Ed., Academic Press, New York, 1983, 77.

105. **Papaj, D. R. and Rausher, M. D.,** Components of conspecific host discrimination behaviour in the butterfly *Battus philenor, Ecology,* 68, 245, 1987.

106. **Pereyra, P. C. and Bowers, M. D.,** Iridoid glycosides as oviposition stimulants for the buckeye butterfly, *Junonia coenia* (Nymphalidae), *J. Chem. Ecol.,* 14, 917, 1988.

107. **Pierce, N. E.,** Lycaenid butterflies and ants: Selection for nitrogen fixing and other protein rich food plants, *Am. Nat.,* 125, 888, 1985.

108. **Pierce, N. E. and Easteal, S.,** The selective advantage of attendant ants for the larvae of a lycaenid butterfly *Glaucopsyche lygdamus, J. Anim. Ecol.,* 55, 451, 1986.

109. **Pierce, N. E. and Mead, P. S.,** Parasitoids as selective agents in the symbiosis between lycaenid butterfly caterpillars and ants, *Science,* 211, 1185, 1981.

110. **Preszler, R. W. and Price, P. W.,** Host quality and sawfly populations: a new approach to life table analysis, *Ecology,* 69, 2012, 1988.

111. **Prokopy, R. J., Averill, A. L., Cooley, S. S., and Roitberg, B. A.,** Behavioral evidence for host races in *Rhagoletis, Oecologia,* 76, 138, 1982.

112. **Prokopy, R. J. and Boller, E. F.,** Artificial egging system for the European cherry fruit fly, *J. Econ. Entomol.,* 63, 1413, 1971.

113. **Prokopy, R. J., Diehl, S. R., and Cooley, S. S.,** Behavioral evidence for host races in *Rhagoletis pomonella* flies, *Oecologia,* 76, 138, 1988.

114. **Prokopy, R. J. and Fletcher, B. S.,** The role of adult learning in the acceptance of host fruit for egglaying by the Queensland fruit-fly, *Dacus tryoni, Entomol. Exp. Appl.,* 45, 259, 1987.

115. **Prokopy, R. J., Papaj, D. R., Opp, S. B., and Wong, T. T. Y.,** Intra-tree foraging behavior of *Ceratitis capitata* flies in relation to host fruit density and quality, *Entomol. Exp. Appl.,* 43, 159, 1987.

116. **Ramaswamy, S. B.,** Host finding by moths: sensory modalities and behaviours, *J. Insect Physiol.,* 34, 235, 1988.

117. **Ramaswamy, S. B., Ma, W. K., and Baker, G. T.,** Sensory cues and receptors for oviposition by *Heliothis virescens, Entomol. Exp. Appl.,* 43, 159, 1987.

118. **Raupp, M. J. and Denno, R. F.,** Leaf age as a predictor of herbivore distribution and abundance, in *Variable Plants and herbivores in Natural and Managed Systems,* Denno, R. F. and McClure, M. S., Eds., Academic Press, New York, 1983, 91.

119. **Raupp, M. J. and Denno, R. F.,** The suitability of damaged willow leaves as food for the leaf beetle, *Plagiodera versicolora, Ecol. Entomol.,* 94, 443, 1984.

120. **Rausher, M. D.,** Search image for leaf shape in a butterfly, *Science,* 200, 1071, 1978.

121. **Rausher, M. D.,** Ecology of host-selection behavior in phytophagous insects, in *Variable Plants and Herbivores in Natural and Managed Systems,* Denno, R. F. and McClure, M. S., Eds., Academic Press, New York, 1983, 223.

122. **Rausher, M. D.,** Tradeoffs in performance on different hosts: evidence from within- and between-site variation in the beetle *Deloyala guttata, Evolution,* 38, 582, 1984.

123. **Rausher, M. D.** Variability for host preference in insect populations: mechanistic and evolutionary models, *J. Insect Physiol.,* 31, 973, 1985.

124. **Rausher, M. D.,** Is coevolution dead? *Ecology,* 69, 898, 1988.

125. **Rausher, M. D. and Englander, R.,** The evolution of habitat preference. II. Evolutionary genetic stability under soft selection, *Theor. Pop. Biol.,* 31, 116, 1987.

126. **Regelmann, K.,** Competitive resource sharing: a simulation model, *Anim. Behav.,* 32, 226, 1984.

127. **Renwick, J. A. A.,** Plant constituents as oviposition deterrents to lepidopterous insects, in *Biologically Active Natural Products,* Cutler, H. G., Ed., American Chemical Society, Washington, D.C., 1988, 378.

128. **Renwick, J. A. A. and Radke, C. D.,** Chemical stimulants and deterrents regulating acceptance or rejection of crucifers by cabbage butterflies, *J. Chem. Ecol.,* 13, 1771, 1987.

129. **Rhoades, R. F. and Cates, R. G.,** Toward a general theory of plant antiherbivore chemistry, *Recent Adv. Phytochem.,* 10, 168, 1976.

130. **Roininen, H. and Tahvanainen, J.,** Host selection and larval performance of two willow-feeding sawflies, *Ecology,* 70, 129, 1989.

131. **Root, R. B.,** Organization of plant-arthropod associations in simple and diverse habitats: the fauna of collards *(Brassica oleracea), Ecol. Monogr.,* 43, 95, 1973.

132. **Rossiter, M. C.,** Genetic and phenotypic variation in diet breadth in a generalist herbivore, *Evol. Ecol.,* 1, 272, 1987.

133. **Sachdev-Gupta, K., Renwick, J. A. A., and Radke, C. D.,** Isolation and identification of oviposition deterrents to the cabbage butterfly, *Pieris rapae,* from *Erysimum cheiranthoides, J. Chem. Ecol.,* in review.

134. **Schultz, J. C.,** Many factors influence the evolution of herbivore diets, but chemistry is central, *Ecology,* 69, 896, 1988.

135. **Schultz, J. C., Nothnagle, P. J., and Baldwin, I. T.,** Seasonal and individual variation in leaf quality of two northern hardwood tree species, *Am. J. Bot.,* 69, 753, 1982.

136. **Scriber, J. M. and Slansky, F., Jr.,** The nutritional ecology of immature insects, *Annu. Rev. Entomol.,* 26, 183, 1981.

137. **Shapiro, A. M.,** Ecological and behavioral aspects of coexistence in six crucifer-feed-

ing pierid butterflies in the central Sierra Nevada, *Am. Midl. Nat.,* 93, 424, 1975.

138. **Singer, M. C.,** Evolution of food-plant preferences in the butterfly *Euphydryas editha, Evolution,* 25, 383, 1971.

139. **Singer, M. C.,** Determinants of multiple host use by a phytophagous insect population, *Evolution,* 37, 389, 1982.

140. **Singer, M. C.,** Quantification of host preferences by manipulation of oviposition behavior in the butterfly *Euphydryas editha, Oecologia,* 52, 224, 1983.

141. **Singer, M. C., Ng, D., and Thomas, C. D.,** Heritability of oviposition preference and its relationship to offspring performance within a single insect population, *Evolution,* 42, 977, 1988.

142. **Smiley, J. T.,** Plant chemistry and the evolution of host specificity: new evidence from *Heliconius* and *Passiflora, Science,* 201, 745, 1978.

143. **Smiley, J. T.,** Are chemical barriers necessary for evolution of butterfly-plant associations? *Oecologia,* 65, 580, 1985.

144. **Stadler, E.,** Contact chemoreception, in *Chemical Ecology of Insects,* Bell, W. J. and Carde, R. T., Eds., Sinaeur, Sunderland, MA, 1984, 3.

145. **Stanton, M. L.,** Searching in a patchy environment: foodplant selection by *Colias p. eriphyle* butterflies, *Ecology,* 63, 839, 1982.

146. **Stanton, M. L.,** Spatial patterns in the plant community and their effects upon insect search, in *Herbivorous Insects: Host-Seeking Behavior and Mechanisms,* Ahmad, S., Ed., Academic Press, New York, 1983, 125.

147. **Strong, D. R.,** Insect host range, *Ecology,* 69, 885, 1988.

148. **Strong, D. R., Lawton, J. H., and Southwood, T. R. E.,** *Insects on Plants: Community Patterns and Mechanisms,* Harvard University Press, Cambridge, MA, 1984.

149. **Tabashnik, B. E.,** Host range evolution: the shift from native legume hosts to alfalfa by the butterfly, *Colias philodice eriphyle, Evolution,* 37, 150, 1983.

150. **Taylor, C. E.,** Habitat selection within species of *Drosophila:* a review of experimental findings, *Evol. Ecol.,* 1, 389, 1988.

151. **Thomas, C. D.,** Behavioral determination of diet breadth in insect herbivores — the effect of leaf age on choice of host species by beetles feeding on *Passiflora* vines, *Oikos,* 48, 211, 1987.

152. **Thompson, J. N.,** Evolutionary ecology of the relationship between oviposition preference and performance of offspring in phytophagous insects, *Entomol. Exp. Appl.,* 47, 3, 1988a.

153. **Thompson, J. N.,** Variation in preference and specificity in monophagous and oligophagous swallowtail butterflies, *Evolution,* 42, 118, 1988b.

154. **Thompson, J. N.,** Coevolution and alternative hypotheses on insect/plant interactions, *Ecology,* 69, 893, 1988c.

155. **Tinbergen, N.,** On aims and methods of ethology, *Z. Tierpsychol.,* 20, 87, 1963.

156. **Traynier, R. M.,** Long-term changes in the oviposition behavior of the cabbage butterfly, *Pieris rapae,* induced by contact with plants, *Physiol. Entomol.,* 4, 87, 1979.

157. **Turelli, M. and Hoffman, A. A.,** Effects of starvation and experience on the response of *Drosophila* to alternative resources, *Oecologia,* 77, 497, 1988.

158. **Vacek, D. C., East, P. D., Barker, J. S. F., and Soliman, M. H.,** Feeding and oviposition preferences of *Drosophila buzzatii* for microbial species isolated from its natural environment, *Biol. J. Linn. Soc.,* 24, 175, 1985.

159. **Via, S.,** The quantitative genetics of polyphagy in an insect herbivore. II. Genetic correlations in larval performance within and among host plants, *Evolution,* 38, 896, 1984.

160. **Wallner, W. E. and Walton, G. S.,** Host defoliation: a possible determinant of gypsy moth population quality, *Ann. Entomol. Soc. Am.,* 72, 62, 1979.

161. **Ward, S. A.,** Optimal habitat selection in time-limited dispersers, *Am. Nat.,* 129, 568, 1987.

162. **Wasserman, S. S.,** Partial paternal inheritance of realized fecundity in a bruchid beetle, *Callosobruchus maculatus, Behav. Genet.,* 18, 193, 1988.

163. **Watanabe, M.,** Multiple matings increase the fecundity of the yellow swallowtail butterfly, *Papilio xuthus* L., in summer generations, *J. Insect Behav.,* 1, 17, 1988.

164. **Watt, K. E. F.,** Comments on fluctuation of animal populations and measures of community stability, *Can. Entomol.,* 96, 1434, 1964.

165. **Watt, K. E. F.,** Community stability and the strategy of biological control, *Can. Entomol.,* 97, 887, 1965.

166. **Wiklund, C.,** The evolutionary relationship between adult oviposition preferences and larval host plant range in *Papilio machaon* L., *Oecologia,* 18, 185, 1975.

167. **Wiklund, C.,** Generalist vs. specialist oviposition behavior in *Papilio machaon* (Lepidoptera) and functional aspects of the hierarchy of oviposition preferences, *Oikos,* 36, 163, 1981.

168. **Williams, E. H. and Bowers, M. D.,** Factors affecting host plant use by the montane butterfly *Euphydryas gillettii* (Nymphalidae), *Am. Midl. Nat.,* 118, 153, 1987.

169. **Williams, K. S. and Myers, J. H.,** Previous herbivore attack of red alder may improve food quality for fall webworm larvae, *Oecologia,* 63, 166, 1984.

170. **Williams, L. H.,** The feeding habits and food

preferences of the Acrididae and the factors which determine them, *Trans. R. Entomol. Soc. London,* 105, 423, 1954.

171. **Wood, T. K.,** Divergence in the *Enchenopa binotata* Say complex (Homopter: Membra-cidae) effected by host plant adaptation, *Evolution,* 34, 147, 1980.

172. **Worthen, W. B.,** Effects of resource density on mycophagous fly dispersal and community structure, *Oikos,* 54, 145 (Table 1), 1989.

INDEX

A

Abiotic factors, 78—80
Acrolepiopsis assectella, 45, 56
Acyrthosiphon pisum, 48, 126, 134
Adaptation, see also Adaptive syndromes
  experience-induced behavioral changes and,
    60—64
  to phenolglucosides, 87
  physiological, 149
Adaptive advantage, 64
Adaptive syndromes, 2, 17—27, 30, see also
    Adaptation
  defined, 17
  in eruptive species, 20—27
  in latent species, 18—20, 25—27
Adaptive value, 60
Aedes taeniorhynchus, 121
Age, 11
Aglais urticae, 100
Agonistic effects of phenolglucosides, 88—90
Agraulis vanillae, 100
Agria affinis, 124, 133
Agropyron spp., 142
Air pollution, 79
Alanine, 119, 121
Aldehyde, 89
Alkaloids, 51, 149, see also specific types
Allelochemicals, 45, 147—150, see also specific
    types
Altitude, 84
Amino acid feedback, 118
Amino acids, 120, 133, 134, see also specific
    types
  in blood, 119
  essential, 119
  free, 118, 143
  limitation of, 143
  mix of, 140
  unwanted, 147
  utilization efficiencies for, 139
Ancestral population, 179
Antagonistic effects of phenolglucosides, 86—
    87
Antheraea polyphemus, 48
Anticarsia gemmatalis, 114, 147, 150
Antifeedant activity, 54
Ants, 45, 91, see also *Prenelopis* spp.; specific
    types
Aphididae, see Aphids
Aphids, 13, 25, 81, 91, see also specific types
Aphis fabae, 134
Apis mellifera, 96, 107, 136
Apple maggot fly, 57
Aquatic program, 40
Argyrotaenia velutinana, 139
Aristochia
  serpentaria, 59
  spp., 58, 59

Aristochrysa prasina, 90
Arroyo willow, see *Salix lasiolepis*
Arroyo willow shoot galler, 23
Artificial diets, 51, 114—116
  defined, 117
  dietary selection and, 130—133
  dilution of, 114—116
Asclepias curassavica, 148
Asobara spp., 42
Associative learning, 40, 43, 55, 58, 135—137
Atropa belladonna, 52
Australian sheep blowfly, see *Lucilia cuprina*
Autochthonously produced secretion, 89
Aversion, 47
Aversion learning, 53, 54, 104, 135, 136
Azadirachtin, 44

B

Balsam fir, 5
Bark beetles, 5, 8, 28—30, see also specific
    types
Battus philenor 57, 59, 63, 103, 164, 169
Bees, 97, 105, see also specific types
  bumble, 97, 107
  honey, see *Apis mellifera*
Beetles, 85, see also specific types
  bark, 5, 8, 28—30
  Colorado potato, 167
  confused flour, see *Tribolium confusum*
  leaf, see Leaf beetles
  water-deprived, 43
Behavior, see also specific types
  experience-induced changes in, see Experi-
      ence-induced behavioral changes
  feeding, see Feeding behavior
  generalist, 60
  predator, 91
  specialist, 60, 64
  speciation and changes in, 64—65
Behavioral host races, 64
Benzaldehyde, 88
Biochemical profile, 51
Biological control of weeds, 51
Biotic interactions, 165
Black blowfly, see *Phormia regina*
Black-headed budworm, 23
Blatella germanica, 114, 121—123, 126
Blood amino acids, 119
Blood composition, 118, 128, 138, 142
Blood nutrients, 143
Blood osmolality, 115, 118, 121, 138
Blowflies, see also specific types
  Australian sheep, see *Lucilia cuprina*
  black, see *Phormia regina*
  carbohydrates in diet of, 121
  nutritional compensation in, 147
  post-feeding inhibition in, 114
  sugar receptors in, 126

Bombyx mori, 117, 140, 143, 145
Brain size, 100
Brassica
  oleracea, 50, 54, 139
  spp., 52—54
Bud galler sawfly, see *Euura mucronata*
Budworms, see also specific types
  black-headed, 23
  eastern spruce, see *Choristoneura fumiferana*
  spruce, 5, 28
Bumble bees, 97, 107
Butterflies, 17, 45, 57, see also specific types
  cabbage white, see *Pieris rapae*
  carbohydrates in diet of, 121
  color and, 100—101
  learning in, 95—108
    color and, 100—101
    deleterious flowers and, 104
    flower constancy and, 97—104
    light intensity learning in, 100—101
    memory and, 102
    nectar extraction and, 104—107
    odor/taste learning, 102—103
    patterns and, 103—104
    resources and, 107
    shapes and, 103—104
    time and, 97, 104—107
    unrewarding flowers and, 104
  light intensity learning in, 100—101
  tiger swallowtail, see *Papilio glaucus*

C

Cabbage white butterfly, see *Pieris rapae*
Cajanus cajan, 58
Caliosamia promethea, 50
Calliphora vomitoria, 41
Callosamia promethea, 48
Callosobruchus maculatus, 58, 179
Calotropis gigantea, 148
Carausius morosus, 50
Carbohydrates, 116, 121—123, see also
    specific types
  concentration of, 133
  deprivation of, 128, 136
  regulation of intake of, 132, 145
Carbon availability, 85
Carboxylesterase, 87
Carrying capacity, 28, 29
Cataglyphis spp., 45
Cecideomylidae, 13, see also specific genuses
Celerio euphorbia, 114, 123
Central excitatory state (CES), 40, 42, 53
Central nervous system (CNS), 41, 42, 44, 125
Centrifugal control, 120
Ceratitis capitata, 57, 121, 134, 166
CES, see Central excitatory state
Chemical cues, 63, 103
Chemical defense, 84, 88—90
Chemical "Gestalt", 48, 62
Chemical legacy hypothesis, 45, 60, 65
Chemical tunnel vision, 62

Chemistry, 163, 180—181
Chemoreceptors, 43
Chemosensory system, 44, 51
Choice index, 48
Choristoneura
  fumiferana, 18, 20—25
  occidentalis, 145
  spp., 20
Chrysomela
  aenicollis, 83, 84, 90
  spp., 88, 89
  tremulae, 88, 89
  vingitipunctata, 81
Chrysomelidae, 81, see also specific genera
Chrysomelina, 88
Cicer arietinum, 58
Cimicifuga simlex, 103
Classical conditioning, 40, 43
Climatic release, 23, 29
CNS, see Central nervous system
Cockroaches, 121, see also specific types
  nutritional compensation in, 147
Coleoptera spp., 58
Colias spp., 45, 63, 100, 167
Colorado potato beetle, 167
Color vision, 100—101
Competition, 26, 27
  among females, 8, 9, 18, 26—29
  among herbivores, 26
  interference, 9, 26
  larval, 8
  life tables and, 29
Conditioned stimulus, 40, 41, 52, 53, 57
Conditioning, 56, see also specific types
  classical, 40, 43
  instrumental, 40, 55, 57, 58
  olfactory, 43
  operant, 43
  second order, 54
Confused flour beetle, see *Tribolium confusum*
Corn earworm, see *Heliothis zea*
Corpora cardiaca, 44, 115, 118
Cottonwood, 25
Crataegus
  mollis, 57
  spp., 57
Crop emptying rate, 115
Cross-induction, 166
CS, see Conditioned stimulus
Cues, see also specific types
  chemical, 63, 103
  olfactory, 181
  recognition, 97, 103
  taste, 181
  volatile chemical, 63
Curculionidae, 13, see also specific species
Cynipidae, 13, see also specific genuses
Cynipid gall wasps, see Cynipidae

D

Dacus tryoni, 166

Danaus chrysippus, 148
Datana ministra, 117, 139
Decision making, 104
Defenses, see also Resistance; specific types
  chemical, 84, 88—90
  phenolic, 10
  strategies for, 162
Delia radicum, 165
Density, see Population density
Derived population, 179
Deterrent receptors, 44
Deterrents, 44, 115, 125, 148, see also specific
    types
  feeding, 61, 150
  high levels of, 133
Detoxifying enzymes, 149, see also specific
    types
Developmental time, 81
Developmental variation, 79
Diacrisia virginica, 52
Diet, see also Nutrients
  artificial, see Artificial diets
  breadth of, 162
  dilution of, see Dietary dilution
  imbalance in, 133, 136
  selection of, see Dietary selection
  water in, 115
Dietary dilution, 114—116
  nutrient response to, 116—123
Dietary selection, 54—55, 129—138
  artificial diets and, 130—133
  mechanisms of, 133—138
  natural foods and, 130
Dietary specialization, 80
Digestive enzymes, 142, 148, see also specific
    types
Dioryctria albovitelia, 25
Diplacus aurantiacus, 148
Diprionid sawflies, 18, 22
Diptera spp., 64, 166
Direct feedbacks, 137—138
Discovery time, 97
Diseases, 20, 22, 30, see also specific types
Dispersal, 20, 22
Drosophila
  busckii, 175, 177
  magnaquinaria, 167, 168, 174
  melanogaster, 61, 166, 175
  mojavensis, 167, 168
  quinaria, 167, 176
  spp., 42, 45, 60, 61, 165, 169, 177
  suboccidentalis, 168, 177
  tripunctata, 65, 177, 179
Dual-process theory, 41

E

Early adult experience, 60, 64, 166
Early-imaginal experience, 59—60, 64
Eastern spruce budworm, see *Choristoneura
    fumiferana*
Ecological imprinting, 42

Ecology, 2, 27, 107
  nutritional, 112
Efferent nerves, 120
Egglaying, see Oviposition
Elevation, 79
Emergent properties, 17—27, 30
  defined, 17
  in eruptive species, 20—27
  in latent species, 18—20, 25—27
Emigration, 8, 20, 22
Endemic populations, 3, 7
Endocrine control of receptor sensitivity, 44
Endopterygotes, 64, 132, see also specific
    types
Enemy-free space, 169
Environmental stress, 79, 85, see also specific
    types
Environmental variation, 164—165
Enzymes, see also specific types
  detoxifying, 149
  digestive, 142, 148
Epidemic disease, 20, 22
Epidemic populations, 3, 7
Epigenetic factors, 61, 62, 64
Epilachna varivestis, 127
Epistatic interaction, 177
Eruptive species, 2, 4, 17, see also specific
    types
  adaptive syndromes in, 20—27
  competition in, 29
  defined, 3
  ermergent properties in, 20—27
  latent species vs., 25—27
  mortality factors and, 30
  permissive system for, 23
  phylogenetic constraints in, 20—27
  population dynamics of, 27
Essential amino acids, 119, see also specific
    types
Estigmene congrua, 52
Euchaetias egle, 63
Euphydryas
  chalcedona, 148, 149
  editha, 168, 173
Eusociality, 96, 107
Euura
  exiguae, 8
  lasiolepis, 2, 8, 11, 26
    adaptive syndromes in, 18—20
    emergent properties in, 18—20
    parasitoids and, 30
    phylogenetic constraints in, 18—20
  mucronata, 8
  spp., 11, 84
Evolution, 182
  life history, 20, 27
  life tables and, 27—28
  of nervous system, 64
Exopterygotes, 132, see also specific types
Experience, see also specific types
  early adult, 60, 64, 166
  early-imaginal, 59—60, 64

host selection and, 166, 181—182
pre-imaginal, 59—60
Experience-induced behavioral changes, 43—45
  adaptive significance of, 60—64
  feeding-type, see Feeding behavior
  oviposition-type, 55—59
  speciation and, 64—65
Experiential factors, 174—175
Experimental methods, 7
Experimental problem in life tables, 7—8
Exposure heterogeneity, 101
Extragenetic factors, 60

F

Fabaceae, 52, see also specific genera
Fagaceae, 48—50, see also specific genera
Fall webworm, 29
Fecundity, 85
Feeding behavior
  blood composition and, 118
  changes in, 46—55
    dietary self-selection and, 54—55
    food-aversion learning and, 52—54
    food finding and, 55
    food-related stimuli and, 45—47
    habituation and, 45—47
    host range and, 51—52
    IFP and, 47—52
  optimality of, 112
Feeding deterrents, 61
Feeding experience, 45
Feeding inhibitors, 46
Feeding patterns, 115, 141
Feeding preference, see Food preference
Firs, 5
First-instar larvae, 17
Fitness, 40, 61, 62
  larval, 179
Flies, see also specific types
  apple maggot, 57
  blow, see Blowflies
  house, see *Musca domestica*
  meal size in, 122
  Mediterranean fruit, see *Ceratitis capitata*
  sarcophagid, see *Agria affinis*
  saw, see Sawflies
Flight capacity, 22
Flightless females, 25
Flower handling, 97
Flow-through respirometer, 139
Food-aversion learning, 43, 52—54
Food-avoidance learning, 52, 53
Food choice, 133
Food deprivation, 124—138
Food finding, 55, 63
Food imprinting (induction of food preference (IFP), 48—52

Food preference, 42
  induction of (IFP), 47—52
Food quality, 8
Food-related stimuli, 46—47
Food seeking behavior, 58
Food selection behavior, 58
Foraging, 17, 58, 59, 62, 96, 107
Fragrance, 103
Free amino acids, 118, 143
Fruit flies, Mediterranean, see *Ceratitis capitata*

G

Galerucinae, 88, see also specific genera
Gallerucella lineola, 81, 85, 89, 90
Galling sawflies, 8—11, 13—14, 17, see also specific types
Gall midges, see Cecidomylidae
Gaultheria spp., 52
Gene flow, 64, 65
Generalists, 60, 64, 96, 97, 107, 169, 173, see also specific types
Genetic background, 60, 61, 65
Genetic correlations, 177
Genetics, 176—179, 182
Genetic variance, 168, 177
Genotypes, 61
Glucose, 89, 122, 124, 133
beta-Glucosidase, 86, 87, 89
Glucoside, 89
Glutamine, 121
L-Glutamine, 119
Glycine, 121
Glycosides, 10, 89, see also specific types
"Good learners", 61
Grasshoppers, 55, 63, see also specific types
Gray larch tortrix, see *Zeiraphera griseana*
Green leaf volatiles, 51
Gurania spp., 104
Gustatory receptors, 115, 119, 122
Gustatory sensitivity, 120
Gut emptying rates, 118, 128
Gut lesions, 86
Gut volume, 141

H

Habitats
  heterogenous, 81
  selection of, 42, 174
Habituation, 40, 41, 46—47, 56, 149
Haltica spp., 64
Helianthus annuus, 103
Heliconius spp., 17, 58, 59, 97, 167
Heliothis zea, 48, 50, 54, 130, 131, 136, 140
  detoxification enzymes in, 50, 149
  nutritional requirement changes in, 142
Hemimetabola, 64
Hemolymph osmolality, 142

Herbivores, see also specific types
  abundance of, 82
  antagonistic effects of phenolglucosides on,
      86—87
  competition among, 26
  distribution of, 80—85, 91
  opportunistic, 23
  population dynamics of, 28
  stealthy, 23
Heterogenous habitats, 81
HHSP, see Hopkins Host Selection Principle
Hierarchy-threshold model, 171, 181
Hippodamia convergens, 90
Histidine, 121
History effect, 99
Honey bees, see *Apis mellifera*
Hopkins Host Selection Principle (HHSP), 60
Hormones, 44, 114, 118, 126, 138, 140, see
      also specific types
Hornworms, tobacco, see *Manduca sexta*
Host plants, 45, 165, see also plants
  acceptability of, 169, 174
  choice of, 162, 165, 166, 170, 171
  defense strategies of, 162
  finding of, 62, 63
  imprinting of, 48
  induction of, 137
  quality of, 2, 8, 18, 27—30
  range of, 51—52
  resistance of, 5, 10, 25, 29
  selection of, see Host selection
  shifts in, 167
  specialization of, 64
  stress in, 23
  suitability of, 169
  susceptibility of, 29
Host-plant searching behavior, 63
Host-plant seeking behavior, 58, 62
Host selection, 42, 161—182
  categories of plants as related to, 45—46
  chemistry and, 180—181
  evolution and, 182
  experience and, 181—182
  experiential factors in, 174—175
  genetics and, 176—179, 182
  learning and, 181—182
  mechanistic factors in, 181
  ontogenetic factors in, 165—166, 174—175
  phylogenetic factors in, 166—168
  phylogenic factors in, 176—179
  proximate factors in, 163—165, 174
  theories of, 169—173
  ultimate factors in, 168—169, 179—180
  unanswered questions in, 181—182
Housefly, see *Musca domestica*
Hunger, 136, 145
6-Hydroxycyclohexenane, 86
Hymenoptera, 40, see also specific genera
Hyperdispersed resources, 107

Hyphantria cunea, 48

I

ICP, see Induced oviposition preference
IFP, see Induction of food preference
Imaginal experience, 59—60, 64
Immigration, 8
Imprinting, 40, 42—43, 48, see also specific
      types
  food (induction of food preference (IFP), 48—
      52
  host-plant, 48
  oviposition site, 56
Individual variability, 61, 63
Induced feeding preference, 42
Induced oviposition preference (ICP), 55, 56
Induced preference, 41
Induction of food preference (IFP), 48—52
Induction index, 48
Information acquisition, 104
Ingestion rate, 125
Inheritance, 177
Initial choice of food, 133
Instrumental conditioning, 40, 55, 57, 58
Interference, 100
Interference competition, 9, 26
Intermeal intervals, 114—115, 118, 121, 122,
      124, 127
Interplant variation in quality, 8
Interspecific variation, 77—78
Intraplant variation in quality, 8
Intraspecific discrimination, 164
Intraspecific variations, 78—80, 83
Isoxazolinone glycosides, 89, see also specific
      types

K

Key factors, defined, 7
K-selection, 4

L

Landmarks, 107
Larvae
  competition in, 8
  feeding of, 28
  first-instar, 17
  fitness of, 179
  performance of, 9, 90, 176
  survival of, 11, 81, 90
Latent species, 2, 4, 8—11, 13—14, 17, see
      also specific types
  adaptive syndromes in, 18—20, 25—27
  defined, 3
  emergent properties in, 18—20, 25—27
  eruptive species vs., 25—27
  mortality factors and, 30

phylogenetic constraints in, 18—20, 25—27
population dynamics of, 27
restrictive system for, 23
weather and, 29
Leaf age, 84—85
Leaf beetles, 81, see also Chrysomelidae,
    specific genera
phenolglucosides and natural enemies of,
    90—91
Leaf consumption, 91
Learned hunger, 136
Learned oviposition preference, 56, 58, 60, 63,
    64
Learned preference, 135
Learned specific appetite, 136
Learning, see also specific types
  associative, 55, 58, 135—137
  aversion, 53, 54, 104, 135, 136
  in butterflies, see under Butterflies
  in dietary selection, 135—137
  flower constancy and, 97—104
  food-aversion, 43, 52—54
  food-avoidance, 52, 53
  in food finding, 55
  by generalists, 96
  host selection and, 166, 181—182
  light intensity, 100—101
  location, 96
  nectar extraction, 104—107
  nonassociative, 135
  odor/taste, 102—103
  one-trial associative, 58
  pattern, 103—104
  relative payoff sum (RPS), 169
  shape, 103—104
  spatial, 96
  time and, 97, 104—107
  time-and-place, 107
  trial-and-error, 43, 55
  types of, 40—43
Learning curve, 97
Lepidoptera, 13, 62, 64, 99, 166, see also
    specific genuses
Leptinotarsa decemlineata, 42, 45, 46, 48, 51,
    63, 149
Lesions in gut, 86
Leucine, 119
Life history, 17, 20, 27
Life tables, 2, 4—9, 27
  competition and, 29
  evolutionary perspective on, 27—28
  experimental problem in, 7—8
  mortality-natality problem in, 8
  natural enemies and, 29—30
  plot problem in, 5—7
  reinterpretation of, 27—30
  sampling problem in, 4—5
  weather and, 29
Light intensity, 85, 100—101
Limeitis
  archippus, 48
  astyanax, 48
  rubidus, 48

Linoleic acid, 51
Linolenic acid, 51
Lipids, 51, 124, 131, 142, 144, see also specific
    types
Location learning, 96
Lochmea caprea, 81
Locust, 43
Locusta
  migratoria, 44, 47, 53, 114, 123, 132, 133
    carbohydrates and, 137
    control of feeding in, 121
    food deprivation in, 124, 125
    intermeal intervals in, 124
    learning in, 136
    nutrient deprivation in, 128
    nutritional compensation in, 117—121
    nutritional requirement changes in, 141—143
  spp., 126, 128, 150
Locusts, 114, see also specific types
Long-term memory, 42, 44, 102
LTM, see Long-term memory
Lucilia
  cuprina, 116, 121, 146
  spp., 123
Lupinus spp., 149
Lymantria dispar, 48
Lysine, 119

M

Macrosiphon euphorbiae, 134
Malacosoma americanum, 51
Malaise, 54, 136, 138
Mamestra brassicae, 47, 61
Manduca sexta, 42, 45, 48, 51, 52, 115, 116
  allelochemicals and, 42, 149
  food deprivation in, 125, 127
  intermeal intervals in, 124
  nutritional requirement changes in, 116, 141,
    143
  post-ingestive compensation in, 138
  taste receptors in, 120
Maxillary palps, 119
Maze, 43
Meal duration, 116, 125
Meal size, 115—116, 121, 122, 125, 141
Mechanistic studies, 7
Mediterranean fruit fly, see *Ceratitis capitata*
Melanoplus
  bivittatus, 114
  sanguinipes, 55, 114
Memory, 44
  capacity for, 97
  constraints on, 97
  long-term, 42, 44, 102
  short-term, 45, 102
Methionine, 119, 134
Midges, gall, see Cecidomylidae
Midgut ceca, 142
Midgut-beta-glucosidase activity, 86
Migration, 8, 141
Monoterpenes, 88, 89, 91, see also specific
    types

Morphological characters of plants, 164
Mortality, 8, 20, 30
Mosquito, see *Aedes taeniorhynchus*
Motivation, 172
Musca domestica, 43, 45
Myzus persicae, 134

N

Narrowleaf cottonwood, 25
Natality factors, 8, 30
Natural enemies, 11, 29—30, 90—91
Natural foods, 130
Natural selection, 60, 168—169
Nectar collection, 96
Nectar extraction, 104—107
Neem extract, 46
Nemeritis canescens, 42
Neo-Hospkins host selection principle, 60
Neophobia, 135, 137
Nepeta cataria, 104
Nerium oleander, 52
Nervous system, 41, 42, 44, 64, 125
Neural mechanisms, 43—45
Neurotransmitters, 137, see also specific types
Nicotine, 44
Nicotine hydrogen tartrate, 47
Nitrogen, 116, 138, 139
Nitrogen fertilization, 79
Nonassociative learning, 40, 135
Non-host plants, 45, 52
Nutrients, see also Diet; Nutritional compensation; specific types
  artificial, 51
  in blood, 143
  deprivation of, 128—129
  effects of, 128
  imbalances in, 123—124
  peripheral sensitivity and, 129
  response of to dietary dilution, 116—123
Nutrient stress, 22, 79
Nutritional compensation, 111—151
  allelochemicals and, 147—150
  artificial diet dilution and, 114—116
  carbohydrates and, 121—123
  for changes in nutritional requirements, 141—146
  contraints on, 146—148
  deprivation of specific nutrients and, 128—129
  dietary selection and, see Dietary selection
  in *Locusta migratoria*, 117—121
  meal size and, 115—116
  metabolic costs of, 146
  nutrient imbalances and, 123—124
  nutrient response to dilution and, 116—123
  post-ingestive, 138—140
  previous food deprivation and, 124—128
  proteins and, 116—121
Nutritional deficiency, 136
Nutritional ecology, 112
Nutritional feedbacks, 115, 116, 120, 122, 123, 132, 135
Nutritional inadequacy recognition, 133—135

Nutritional requirement changes, 141—146
Nutritive value, 85
Nymphalidae, 100, see also specific genera
Nymphalis antiopa, 91

O

Octopamine, 125, 137
Odor, 54
Odor/taste learning, 102—103
Olfaction, 58
Olfactory conditioning, 43
Olfactory cues, 181
Oligophagous species, 47, 52, 53, 57, 61
Oligophagy, 50, 52
Oncopeltus fasciatus, 120, 127
One-trial associative learning, 58
Ontogenetic factors, 165—166, 174—175
Operant conditioning, 43
Opportunistic herbivores, 23
Optimal food, 123
Optimality of feeding behavior, 112
Optimality theory, 62
Osmotic effect, 118, 128
Ovaries, 20, 145
Oviposition, 62, 65
  learned, 63
  learning and, 96
  in living plant tissue, 25
  on twigs, 25
  in winter, 25
Oviposition behavior, 2, 9, 17, 25, 168
  experience-induced changes in, 55—59
Oviposition marking substances, 58
Oviposition preference, 9, 55, 81
  induced (ICP), 55, 56
  learned, 56, 58, 60, 63, 64
  tests of, 60
Oviposition sites, 8
  finding of, 43, 55
  imprinting of, 56
  plant chemistry and, 163
  selection of, 57
Oviposition substrate, 57
Oxidase, 89
Ozone stress, 79, 85

P

Papilio
  glaucus, 86—87
  machaon, 48, 50, 52, 59
  oregonius, 167
Pararge aegeria, 100
Parasitoids, 20, 22, 29, 30, 60, see also specific types
Passiflora spp., 59
Pattern learning, 103—104
Pemphigus betae, 25, 26, 81
Peripheral gustatory sensitivity, 120
Peripheral sensitivity, 129
Periplaneta americana, 43, 121, 124, 126, 129, 135

Peritrophic membranes, 149
Permissive system, 23
Petunia spp., 52
Phagoletis pomonella, 57, 63
Phagostimulants, 115, 133, 134, 148
Phagostimulation, 81, 133, 150
Phagostimulatory excitation, 122
Pharmacophagy, 148
Phaseolus radiatus, 58
Phasmids, 166
Phenolglucosides, 76—80, see also specific
    types
  adaptation to, 87
  agonistic effects of, 88—90
  altitude and, 84
  analysis of, 76—77
  antagonistic effects of, 86—87
  chemical structure of, 76—77
  detoxification of, 86
  distribution of, 77—80
  field studies of, 82—84
  herbivore distribution patterns and, 80—85
  in host plant leaves, 80—85
  laboratory studies of, 81—82
  leaf age and, 84—85
  natural enemies of leaf beetles and, 90—91
  specialization on leaves rich in, 86—90
  toxicity of, 86
Phenolic defenses, 10
Phenolic glycosides, 10, see also specific types
Phenolics, see also specific types
Phenylalanine, 119
Phoebis sennae, 100
Phormia
  regina, 41—43, 121, 122, 129, 145
  spp., 61, 123
Phratora
  laticollis, 82
  spp., 88
  vitellinase, 48, 81—83, 85, 89, 90
  vulgatissima, 82
Phryganidia californica, 138
Phyllocolpa spp., 84
Phyllodectina, 88
Phylogenetic constraints, 2, 17—27, 30
  defined, 17
  in eruptive species, 20—27
  in latent species, 18—20, 25—27
Phylogenetic factors, 166—168
Phylogeny, 176—179
Physiological adaptation, 149
Physiological factors, 165
Phytochemicals, 51, 61, see also specific types
Pieridae, 100, see also specific genera
Pieris
  brassicae, 44, 46—48, 50
    carbohydrates in diet of, 121
    color and, 100
    intermeal intervals in, 124
    light intensity learning in, 100
  rapae, 44, 48, 56, 57, 96, 139
    flower constancy and, 99
    physiological factors and, 165
    post-ingestive compensation in, 139
    switching rates in, 103
    unrewarding flowers and, 104
  spp., 45
  virginiensis, 164
Pilosity, 81
Pine, 18, see also specific types
Plagiodera
  versicolora, 81, 83—85, 89—91
  vitellinae, 90
Plants
  architecture of, 28
  chemistry of, 163
  growth of, 28
  host, see Host plants
  morphological characters of, 164
  nitrogen in, 116, 138, 139
  non-host, 45, 52
  sex of, 78
  stress in, 14, 17, 23, 28, 29
  surface of, 53
  susceptibility of, 29
Pleiotropic gene action, 177
Plot problem in life tables, 5—7
Pollen feeding, 97
Pollution, 79
Polygenic inheritance, 177
Polygonia interrogationis, 48
Polyphagous species, 22, 47, 52—54, 57, 61,
    see also specific types
Polyphagy, 50—52
Pontania
  proxima, 82, 83
  spp., 84
Population density, 3, 7, 27
  fluctuations in, 23
  mortality and, 20
  steady, 3
Population dynamics, 2, 5, 8, 20, 27
  of eruptive species, 27
  female behavior role in, 10
  herbivore, 28
  of latent species, 27
Population regulation, 7
Population stability, 11
Populin, 81
Populus
  alba, 82
  deltoides, 79, 85
  nigra, 82
  spp., 11, 76, 78, 82
  tremuloides, 78, 79, 85, 86
Post-ingestive compensation, 138—140
Post-ingestive effect, 51, 52
Post-ingestive malaise, 54
Post-prandial quiescence, 114, 119, 122
Predators, 20, 22, 30, 84, 90, see also specific
    types
  behavior of, 91

susceptibility to, 97, 100
Preference, 99, see also specific types
  associatively learned, 136
  food, see Food preference
  induced, 41, 42
  learned, 135
  oviposition, see Oviposition preference
  relative, 52
Pre-imaginal experience, 59—60
Preingestive effect, 51
Prenelopis spp., 91
Previous food deprivation, 124—138
Process studies, 7
Proovigenesis, 22
Protein hunger, 145
Proteins, 116—121, 123, 131, see also specific
    types
  concentration of, 133
  deprivation of, 128, 136
  precipitation of, 148
  retention time of in gut, 142
Proximate factors, 27, 163—165, 174
Pseudoconditioning, 40—42
Pteromalus spp., 11
Pterostichus melanarius, 43
Pupal size, 81

Q

Quiescence, 114, 119, 122

R

Raffinose, 122
Ramet age, 11
Ranunculaceae, 103, see also specific genera
Receptor sites, 44
Recognition cues, 97, 103
Regulation of population, 7
Reinforcement, 42, 43, 58, 59
Relative payoff sum (RPS), 169
Relative preference, 52
Reproductive isolation, 64, 65
Reproductive strategy, 18
Resistance, 5, 10, 25, 29, see also Defenses;
    specific types
Resource concentration hypothesis, 165
Resource quality, 18, 20, 25, 27—29
Resource quantity, 28—29
Resource regulation, 11
Respirometer, 139
Restrictive system, 23
Reusable resources, 107
Rhagoletis
  pomonella, 57, 166
  spp., 167
Risk concentrators, 18, 23
Risk-prone reproductive strategy, 18
Risk spreaders, 23
Rolling fulcrum model, 170
RPS, see Relative payoff sum

R-selection, 4

S

Salicaceae, 76, 77, 79, 86, 88, see also specific
    genera
Salicin, 76, 81, 82, 86, 88, 89
Salicortin, 76, 82, 86, 89
Salicylaldehyde, 88, 90, 91
Saligenin, 89
Salix
  alba, 81, 82, 85
  babylonica, 84, 85, 90
  caprea, 81, 82
  carea, 89
  cinerea, 8, 28, 81, 82
  crestera, 83, 84
  dasyclados, 78, 79, 81, 85
  exigua, 8
  fragilis, 78, 81, 82, 85
  lasiolepis, 3, 8, 29, 79, 83, 84
  nigricans, 81
  orestera, 79
  planifolia, 79
  purpurea, 81
  spp., 11, 76, 78, 80, 81
  viminalis, 78, 81, 85
Sampling problem in life tables, 4—5
Sarcophagid fly, see *Agria affinis*
Satyridae, 100, see also specific genera
Sawflies, 8, 84, 90, see also specific types
  adaptive syndromes in, 18
  bud galler, see *Euura mucronata*
  densities of, 11
  diprionid, 18, 22
  galling, 8—11, 13—14, 17
  pattern of attack by, 11
  phylogenetic constraints in, 18
  shoot-galling, see *Euura*
  Swaine jack pine, 23, 28
  vulnerability of to parasitoids, 29
Schistocerca
  americana, 54, 135, 142
  gregaria, 41, 47, 53, 114, 119, 140
    allelochemicals and, 149
    food deprivation in, 125
  spp., 150
Scotts pine, 18
Scrophulariaceae, 51, see also specific genera
Searching, 63
Searching image, 55, 58, 62
Search rate, 58
Seasonal variations, 78, 79
Second order conditioning, 54
Secretogogue mechanism, 140
Seeing, 58
Selection, 61, 64, see also specific types
  dietary, see Dietary selection
  natural, 60, 168—169
Selective advantage, 61
Senecio spp., 53

Sensitization, 40—42
Serine, 119
Serotonin, 136, 137
Sex allocation, 20
Sex of plant, 78
Sex ratio, 8, 22
Shape learning, 103—104
Shoot-galling sawfly, see *Euura*
Short-term memory, 45, 102
Sinigrin, 44, 56, 57
Solanaceous alkaloids, 51
Sorghum spp., 53
Spatial density, 11
Spatial learning, 96
Specialists, 60, 64, 84, 97, 107, 169, see also
    specific types
  generalists vs., 173
Specialization
  dietary, 80
  on leaves rich in phenolglucosides, 86—90
Speciation, 64—65
Specific amino acid feedback, 118
Specificity, 20
Specific nutrient effects, 128
Specific nutrient feedback, 120, 123
Spiders, 90, see also *Xysticus* spp.; specific
    types
Spinacea
  oleracea, 54
  spp., 54
Spodoptera
  eridania, 50, 114, 149
  exempta, 53, 120
  exigua, 140, 142, 148
  littoralis, 134, 137, 143
    dietary selection and, 132
    food deprivation in, 125
    nutrient deprivation in, 128
    taste receptors in, 120
  spp., 44, 128
Spruce budworms, 5, 28
  eastern, see *Choristoneura fumiferana*
Stable population, 25
Stable resources, 107
"Starving-to-death-at-Lucullian-banquets" phe-
    nomenon, 50, 62
Steady populations, 3, 20, 27
Stealthy herbivores, 23
Stimuli, see also specific types
  conditioned, 40, 41, 52, 53, 57
  food-related, 45—47
  unconditioned, 40, 41, 52, 53
  visual, 55
STM, see Short-term memory
Stress, see also specific types
  environmental, 79, 85
  host-plant, 23
  nutrient, 22, 79
  ozone, 79, 85
  plant, 14, 17, 23, 28, 29
  water, 22, 79

Stretch-induced hormonal mechanism, 140
Stretch receptors, 114, 125, 142, 145
Strychnine, 44
Substrate, see also specific types
  oviposition, 57
Sucrose, 122
Sugar receptors, 126
Sugars, 122, see also specific types
Supella longipalpa, 132, 143
Surfactants, 149, see also specific types
Susceptibility of plants, 29
Swaine jack pine sawfly, 23, 28
Switching rates, 99, 103
Synovigenic ovaries, 20

T

Tannic acid, 149, 150
Taste cues, 181
Taste learning, 102—103
Taste receptors, 125
Taste sensilla, 119, 126
Tephritidae, 177, see also specific genera
Tiger swallowtail butterfly, see *Papilio glaucus*
Tilia americana, 117, 139
Time-and-place learning, 107
Tobacco hormworm, see *Manduca sexta*
Tortrix, gray larch, see *Zeiraphera griseana*
Trapline, 107
Trehalose, 122
Tremulacin, 86
Trial-and-error learning, 43, 55
Tribolium confusum, 54, 130
Trichomes, 89
Tropaeolum majus, 50, 52
Tropaseolaceae, 52, see also specific genera
Trypsin, 140, 142
Tunnel vision, 64

U

Ultimate factors, 27, 168—169, 179—180, see
    also specific types
Umbelliferae, 50, see also specific genera
Unconditioned stimulus, 40, 41, 52, 53
Unifolium spp., 52
Unstressed plants, 17
US, see Unconditioned stimulus

V

Valine, 119
Verbascum thapsus, 51
Vigna
  sinensis, 52
  unguiculata, 58
Vigor, 11
Vision, 58
Visit rates, 99
Visual cues, 55, 57, 59, 63
Visual stimuli, 55

Vitamins, 124, 131, see also specific types
Vitellogenesis, 129, 145
Volatile chemical cues, 63
Volatile substances, 51, 53, see also specific
    types
Volumetric feedback, 115, 127
Vulnerability, 29

W

Wasps, cynipid gall, see Cynipidae
Water in diet, 115
Water stress, 22, 79
Weather, 29
Webworms, fall, 29
Weed control, 51
Weevils, see Curculionidae
Wheat germ diet, 51

Willows, 81, see also specific types
    arroyo, see *Salix lasiolepis*
    galling sawflies on, 8—11, 13—14, 17
    heterogeneity of, 11
Windows of vulnerability, 29
Winter oviposition, 25

X

Xysticus spp., 90

Z

Zeiraphera griseana, 23, 30
Zerynthia rumina, 174
Zonocerus variegatus, 46
    orestera, 79

9 781138 560321